TINOCO

ORGANIC CHEMISTRY OF NUCLEIC ACIDS

Part A

Contributors:
N. K. KOCHETKOV, E. I. BUDOVSKII, E. D. SVERDLOV, N. A. SIMUKOVA,
M. F. TURCHINSKII, AND V. N. SHIBAEV

ORGANIC CHEMISTRY OF NUCLEIC ACIDS
Part A

Edited by N. K. Kochetkov and E. I. Budovskii

Translated from Russian by Basil Haigh
Translation edited by Lord Todd and D. M. Brown

ℚ PLENUM PRESS • London and New York • 1971

Plenum Publishing Company Ltd.
Davis House
8 Scrubs Lane
Harlesden
London NW10 6SE
Tel . 01 - 969 - 4727

U.S. Edition published by
Plenum Publishing Corporation
227 West 17th Street
New York, N.Y. 10011

Library of Congress Catalog Card Number 77-178777
SBN 306-37531-1

The original Russian text was first published by Khimiya Press in Moscow in 1970. The present translation is published under an agreement with Mezhdunarodnaya Kniga, the Soviet book export agency.

Н. К. Кочетков,   Э. И. Будовский,

ОРГАНИЧЕСКАЯ ХИМИЯ НУКЛЕИНОВЫХ КИСЛОТ

ORGANICHESKAYA KHIMIYA NUKLEINOVYKH KISLOT

Set in cold type by Plenum Publishing Company Ltd
Printed in Great Britain by Page Bros. Ltd., Norwich

Foreword

The study of nucleic acids is one of the most rapidly developing fields in modern science. The exceptionally important role of the nucleic acids as a key to the understanding of the nature of life is reflected in the enormous number of published works on the subject, including many outstanding monographs and surveys. The pathways of synthesis and metabolism of nucleic acids and the many and varied biological functions of these biopolymers are examined with the utmost detail in the literature. Nearly as much attention has been paid to the macromolecular chemistry of the nucleic acids: elucidation of the size and shape of their molecules, the study of the physicochemical properties of their solutions, and the appropriate methods to be used in such research.

The surveys of the chemistry of nucleic acids which have been published so far deal almost entirely with their synthesis and, in particular, with the synthetic chemistry of monomers (nucleosides and nucleotides); less attention has been paid to the synthesis of polynucleotides. There is yet another highly important aspect of the chemistry of nucleic acids which is still in the formative stage, the study of the reactivity of nucleic acid macromolecules and their components. This can make an important contribution to the determination of the structure of these remarkable biopolymers and to the correct understanding of their biological functions. Research in this direction has begun to make rapid progress in recent years and its scope has increased enormously. Nevertheless, this aspect of nucleic acid chemistry has not yet been adequately reflected in those surveys which have been published in this field, with the exception of a few publications devoted to more or less specialized problems.

The authors of this present monograph have tried to remedy this deficiency, while recognizing the difficulty of the task they have undertaken. The book deals with reactions of the nucleic acids and their components; these reactions lead to structural changes in nucleic acids; to their chemical modification

Although the ultimate purpose of the book is to familiarize the reader with the chemical reactions of the polynucleotides, most of the material is nevertheless concerned with chemical conversions of the

v

monomeric components of nucleic acids ; nucleosides and nucleotides. The main reason for this is that a proper understanding of the reactivity of polymers is unthinkable without knowledge of the reactions of their monomeric components. Moreover, the bulk of the research at the present time is still being undertaken at the monomer level, and only a relatively few studies of the chemical modification of the biopolymers themselves have been made. Nevertheless, the facts which are known concerning reactions of polymers, especially in cases where the problem has been studied in considerable depth and detail, can shed light on the structure of nucleic acids. What is more, they can point the way to a rational use of the corresponding reactions in the study of the biological function of nucleic acids.

The material of this monograph is divided into two parts. Part A (Chapters 1–4) deals with general aspects of the organic chemistry of nucleic acids. The conformation and electronic structure of the monomers, the reactivity of the heterocyclic bases (looking at the problem also from the standpoint of quantum chemistry), and the important question of noncovalent interactions in the polymer chain of nucleic acids all receive attention. It was thought advisable to begin the first part of the book with a brief survey of the classification and distribution of nucleic acids and the principles governing the establishment of their primary structure.

Part B (Chapters 5–12) is concerned with the special organic chemistry of the nucleic acids. The various types of reactions of heterocyclic rings, and reactions of the carbohydrate residue and the phosphate group are examined. A separate chapter is devoted to a brief account of the photochemistry of nucleic acids.

A list of recommended symbols and abbreviations for the poly-nucleotides and their components and derivatives is included.

No special account of the synthetic chemistry of nucleic acids and of their monomeric components is contained in this book. Because of the existence of a number of monographs on this subject, and to keep the size of this book within reasonable limits, it was decided to omit an examination of the extensive literature on methods of synthesis of nucleosides and nucleotides, and to give a very concise account of the synthesis of polynucleotides only.

Nevertheless many sections of the book border very closely on questions of synthesis, and in the authors' opinion they will be useful to the synthetic chemist. In them he will find information on the reactivity of functional groups in nucleosides and nucleotides and a description of various reactions of the greatest use in synthesis.

The authors are well aware of the weaknesses of their book: the conventional and, sometimes, artificial manner of arrangement of the material, the possibility that some of their personal views expressed in it may be incorrect, and so on. They consider that its publication is justified by the fact, already mentioned, that the organic chemical aspect of the chemistry of nucleic acids has only very recently begun to receive its due measure of attention, and its study is virtually in its infancy. For this reason the authors hope that their generalization of the existing data may prove particularly useful. They consider that their task will have been fulfilled if this book helps to foster the further development of research in the organic chemistry of the nucleic acids.

The authors are grateful to the following for their help in the preparation of this book: Corresponding Member of the Academy of Sciences of the USSR D. G. Knorre, Corresponding Member of the Academy of Sciences af the USSR M. A. Prokof'ev, Professor Z. A. Shabarova, Professor Yu. S. Lazurkin, Candidate of Chemical Sciences M. A. Kuz'min, and Candidates of Physico-Mathematical Sciences É. N. Trifonov, M. D. Frank-Kamenetskii and V. I. Danilov.

The Authors

Contents

Part A

Chapter 1

Structure of the Nucleic Acids

Chapter 2

Conformation of Nucleosides and Nucleosides

Chapter 3

Electronic Structure and Reactivity of the Monomer Components of Nucleic acids

Chapter 4

The Secondary Structure of Nucleic Acids

Part B

Chapter 5

Substitutions and Additions in the Heterocyclic Rings of the Bases of Nucleic Acids and their Derivatives

Chapter 6

Reactions of Exocyclic Substituents of the Bases of Nucleic Acids and their Derivatives

Chapter 7

Reactions Involving the Cleavage or Regrouping of Heterocyclic Rings of the Bases of Nucleic acids and their Derivatives

Chapter 8

Hydrolysis of N-Glycosidic Bonds in Nucleosides, Nucleotides and their Derivatives

Chapter 9

Reactions of the Carbohydrate Residues of Nucleic Acids

Chapter 10

Cleavage of Phosphoester Bonds and some other Reactions of Phophate Groups of Nucleic Acids and their Derivatives

Chapter 11

Some Reactions of Minor Components of Nucleic Acids

Chapter 12

Photochemistry of the Nucleic Acids and their Components

Introduction

The group of life sciences occupies a special place largely through the efforts and talents of many of the most distinguished scientists of today; progress in this field is extraordinarily rapid. The wish to understand the nature of life and the most intimate aspects of its phenomena has led investigators to probe into the innermost depths of biological processes, and to interpret them at the molecular level, at which physiological phenomena can be explained ultimately by chemical conversions or physical changes in individual molecules. The concept of molecular biology, as expressed above in the most general terms, has greatly stimulated interest in the study of those chemical substances whose conversions and changes lie at the bases of biological processes. Foremost among these substances are the natural macromolecular compounds: proteins, nucleic acids, polysaccharides, and mixed biopolymers.

Nucleic acids occupy an exclusive place in biological activity, for it is no exaggeration to say that they lie at its source. They are the material substrate which carries genetic information, the basis of the whole development of the future organism. Nucleic acids are also the instruments responsible for the synthesis of specific proteins. Even these few remarks are sufficient evidence of the importance of the study of nucleic acids in modern science and they readily account for the extraordinary attention which the investigation of these compounds has received in hundreds of laboratories the world over.

In 1869 Friedrich Miescher isolated from cell nuclei a substance with acid properties, which he called nucleic acid. In the century which has elapsed since that date, the following stages in the study of nucleic acids can be distinguished.

1

1. The preliminary research stage covers the historic work of Miescher, who isolated nucleic acid and showed that it contains carbon, hydrogen, oxygen, nitrogeh, and phosphorus, and also the work of Kossel, who demonstrated the existence of two types of nucleic acids in cells, and finally, the work of Emil Fischer, who studied the purines and pyrimidines, components of nucleic acids.

2. The second stage lasted from the beginning of this century until the end of the 1930s. It was concerned mainly with the study of breakdown products of the nucleic acids. In the course of this work, the monomer components of nucleic acids were isolated and studied. The structure of the carbohydrate residues, nucleosides, and nucleotides was established, by the work principally of Levene and also of Gulland. On the basis of his findings, Levene put forward the tetranucleotide hypothesis of nucleic acid structure, but this has not been subsequently confirmed. The development of research in this field was handicapped by the lack of suitable methods of isolating nucleic acids and of determining their physicochemical and biological properties, as well as by absence of evidence regarding their role in biological processes.

3. At the beginning of the 1940s indirect evidence, followed quickly by direct evidence, of the participation of nucleic acids in the transmission of genetic information was obtained (Avery, McLeod, and McCarty). This acted as a powerful stimulus to the development of the organic chemistry of nucleic acids. As a result of work carried out principally by the Cambridge School under the direction of Todd, the structure of the nucleosides and nucleotides was completely elucidated, methods for their synthesis were developed, and the basic principles governing the structure of nucleic acids as macromolecular compounds were established. These discoveries, with the results of Chargaff's study of polynucleotide composition, formed the bases of Watson and Crick's hypothesis (1953), one of the cornerstones of modern molecular biology.

4. The last stage has seen the rapid development of nucleic acid research in which the results obtained by biologists, physicists, and chemists have been closely interwoven, to their mutual enrichment. The biosynthesis of nucleic acids (Kornberg and Ochoa) and the mechanism of the transmission and realization of genetic information (Crick, Jacob, Monod, Nirenberg) have been studied in this period. The physical and synthetic chemistry of nucleic acids has received great attention (Doty, Khorana). Natural and synthetic oligonucleotides and polynucleotides have been extensively used for research purposes. In this way many of the distinctive structural and functional properties of nucleic acids have been discovered and the code of protein synthesis has been finally elucidated.

5. Finally, in the modern stage, which began in 1965, side by side with the development of biochemical and physical methods of study of the nucleic acids, chemical approaches have received increasing attention. This period has

been marked by the development of methods of establishing the primary structure of low molecular weight ribonucleic acids (Holley, Sanger), by the synthesis of large deoxypolynucleotide templates (Khorana), by the extensive study of the organic chemistry of nucleic acids and their components, and by the use of chemical methods at all stages of investigation: for the isolation of nucleic acids and nucleoprotein complexes and for the study of their structure and functions.

The chemistry of nucleic acids, like the chemistry of other macromolecular compounds, differs in many essential details from the chemistry of the corresponding monomer components. Nucleosides and nucleotides are themselves polyfunctional compounds, although differences in the reactivity of the groups occurring in the four usual types of nucleotide components are comparatively small. Polynucleotides are giant molecules with many reaction centres.

Some aspects of nucleic acid chemistry are particulary complex. The reactivity of individual groups in the nucleotide components is dependent not only on the reaction conditions (solvent, pH, temperature, and so on), but also on the presence and character of interaction of the components with each other (in the same chain and on the complementary segment in double-helical double-stranded molecules), and also on interaction with protein molecules, metal ions, and so on. As a rule all these interactions are cooperative, i.e., their changes are a nonlinear function of changes in the reaction conditions. Modification of one component in the polynucleotide chain changes the character and strength of interaction of that component with its neighbours (or with the protein molecule in the case of nucleoproteins). This ultimately influences the reactivity of the components over long segments of the polynucleotide chain.

The reactivity of the nucleotide components, the mechanism and kinetics of the reactions, and the structure and properties of modified components can be studied, as a first approximation, in monomer compounds by the ordinary methods of organic and physical chemistry. However, the analysis of interactions affecting reactivity of individual nucleotide components of a polymer requires the investigation of polymer models: oligonucleotides, single- and double-helical homopolynucleotides and, finally, heteropolynucleotides. The results obtained by work of this type in most cases has enabled rational use to be made of the reaction in the study of the structure and functions of nucleic acids.

When studying structure and functions of polynucleotides it is most convenient to use highly specific reactions leading to the formation of stable modified components with known chemical and functional properties. In fact, however, no reaction which satisfies these requirements completely is known. The importance of a detailed study of the organic chemistry of the nucleic acids will therefore be obvious. Unfortunately, the results obtained by chemical methods often receive only a superficial and primitive interpretation, with the inevitable discrediting of organic chemical approaches to the study of structure and functions of the nucleic acids.

The depth and extent of organic chemical research are naturally determined by the problems for whose solution the reaction will be used. In general,
these problems can be expressed as follows:

1. Purification and fractionation of oligonucleotides and polynucleotides.

2. The study of the primary structure of nucleic acids*.

3. The study of higher structures of nucleic acids and nucleoproteins.

4. Investigation of the functional (biological) properties of nucleic acids.

The chemical methods used to isolate polynucleotides or nucleic acids
can be based on the specific reactivity of minor components or of terminal
groups. These groups may be connected, either directly or after preliminary
chemical modification, with an insoluble carrier or with a molecule which
sharply modifies the physical properties of the polynucleotide (solubility,
partition or sedimentation coefficients, and so on). Such methods have been
used for the isolation and fractionation of transfer RNAs.

There are three possible chemical approaches to the study of the primary structure of nucleic acids: a) successive detachment and determination
of end groups; b) specific cleavage of the polynucleotide chain at particular types of linkages; c) specific modification of nucleoside components of
the polynucleotide and direct electron-microscopic determination of the distribution of these modified components along the polymer chain.

Method a (both when purely chemical methods and when a combination
of chemical reactions and enzymes are used) imposes very rigid and, at the
present time, still practically unattainable demands on the specificity and
quantitativeness of the chemical and enzymic reactions taking place. This
method is therefore used only for the determination of comparatively short
terminal sequences or to establish the structure of oligonucleotides containing not more than ten nucleotide residues.

Method b, the specific splitting of the polynucleotide chain into units,
followed by elucidation of the structure of these units and recreation of the
primary structure of the original polymer, imposes far less rigid demands

---

*Here and later in the book the terms "primary, secondary, tertiary, and quaternary structures of
nucleic acids" are used in the following sense. Primary structure means the sequence of nucleoside
components connected by a phosphodiester bond into a continuous, unbranched polynucleotide chain.
Secondary structure, in the case of single-stranded, mainly homopolynucleotides, means the spatial
arrangement of nucleoside components due to interplanar interaction between bases. In the case of
two complementary chains the secondary structure is a rigid double helix, stabilized both by interplanar interaction between neighbouring bases in the same chain and also by hydrogen bonds between opposite bases in the parallel chains. The tertiary structure is formed by the realization of other
types of fixed arrangement of polynucleotide chains besides the double-helical type. Quaternary
structure implies the spatial arrangement of interacting macromolecules (usually polynucleotides
and polypeptides) in nucleoproteins, such as ribosomes, viruses, and so on.

on the specificity and completeness of the conversions. The chain is split
only once, and the resulting errors are not cumulative. Chemical or combined
chemical and enzymic methods can be used to produce selective splitting of the
polynucleotide chain. In the first case the chemical modification must lead to
labilization of the internucleotide bonds. Reactions of this type, possessing
group specificity (destruction of all pyrimidine bases or detachment of all
purine bases), are widely used to study the distribution of nucleotides in the
DNA molecule. The analogous methods for RNA possess higher specificity.
The second method of splitting the chain, by combined chemical and enzymic
attack, consists essentially of the selective chemical modification of the structure of certain nucleoside components, leading to an increase in the stability
of the corresponding internucleoside bonds to the action of nucleases, and to
the subsequent action of nucleases. For instance, after modification of the
uracil ring with hydroxylamine or a carbodiimide, pancreatic ribonuclease splits
RNA only at the cytidine components, which is equivalent to the use of cytidylyl-
ribonuclease, an enzyme not found in natural sources.

However, the unit method of determination of primary structure, despite
its obvious advantages compared with the method of successive detachment of
terminal fragments, cannot be used in principle to study macromolecular
polynucleotides. As the length of the chain increases, so also does the number of identical units, and this leads to indeterminacy when the original
structure is recreated.

Method c is at present the most promising method of determining the
primary structure of polynucleotides of any type (RNA or DNA) and of any
length. It consists, in principle, of the specific modification (labelling) of
each of the four nucleotides by a definite chemical reaction, as a result of
which the sequence of mononucleotides in the polymer chain can be "examined"
directly under the electron microscope. One of the key problems in this approach — the specific combination of particular bases with the contrasting
group (one or several heavy atoms, causing intensive scatter of electrons) —
can be solved by chemical methods. Attempts to use this principle have been
undertaken, although so far without much success. This is probably because
of an obvious lack of investigation of the specificity and quantitative characteristics of reactions at present used to introduce the contrasting groups.
Nevertheless, if it is remembered that only 10^2-10^3 molecules of a polynucleotide are sufficient for determination of the primary structure by means of
electron-microscopy, the urgency of the discovery of methods of specific introduction of contrasting groups will be obvious.

It was stated above that the reactivity of the nucleotide components is
essentially dependent on the presence of noncovalent interactions with neighbouring components; as a result, chemical methods can be used to study the
secondary structure of nucleic acids. In particular, the influence of complementary interactions between bases on their reactivity is so great that
components of a polynucleotide chain located in single-helical zones can be
selectively modified, and in this way the composition and size of these zones

can be determined. If, therefore, the primary structure of the molecule is known, such single-helical segments can be localized in the chain. Investigations of this type have been widely undertaken in many laboratories using reactions with formaldehyde, acrylonitrile, water-soluble carbodiimides, hydroxylamine, and other agents.

The effect of interplanar (i.e., stacking) interaction on reactivity of the bases is much weaker, although the effect in this case also is quite adequate; it is used to study the structure of single-stranded polynucleotides and of nucleotide coenzymes.

Another promising use of chemical methods is for studying higher structures of nucleic acids and nucleoproteins (enzyme–substrate complexes, viruses, ribosomes, and so on) in a functionally active state. Only the first steps have been made so far in this direction, but the results give good grounds for optimism. When primary and higher structures of nucleic acids are studied in this way, as a rule an attempt is made to secure the maximum (quantitative) degree of modification of components of a certain type, or the kinetics of the main reaction are studied. The mechanism of the reaction, the kinetics of its intermediate stages, and the structure of intermediate products (and in the case of the study of higher structures, the specificity of the reaction) are not of decisive importance. Often a relatively high level of side reactions is also permissible.

An essentially different approach is required for chemical methods used for functional (biological) investigations of nucleic acids. First, in functional investigations as a rule modification of only a very small number of monomer components of the polymer is permitted. For this reason, information concerning the mechanism and kinetics of the main and side reactions, and the structure and properties (including functional properties) not only of the end products, but also of the intermediate products of the reaction, must be available to enable correlation of the chemical and functional changes. Second, since only a few components are modified, it is important to know not only their number, but also their distribution along the chain. Third, the modified components with different structures may have different functional properties, so that even if the velocity of the side reactions is several orders of magnitude below the velocity of the main reaction, they can still make a significant contribution to the changes in functional properties of the polynucleotide, thus making rational interpretation of the results difficult, or sometimes impossible. This is a particularly important fact in the case of functional investigations of the genetic nucleic acids (DNA, virus RNAs). Methods of detection which are used can reveal in this case changes in individual molecules of the polymer contained in the analysed mixtures in very small amounts. On the other hand, in the case of modification of nongenetic nucleic acids (transfer RNA, for example), only the over-all change in functional properties can be observed, and the contribution of each of the modified components is proportional to its content in the mixture.

It is clear that the more reliable and accurate the data concerned with the specificity, mechanism, and kinetics of reactions taking place with the participation of a chemical agent, and also data concerning the structure and chemical properties of the modified components formed in the main and side reactions, the better will be our understanding of the chemical basis of functional specificity of the nucleic acids.

The material described in this section testifies to the enormous possibilities and prospects for further development of the organic chemistry of nucleic acids. A detailed study of the mechanisms and kinetics of reactions between components of the nucleic acids and different reagents, the search for and development of new specific reactions, the investigation of the effect of reaction conditions on reactivity of the components of nucleic acids and nucleoproteins, the determination of the character and nature of noncovalent interactions of the components of nucleic acids with each other, and with protein molecules, metal ions, and so on, are all therefore matters of the utmost importance.

There is no doubt that the harmonious development of the organic and synthetic chemistry, the physical chemistry and biochemistry of nucleic acids in the years to come will give a still deeper insight into the nature of processes taking place in the living cell, and into the chemical mechanisms of biosynthesis and regulation lying at the basis of metabolism.

RECOMMENDED GENERAL BIBLIOGRAPHY

P. Levene and L. W. Bass, Nucleic Acids, Chem. Catalog. Co. , New York (1931).

D. O. Jordan, Chemistry of the Nucleic Acids, Butterworth, London (1960).

R. F. Steiner and R. F. Beers, Polynucleotides, Elsevier, Amsterdam (1961).

F. W. Allen, Ribonucleoproteins and Ribonucleic Acids, Elsevier, Amsterdam (1962).

D. N. Brown and T. L. V. Ulbricht, in: Comprehensive Biochemistry, Vol. 8, M. Florkin and E. H. Stotz (editors), Elsevier, Amsterdam (1963), p. 157.

E. Harbers, Die Nucleinsauren, Thieme Verlag, Stuttgart (1964).

A. Michelson, The Chemistry of Nucleosides and Nucleotides, Academic Press, London – New York (1963).

E. Chargaff and J. N. Davidson (editors), The Nucleic Acids, Vols. 1, 2 (1955); Vol. 3 (1960); Academic Press, New- York – London.

J. N. Davidson and W. E. Cohn (editors), Progress in Nucleic Acid Research, Vols. 1, 2 (1963), Vol. 3 (1964), Vol. 4 (1965), Vol. 5 (1966), Vols, 6, 7 (1967), Vol. 8 (1968), and Vol. 9 (1969), Academic Press, New York – London.

J. N. Davidson, The Biochemistry of the Nucleic Acids (Sixth Edition), Methuen, London (1969).

I. B. Zbarskii and S. S. Debov (editors), The Chemistry of the Nucleic Acids [in Russian], Meditsina, Moscow (1968).

SYMBOLS AND ABBREVIATIONS USED TO DENOTE

POLYNUCLEOTIDES AND THEIR COMPONENTS

AND DERIVATIVES

The authors of any survey must always face the difficult task of producing an orderly system of the symbols and abbreviations used in the many
original works. This is a particularly difficult task in the field of any universal system and the impossibility of strict adherence to any impeccably
logical principle makes this task almost impossible or, at best, highly
vulnerable to criticism, because the concept of "impossible" is highly sub-
jective. However, guided by the recommendations of the Commission on
Nomenclature of the International Union of Pure and Applied Chemistry
(IUPAC) and of the International Union of Biochemists (IUB), the Third All-
Union Working Conference on Chemistry of Nucleotides adopted rules unify-
ing symbols of the nucleic acids, their components, and derivatives. These
rules, which are given below, have been used by the authors when writing
this book. It is to be hoped that in future these rules, with minimal modifications, will become generally accepted. This will make it much easier for
readers to understand not only this book, but also other surveys and original
works published in the USSR and elsewhere.

I. NUCLEOSIDES

1. The usual nucleosides are denoted by the first letter of their name:
A — adenosine; G — guanosine; C — cytidine; U — uridine; T — ribothymidine;
N — any nucleoside.

2. The following symbols are used for some minor components of
nucleic acids: I — inosine; X — xanthosine; Ψ — pseudouridine; $\overset{H}{U}$ — 5,6-di-
hydrouridine; $\overset{S}{U}$ — thiouridine; $\overset{H_2O}{C}$ — 6-hydroxy-5,6-dihydrocytidine.

3. The presence of a 2'-deoxyribose residue in a nucleoside is denoted
by the symbol "d," which is written before the symbol of the nucleoside; in
its absence, the ribonucleotide is implied.

For example: T — ribothymidine (thymine riboside); dT — thymidine
(thymine-2'-deoxyriboside).

4. In oligonucleotides, if all the component nucleosides are of the
deoxy group, the symbol "d" is placed either before the symbol of each
nucleoside or before the symbol of the whole oligomer. In the latter case,
the symbol of the oligonucleotide is included in parentheses.

For example: pdApdTpdCpdG or d(pApTpCpG).

If some component nucleosides in an oligonucleotide are derivatives of
2'-deoxyribose, while others are derivatives of ribose, the symbol "d" is
placed before each deoxyriboside component.

For example: pApTpdGpC.

5. If pentose residues other than ribose and 2'-deoxyribose are present, the symbols for the sugar are placed before the symbol for the nucleoside: a) arabinose, x) xylose, and l) lyxose.

Example: aC — arabinosylcytosine; xU — xylosyluracil; lG — lyxosylguanine.

6. For α-anomers, the symbol "α" is written before the symbol for the nucleoside.

For example: αC — α-ribosylcytosine; αdA — α-2'-deoxyribosyladenine.

7. Where bases of the nucleic acids and nucleosides are mentioned together in the same paper, to avoid misunderstandings which could arise from the use of single-letter symbols, the following three-letter symbols can be used.

Bases: Ade — adenine; Cyt — cytosine; Gua — guanine; Ura — uracil; Hyp — hypoxanthine; Xan — xanthine; Thy — thymine; Oro — orotic acid.

Nucleosides: Ado — adenosine; Cyd — cytidine; Guo — guanosine; Urd — uridine; Ino — inosine; Xao — xanthosine; Thd — ribothymidine; Ord — orotidine; dAdo — 2'-deoxyadenosine, and so on; Puo — any purine nucleoside; Pyo — any pyrimidine nucleoside.

8. For other nucleosides, until general agreement is obtained, appropriate symbols can be introduced subject to the following limitations:

A) The abbreviation must not coincide with one already in accepted use (for other compounds or groups).

B) The abbreviation must not include any hyphens or commas.

C) The meanings of these abbreviations must be qualified in each publication.

II. SUBSTITUTION AT THE SUGAR RESIDUE

1. To denote a phosphoric acid monoester residue

$$-O-\overset{O}{\underset{}{\overset{\displaystyle\|}{P}}}(OH)_2$$

the symbol "p" is used. The same symbol is also used to denote a phosphoric acid diester residue

$$-O-\overset{O}{\underset{O-}{\overset{\displaystyle\|}{P}}}-OH$$

in pyrophosphates and oligonucleotides. In the latter case, the symbol "p" is placed between the symbols of the two nucleoside components joined by the phosphodiester bond.   To denote a phosphodiester residue in oligonucleotides and, in particular, in polynucleotides, a hyphen (-) is also used.

Example:

pApGpUpC   or   pA-G-U-C

···pApCpU···   or   ····A-C-U···

2. The symbol for the substituent (including phosphate groups), if placed to the left of the symbol of the nucleoside, denotes substitution in the 5'-hydroxyl group of the sugar residue.   A symbol written to the right of the nucleoside symbol denotes substitution in the 3'-hydroxyl group.

Example: Ap, adenosine-3'-phosphate; pU, uridine-5'-phosphate; pppdG, 2'-deoxyguanosine-5'-triphosphate; pCp, cytidine-3',5'-diphosphate.

The symbol of a phosphate group located at the 2'-hydroxyl group of ribose of a ribonucleoside is placed to the right of the symbol for the nucleoside, and the position of the substituent is noted between them in parentheses.

Example: G(2')p denotes guanosine-2'-phosphate.

3. A bifunctional phosphate substituent attached at the 2',3'-position of a nucleoside is placed to the right of the nucleoside symbol and separated from it by the sign "> ".

Example: C>p denotes cytidine-2',3'-cyclic phosphate.

If the bifunctional phosphate substituent is attached in the 3',5'- or 2',5'-position, numbers indicating the position of this substituent will be written before the "> " sign in parentheses.

Example: A(3',5')>p denotes adenosine-3',5'-cyclic phosphate.

4. For other substituents than phosphate in hydroxyl groups of the sugar residue of nucleosides, the following symbols can be recommended: Me — methyl; Et — ethyl; Ac — acetyl; Bz — benzoyl (not benzyl); Tr — trityl.

III. SUBSTITUTION IN HETEROCYCLIC BASES

It is recommended that symbols of substituents in heterocyclic bases of monomer components of nucleic acids and oligonucleotides and polynucleotides be placed above the symbol of the corresponding nucleoside.   The position of such a substituent is indicated by a numerical superscript attached to the symbol of the substituent, and the number of substituent groups is shown by a subscript.

Example: $\overset{Me^1}{p}A$ denotes 1-methyladenosine-5'-phosphate (base substitution); . . . A–U–C–C–$\overset{Me_2}{G}$–A. . . denotes a polynucleotide with a dimethylguanosine residue in the chain (base substitution).

IV. POLYNUCLEOTIDE SEQUENCES

1. In oligonucleotides (or fragments of polynucleotides) with a known sequence and with a natural $(3' \to 5')$ type of internucleoside phosphodiester bond, symbols of the nucleosides are separated by the symbol "p" or "-".

Example: pApGpApT or pA-G-A-T. It is preferable to use the symbol "-" for polynucleotides.

Whatever method of abbreviation is used to record polynucleotide sequence, it must be so arranged that the 3'-end of the chain is on the right and the 5'-end on the left. This applies equally to the full sequence of the polynucleotide and to that of its fragments.

Example:

$$5'\text{-end} \qquad\qquad\qquad\qquad 3'\text{-end}$$
$$\text{pG-A-U} \cdots \text{A-U-U-C-G-A-G-U} \cdots \text{C-C-A}$$

2. If the order of the components is unknown, symbols of the nucleosides are put in parentheses and separated by commas.

Example: . . . (A,G,C). . .

When not all the sequence is known, the part which is unknown is placed in parentheses.

Example: . . . -G-C-(A,G)-C. . .

3. Terminal phosphate and pyrophosphate groups in oligonucleotides and polynucleotides are denoted by the symbols "p", "pp" and so on.

Example: pppA-C-G-U-. . .

4. In the case of an unnatural internucleoside phosphodiester bond, the numbers denoting the atoms at which hydroxyl groups forming the phospho-ester bond are located are indicated in parentheses by the side of the symbol for the internucleoside bond.

Example: $Cp(2' \to 5')A$ or $C-(2' \to 5')A$ denotes cytidylyl-$(2' \to 5')$-ade-nosine; $U(5' \to 5')pU$ or $U(5' \to 5')-U$ denotes uridylyl-$(5' \to 5')$-uridine.

V. NATURAL POLYNUCLEOTIDES

1. Natural polynucleotides are denoted by trivial abbreviations; the func-tional characteristics are denoted by a lower case letter in front of the abbre-viation, without any intervening hyphen or comma:

rRNA – ribosomal RNA; mRNA – messenger RNA; tRNA – transfer RNA.

In some cases the sedimentation constant is used to describe RNA. In such cases, the characteristic is separated from the abbreviation by a space, and it is suggested that the following method of writing be used.

Example: 16S RNA denotes RNA with a sedimentation constant of 16 Svedberg units.

2. Individual transfer RNAs are qualified by a superscript to the right of the abbreviation, denoting the corresponding amino acid.

Example: $tRNA^{Val}$ denotes an individual tRNA capable of accepting and transferring a valine residue.

Isoacceptor tRNAs are qualified by an additional subscript on the right. Example: $tRNA_1^{Val}$; $tRNA_2^{Ser}$.

Aminoacylated individual tRNAs are denoted as follows: $Val - tRNA_1^{Val}$; $Ala - tRNA^{Cys}$.

VI. SYNTHETIC POLYNUCLEOTIDES

1. Homopolynucleotides with an unknown number of components are denoted by poly-N or $(N)_n$. If the exact or mean statistical number of components is known, the following method of writing is recommended.

Example: U_{10} denotes decauridylic acid; $U_{\overline{10}}$ denotes oligouridylic acid with a mean number of 10 nucleotide components in the chain.

2. Heteropolynucleotides with a known sequence are written as follows.

Example: poly- (A-U) or $(A-U)_n$ denotes a polymer of adenylyluridylic acid.

If the distribution of nucleotide components along the chain is statistical or unknown, the polymers are written as follows.

Example: poly- (A, C) or $(A,C)_n$ denotes a copolymer of adenylic and cytidylic acids with monomer units in the ratio $1:1$;

Poly- (G_2,C) or $(G_2,C)_n$ denotes a copolymer of guanylic and cytidylic acids with a ratio between monomer units of $2:1$;

$(A_2U)_{\overline{50}}$ denotes a copolymer of adenylic and uridylic acids with a ratio between monomer units of $2:1$ and a mean number of components of 150;

Poly- (dA-dT), or poly-d(A-T), or $(dA-dT)_n$, or $[d(A-T)]_n$ denotes a polymer of deoxyadenylyldeoxythymidylic acid.

When the complete sequence of nucleotide components in a polymer, or in part of a polymer near to one end of the chain, is written the terminal groups are denoted specially (see section V.3).

VII. COMPLEMENTARY PAIRS

1. Complementary base pairs mentioned in the text are separated by a full stop.

Example: ". . . the strength of interaction between the complementary pair guanine·cytosine is greater than between the pair adenine·thymine. . . ."

Complementary pairs of nucleosides are written in the same way.

If the composition of a polynucleotide is written, the monomer components of the unit are separated by a "+" sign.

Example: ". . . an increase in the content of guanine + cytosine in DNA leads to an increase in T_m. . . "

2. When open structures are written, complementary "cross" interactions are denoted by full stops:

Example:

$$\cdots A\text{-}G\text{-}U\text{-}C\text{-}G\cdots$$
$$\cdot\ \ \cdot\ \ \cdot\ \ \cdot\ \ \cdot$$
$$\cdots U\text{-}C\text{-}A\text{-}G\text{-}C\cdots$$

3. In the case of double-helical double-stranded complexes, the abbreviated names of the components of the complex are separated by a full stop.

Example: (poly-A)·(poly-U) or $(A)_n$·$(U)_n$ denotes a double-helical complex of polyadenylic and polyuridylic acids.

Chapter 1
Structure of the Nucleic Acids

I. Introduction

The term nucleic acids describes phosphorus-containing biopolymers made up of residues of nucleosides which, in turn, are N-glycosides of pentoses and derivatives of heterocyclic bases of the purine or pyrimidine series; the nucleoside residues are linked to form a polymer chain by means of phosphodiester bonds. Hydrolysis of nucleic acids gives a high yield of nucleosides or of their phosphoric esters, nucleotides.

The carbohydrate component of nucleic acids may be either D-ribose or 2-deoxy-D-ribose.

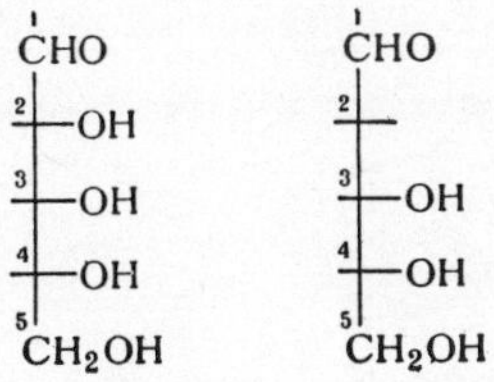

$$\begin{array}{cc}
\overset{1}{C}HO & \overset{1}{C}HO \\
{}^{2}\!\!-\!\!OH & {}^{2} \\
{}^{3}\!\!-\!\!OH & {}^{3}\!\!-\!\!OH \\
{}^{4}\!\!-\!\!OH & {}^{4}\!\!-\!\!OH \\
{}^{5}CH_2OH & {}^{5}CH_2OH
\end{array}$$

Two types of nucleic acids can accordingly be distinguished: ribonucleic acids (RNA) and deoxyribonucleic acids (DNA).

The nucleosides are β-N-pentofuranosides of heterocyclic bases, and their structures are shown in Formulae I-IV. Usually residues of at least four nucleosides are present in the RNA or DNA molecule. The usual components of RNA are adenosine (Ia), guanosine (IIa), cytidine (IIIa), and uridine (IVa); of the four nucleosides most commonly found in DNA, three — deoxyadenosine (Ib), deoxyguanosine (IIb), and deoxycytidine (IIIb) — contain the same bases as the ribonucleosides, while the fourth — thymidine (IVb) — differs from the corresponding ribonucleoside (uridine) by possessing an additional methyl group in the heterocyclic base.

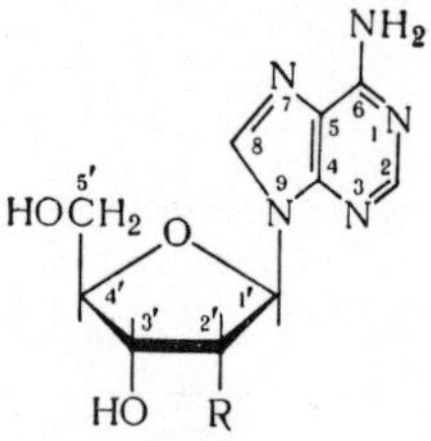

Ia (R=OH)
Ib (R=H)

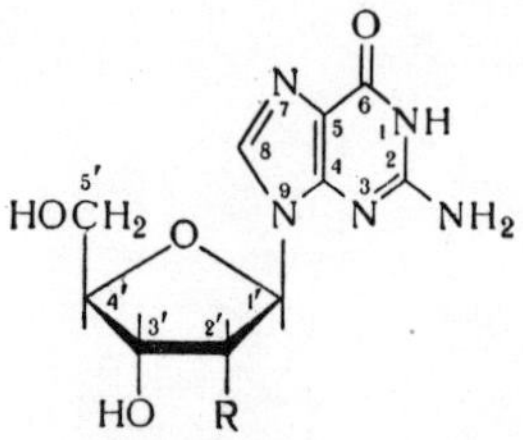

IIa (R=OH)
IIb (R=H)

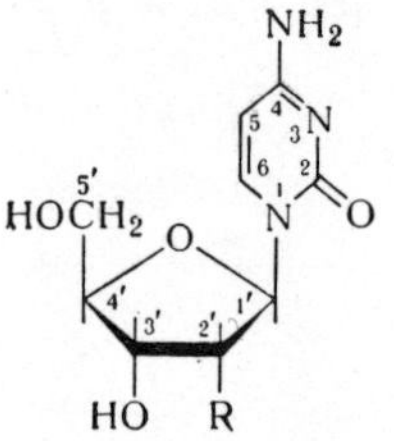

IIIa (R=OH)
IIIb (R=H)

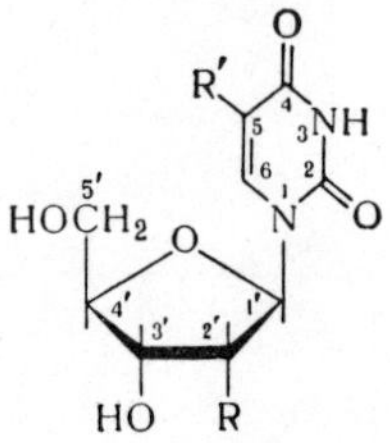

IVa (R=OH, R'=H)
IVb (R=H, R'=CH₃)

The monomer components of nucleic acids, or nucleosides, are linked together by phosphodiester bonds to form a polynucleotide chain. The phosphodiester group links the 3'-hydroxyl group of one nucleoside residue with the 5'-hydroxyl group of the next nucleoside residue. The polynucleotide chain of the nucleic acids is thus a linear structure in which nucleosides are linked together by 3',5'-phosphodiester bonds, and the nucleotides are arranged in the chain in a strictly definite order for each nucleic acid. The general structure of RNA and DNA is illustrated below in Formula V:

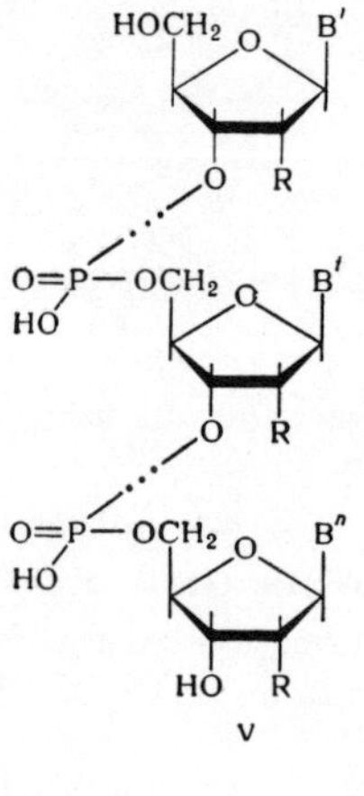

B'....B'....B''- base residues

R= H (for DNA)

R= OH (for RNA)

V

It will be clear that none of the monomer components of a deoxyribonucleic acid, except those at the ends of the chain, contain free hydroxyl groups; in ribonucleic acids, on the other hand, the monomer components of the polynucleotide chain have a free hydroxyl group at the C-2' position, next to the phosphodiester group. This difference in structure determines the marked difference in physicochemical properties of RNA and DNA.

The explanation of the fact that specific covalent interaction can take place between particular base pairs in the nucleoside residues of the polynucleotide chain shed considerable light on the biological functions of nucleic acids. This interaction, through hydrogen bonds, takes place between "complementary" base pairs; these are pairs of the natural bases adenine – thymine (VIa) and guanine – cytosine (VIb).

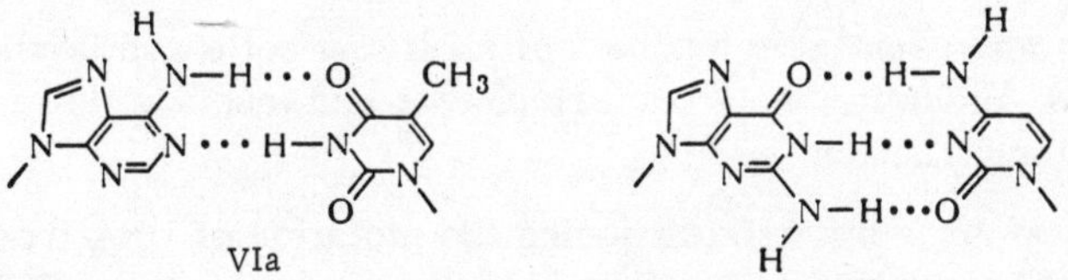

VIa   VIb

The first suggestion of interaction of this type was made during the study of the macromolecular structure of DNA (the Watson – Crick hypothesis)*; it has subsequently been confirmed experimentally (see Chapter 4), and is now one of the cornerstones of molecular biology and molecular genetics.

Together with proteins, polysaccharides, and lipids, nucleic acids are essential components of all living cells, which usually contain both RNA and DNA. Nucleic acids are also found in simpler, parasitic forms of life, namely in viruses. Virus particles often consist only of protein and DNA or RNA.

Histochemical and cytochemical methods of detection of nucleic acids in living cells [1-3] are usually based on the characteristic UV-absorption of nucleic acids and specific colour reactions due to the liberation of reducing groups of 2-deoxyribose residues during mild acid hydrolysis (the Feulgen reaction for DNA) or to the ability of polynucleotides to form complexes with basic dyes (Brachet's reaction for RNA; luminescence methods based on interaction with Acridine Orange). A most important sign, enabling a substance giving these reactions in a tissue section to be reliably identified as DNA or RNA, is the disappearance of the characteristic UV-absorption or of the cytochemical reaction after treatment of the section with nucleases —enzymes

---

*DNA exists in solutions characteristically as complexes formed of two polynucleotide chains, stabilized through interaction between complementary bases of different chains and hydrophobic interaction between bases in the same chain. These complexes are described as double-stranded.

catalysing the hydrolysis of nucleic acid polymers to compounds of low molecular weight*.

This phenomenon, the hydrolysis of a polymer to compounds of low molecular weight by the action of ribonuclease or deoxyribonuclease (nucleases splitting RNA and DNA respectively) can also be used to identify a polymer isolated from a cell as a nucleic acid. Other characteristic properties of nucleic acids include ultraviolet absorption with a maximum at about 260 nm and the presence of phosphorus and ribose or 2-deoxyribose, which can easily be demonstrated by appropriate colorimeteric reactions [5, 494].

II. Methods used to isolate DNA and to determine its properties.

The principal types of DNA

DNA is an essential component of all living cells and is also found in many viruses, including most bacteriophages and insect viruses and nearly all known animal viruses.

The chief problem arising during the isolation of DNA from natural sources is its separation from other biopolymers – protein, RNA, and polysaccharides.

In modern methods of isolation of DNA [6–8], the principal stage is usually extraction of the biological material with phenol [9]. After separation into layers, the DNA is either found in the aqueous phase or it is left as a residue in the interphase, while most protein is denatured and goes into the phenolic phase. To remove protein from DNA preparations, they can be treated with detergents [10] or with a mixture of chloroform and isoamyl (or octyl) alcohol [11, 12], or incubated with proteolytic enzymes, such as pronase [13]. RNA is separated from DNA by fractional precipitation with alcohol and treatment of the product with ribonucleases. The greater part of the polysaccharide is usually removed during fractional precipitation with ethanol or isopropanol; in some cases additional purification is necessary (extraction with methylcellulose, fractionation of cetavlon salts, electrophoresis).

A specific problem in DNA isolation is the extremely high sensitivity of these long linear molecules to hydrodynamic shearing forces. During mixing or filtration of the solution and its aspiration into a pipette, stresses arise in long linear molecules and they may be sufficient to break the covalent bonds [14], with consequent fragmentation of the DNA molecules. Because of this, in modern methods of DNA isolation these operations are reduced to the minimum, even though this may mean a smaller yield of DNA and less

---

*The specificity and mechanism of action of the nucleases are examined in more detail in the section on determination of the nucleotide sequence in nucleic acids (page); a discussion of other matters concerning nucleases will be found in a recently published monograph [4].

satisfactory deproteinization [15]. The resulting DNA preparations are usually described by their molecular weight and their base composition.

The methods most widely used to determine the molecular weight of
DNA [7, 16, 17] are those based on measurement of the sedimentation rate of
macromolecules. This determination can be made in several ways; the one
which has recently achieved the greatest popularity is that based on zonal
centrifugation in a sucrose density gradient in a preparative ultracentrifuge
[18, 21]. In this case the distribution of substances relative to their sedimentation rate can be determined by radioactive labelling, thus giving the
method high sensitivity. Another advantage is that the method can be used virtually unchanged for the preparative fractionation of nucleic acids. A number of
empirical equations relating the sedimentation rate of the double-stranded
DNA complex with its molecular weight, determined by independent methods,
have been suggested [19-24]. The last of these equations [24] covers the
range of molecular weights from $0.2 \cdot 10^6$ to $130 \cdot 10^6$.

A similar empirical equation has been suggested for the relationship
between viscosity and molecular weight for double-stranded DNA complexes
[24]. Light scattering methods are of limited value for determining molecular weight [7], for they are suitable only up to molecular weights of about
$25 \cdot 10^6$.

The only reliable and absolute method of determination of molecular
weight which can be used for high-polymer DNA is that based on autoradiography of DNA molecules labelled with ^{32}P [20]; by counting the "stars" on
the photographic emulsion the number of phosphorus atoms in the investigated
molecule can be determined. However, this method is very laborious and it
is used only occasionally. A more widely used method is that based on direct
observation of DNA molecules and measurement of their length under the
electron microscope [25, 26]. X-ray structural investigations of double-
helical DNA complexes have shown that a segment of double-stranded complex measuring 1 Å in length corresponds to a molecular weight of 192, and
on this basis the molecular weight of a prepared sample can be determined.
Finally, methods of determination of molecular weight based on chemical
determination of terminal groups have recently begun to be used (see page
31).

The nucleotide composition of DNA is usually determined by quantitative chromatography of the purine and pyrimidine bases or of nucleotides
obtained after hydrolysis of the polymer (see page 43 for details). The
nucleotide composition of DNA is directly related to two physical properties
of double-stranded complexes which are frequently used to characterize
prepared specimens [27, 28]. One of these is the melting temperature (T_m),
the temperature at which the double-stranded complex breaks up into single-
stranded molecules; this process can easily be observed from changes in the
UV-absorption or optical rotation of the solution (see Chapter 4 for details).

Another characteristic constant of DNA, its buoyant density (ρ) can be determined from the results of equilibrium centrifugation [29]. This type of centrifugation is usually carried out in salt solutions of high density; most frequently caesium chloride or sulphate is used. During prolonged centrifugation, a density gradient of the solution is established, and the DNA collects in the narrow zone of equilibrium between the centrifugal force and the centripetal force, which forces are determined by the difference in density between the sedimented substance and the salt solution used in that particular zone. Equilibrium centrifugation in a CsCl density gradient can be used not only as an analytical method for the identification of a DNA preparation, but also as a useful preparative method for separating DNAs of differing nucleotide composition. Preparative ultracentrifugation in a sucrose density gradient is similarly used for the separation of DNA molecules of differing sedimentation rate.

Chromatographic methods can also be used for DNA fractionation. The most popular method is that of chromatography on columns packed with methylated albumin on kieselguhr (MAK) [30] or with calcium phosphate (hydroxylapatite) [31]. Countercurrent distribution methods based on distribution between nonmiscible aqueous solutions of dextran and polyethylene glycol, have recently begun to be used for DNA fractionation [32].

DNA of viruses and phages [33-36, 495]. Isolation of DNA from viruses is a relatively simple matter, because virus particles can be obtained in a highly purified state, and in many cases all that is necessary to obtain DNA is to separate the protein. In such cases, the mildest possible methods of isolation can be used [37], such as deproteinization by concentrated sodium perchlorate solution [38]. Operations producing hydrodynamic shearing forces can be reduced to the minimum, and virus DNA can be obtained in a highly polymerized state. The process of isolation of virus DNAs, and their degree of nativeness can easily be controlled, because these compounds possess biological activity which is very easily measured quantitatively (infectivity).

Most viruses and phages contain DNA as a double-stranded complex with molecular weight $(30-200) \cdot 10^6$. The favourite sources for isolation of phage DNAs are the T bacteriophages infecting Escherichia coli; much work has also been done on the study of DNAs of bacteriophages λ, α, and SP8. Of the animal viruses, those which have been studied in greatest detail are the viruses of herpes simplex, vaccinia, polyoma, and papilloma, and the adenoviruses [the DNAs of the last of these have a much smaller molecular weight, of the order of $(4-5) \cdot 10^6$].

One group of small bacteriophages (ϕX174, fd, S13, etc.) contains a single-stranded DNA of low molecular weight. A value of $1.7 \cdot 10^6$ has been found for the molecular weight of the DNA of bacteriophage ϕX174.

A double-stranded DNA complex with twice the molecular weight [39], the "replicative form" of DNA of phage ϕX174, can be obtained, along with the single-stranded DNA present in the virus particles, from cells infected with phage ϕX174 by centrifugation in a density gradient and chromatography on MAK. The presence of forms of virus DNA differing in structure from the DNA of the virus particles, in a cell infected by the virus, has also been demonstrated for many other viruses.

Determination of the DNA content per virus particle shows that usually such a particle contains only one DNA molecule or one double-stranded complex. It is natural, therefore, to expect that DNA samples isolated from viruses will, chemically speaking, be individual DNA molecules or, at least, complexes of two individual molecules. This hypothesis is valid in many cases, and sometimes after destruction of the double-stranded complex, it is possible to separate the individual polynucleotide chains by equilibrium centrifugation.

However, things are not always so easy. In some cases the DNA macromolecule of the phage particle consists, not of two long polynucleotide chains, but of several shorter chains (DNA of phage T5 for example). In other cases (notably the DNA of phage T2), DNA isolated from the virus particles contains a mixture of molecules which differ in their chemical structure because of cyclic permutation of the fragments, as illustrated below

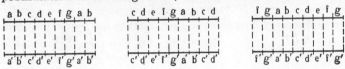

where a, b, . . . , g represent certain segments of one polynucleotide chain, and a', b', . . . , g' represent the complementary segments of the other chain.

DNA of the cell nucleus [7, 17, 40-42]. DNA in the cell nucleus of bacteria and other microorganisms can be observed as filamentous structures whose thickness, as shown by electron microscopy, corresponds to the thickness of the double-helical DNA complex [43].

Genetic observations suggest that the bacterial cell nucleus contains one DNA molecule of extremely high molecular weight. After destruction of bacteria under very mild conditions, DNA structures corresponding in length to a molecular weight of $\sim2800 \cdot 10^6$ can be observed by autoradiography [44]. Such large molecules cannot be obtained in an intact state by modern methods of isolation of nucleic acids. Even with the most careful isolation, the samples of DNA obtained have a much lower molecular weight. For instance, a DNA preparation can be obtained from Haemophilus influenzae [45, 46], whose behaviour on ultracentrifugation and whose length under the electron microscope correspond to a molecular weight of $400 \cdot 10^6$, while from Bacillus subtilis and Escherichia coli specimens with a molecular weight of $\sim250 \cdot 10^6$ have been obtained [13]. These specimens contain large quantities of protein and RNA; after more effective deproteinization and removal of RNA the molecular weight falls [47, 48] to $120 \cdot 10^6$. In the usual methods of DNA preparation [8, 12, 37, 49, 50], a double-stranded component with molecular weight

$(10-30)\cdot 10^6$ is isolated (this corresponds to approximately one-hundredth of an intact molecule of bacterial DNA). It is natural to suppose that these specimens of bacterial DNA are mixtures of many different components and the isolation of individual compounds from these mixtures is an extremely difficult procedure with present-day methods of fractionation.

The results of genetic investigations have shown that, besides the main mass of bacterial DNA, the cells also contain elements of genetic material which can function independently. These elements are known as "episomes" or "plasmids." Some of them have been isolated and identified as double-stranded DNA complexes of relatively low molecular weight [51-55]. It seems probable that the specimens of episomal DNA thus obtained are chemically individual compounds.

DNA in the cell nuclei of higher animals and plants is a component of a complex morphological structure, the chromosome [56, 57]. Besides DNA, the chromosome also contains basic proteins, or histones, and small quantities of RNA and nonhistone protein. The molecular organization of chromosomes is complex and has not yet been fully elucidated; however, it has been shown that a single cell contains an assortment of heterogeneous DNA molecules. The size of the intact DNA molecules within the chromosome has not yet been settled. Autoradiographic studies have revealed [58] strands of DNA reaching a length of 1.8 mm, or about twice the length of molecules of intact bacterial DNA [44]. In another investigation [59], strands over 2 cm in length have actually been obtained. Be that as it may, existing methods of isolation of DNA from animal tissues [37, 60] yield specimens of much lower molecular weight [$(50-60)\cdot 10^6$].

Although specimens of DNA so far available from bacteria, animals, and plants are undoubtedly mixtures of several components, their effective fractionation is impossible with existing methods. In a few cases, however, the results indicate that the total DNA specimen contains polynucleotides which differ in their properties from the main mass of DNA in the cell nucleus, and which are thus special varieties of DNA. For instance, by fractionation using countercurrent distribution, centrifugation in a sucrose gradient, and chromatography on calcium phosphate or MAK columns, fractions with properties corresponding to those of single-stranded DNA have been obtained from specimens of bacterial DNA [61-65]. It is postulated that this DNA is an intermediate product during reproduction of DNA in the cell.

By equilibrium centrifugation of DNA specimens from many animals and plants in a CsCl density gradient, besides the main DNA peak, a small "satellite" peak is also found [66, 67]. In some cases it has been shown that the appearance of a satellite peak is due to the presence of extranuclear DNA in the specimen (see below). In other cases, principally in DNA from mammalian tissues, the satellite DNA has been shown to come from the cell nucleus. Two of these polynucleotides have received the most study: satellite DNA from mouse liver [66-70, 498] and satellite DNA from the tissues of some species of crab [71-73]. In quantity they account for between 10 and 30% of the total DNA of the cell nucleus; the molecular weight of mouse satellite DNA is $40 \cdot 10^6$. These polynucleotides exist as double-stranded complexes; a characteristic property [74] is the ease with which they are renatured, i.e., the ease with which the double-stranded complex is restored after being broken.

Extranuclear DNA [75-78, 503]. Although the cytochemical evidence suggests that most DNA is concentrated in the cell nucleus, DNA has also been shown to exist in other subcellular particles [79]. This cytochemical conclusion is confirmed by the isolation of DNA from mitochondria and chloroplasts.

Isolation of DNA from mitochondria [80-82] is carried out by the same methods as are used to isolate DNA from the cell nucleus; in this case, however, the corresponding subcellular particles are first isolated, and before disintegration they are treated with deoxyribonuclease, so that any possible nuclear DNA present as an impurity can be removed. Investigation of the physical properties of mitochondrial DNAs isolated from different sources shows that in many cases this type of DNA differs from nuclear DNA in its buoyant density and melting temperature and, therefore, that it differs in its nucleotide composition. Mitochondrial DNA has a molecular weight [83-87] of about $10 \cdot 10^6$ and it exists as a double-stranded complex which readily undergoes renaturation.

To isolate DNA from chloroplasts [88-90], the same method is used as for isolation of mitochondrial DNA: treatment of the subcellular particles with deoxyribonuclease before disintegration. In this case, however, it is impossible to remove all the nuclear DNA, and the DNA of the chloroplasts is further purified by equilibrium centrifugation in a CsCl density gradient. DNA from chloroplasts of the green algae Euglena gracilis has been investigated most thoroughly [91]. It differs from nuclear DNA in its base composition and its exists as a double-stranded complex with molecular weight $10 \cdot 10^6$.

The facts described above show that several types of polydeoxyribonucleotides, differing in their composition and properties, exist in each living cell. The biological function of all these types of DNA has not yet been completely elucidated, although in the modern view it is analogous in every case: to provide for self-reproduction of the cell (or subcellular particle), for storage and transfer of genetic information, and finally, for synthesis of the RNA which subsequently participates in protein synthesis in the cell.

III. Methods used to isolate RNA and to determine its properties.

The principal types of RNA

Like DNA, RNA is an essential component of all living cells. In addition, it is found in all viruses of plants so far investigated, and also in several bacteriophages, and many viruses of insects and animals. Whereas the biological functions of different types of DNA are evidently similar, different types of RNA have different biological functions. The functions of virus RNA are evidently analogous to those of virus DNA, i.e., this type of RNA contains the genetic information necessary for construction of the virus particle, transmits it from generation to generation and, in addition, it serves for the synthesis of enzymes and proteins of the virus membrane necessary for construction of virus particles. As is well known, at least three types of RNA are distinguished in the living cell, and each has its own biological function. First, there is ribosomal RNA, a structural component of the subcellular particles where protein is synthesized. The second type is an RNA of relatively low molecular weight, capable of combining with amino acids, and the aminoacyl derivatives thus formed serve as original material for protein biosynthesis; this RNA fraction is called transfer RNA *. Finally, the third type of RNA which has been discovered is messenger RNA†, in which the nucleotide sequence determines the sequence of amino acids in protein.

Methods of isolation of RNA [6, 8] are, in general, similar to those used to isolate DNA. To obtain RNA from cells or subcellular particles, they are usually extracted with phenol [92-95]; the residual protein is often removed by treatment with detergent or a mixture of chloroform and octyl alcohol. Separation of DNA from RNA usually takes place at the stage of extraction, because the conditions can be so arranged that RNA migrates to the aqueous phase while DNA remains in the interphase. Fractional precipitation is frequently used for this purpose, and treatment of the specimen with deoxyribonuclease less frequently. The various types of cellular RNA can be separated by fractional precipitation, by centrifugation in a sucrose density gradient [96, 97], by chromatography [30, 98, 99], and by electrophoresis in gel [100-103].

Since RNA usually has a much smaller molecular weight than DNA, the problem of degradation through hydrodynamic shearing forces during isolation of the specimens is much less acute in this case. By contrast, the problem of enzymic degradation of RNA by the action of ribonucleases, both intracellular and exogenous (for example, derived from the experimenter's fingers [104]), is very serious. Ribonucleases are usually very resistant enzymes, not inactivated by heating or by treatment with phenol. Accordingly, during isolation of RNA it is recommended that inhibitors of ribonucleases, such as bentonite (aluminosilicate), polyvinyl sulphate, or iodoacetamide, be added constantly.

The basic characteristics used to describe an RNA, just as in the case of DNA, are its molecular weight and nucleotide composition. The molecular weight of RNA can be determined by the light scattering method [105], or chemically, by determination of terminal groups (see page 31); ultracentrifugation in a sucrose density gradient is most frequently used for this

---

*Sometimes referred to as "soluble RNA" or "adaptor RNA."

†The Russian name for this type is "informational RNA."

purpose. A number of empirical equations connecting the sedimentation con-
·stant* with the molecular weight under certain conditions have been suggested
[105-107]; the equation put forward by Spirin [107] is the one most frequently
used. The nucleotide composition of RNA is usually determined by quantitative analysis of the nucleotides formed after hydrolysis of the polymer (see
page 43).

In many cases it has been suggested that an RNA preparation can be
characterized, not by its molecular weight, but by its sedimentation constant
or electrophoretic mobility, which correlates closely with the sedimentation
constant [108].

Virus RNA [109-112]. Virus RNA is usually isolated from purified
virus particles by the phenol or detergent methods (typical methods are described in [113-116]). As a rule, the nucleic acids of viruses exist as single-
stranded molecules with molecular weights of $(1-3) \cdot 10^6$, although examples
with much lower molecular weights are known. For instance, the RNA from
one variety of phage Q_β has a molecular weight of $\sim 2 \cdot 10^5$, which corresponds
to a content of 550 nucleotide residues [117]. Favourite materials for study are
RNAs from tobacco mosaic virus and turnip yellow mosaic virus, from polio-
viruses of animals, and influenza viruses. Finally, many investigations have
been made of bacterial viruses, or bacteriophages, notably f2, MS2, R17,
M12, fr, and Q_β. The degree of nativeness of the isolated virus RNA can be
determined by means of a biological test based on its ability to infect the organisms which are the host for the virus concerned.

There is a small group of viruses whose RNA exists as a double-stranded complex; isolation of the RNA usually yields several components. This
group of viruses includes reovirus [118], wound tumour virus of plants [119],
and rice dwarf virus [120].

The double-stranded complex composed of the original (+)-strand of
virus RNA and its complementary (-)-strand of RNA synthesized in the host
organism during virus infection is an intermediate product for virus reproduction in the cell (the "replicative form"). Such a complex is resistant to
the action of small concentrations of ribonuclease, and it can be isolated and
separated from other forms of RNA by centrifugation in a density gradient or
by chromatography on MAK [121].

Ribosomal RNA [40, 122-124]. Most RNA of the living cell is concentrated in the cytoplasm, in ribonucleoprotein granules called ribosomes.
These granules may exist in the cytoplasm in a free state; in the cells of
animals and plants they are usually joined to membranes of the endoplasmic
reticulum, forming the "rough endoplasmic reticulum." Ribosomes can be
isolated preparatively by ultracentrifugation, and used as the raw material

---

*To abolish the effect of secondary structure, it is suggested that the sedimentation constant be
determined in dimethyl sulphoxide [496] or in the presence of formaldehyde [497].

for the preparation of ribosomal RNA. Preliminary isolation of the ribosomes, however, is not essential, because ribosomal RNA accounts for up
to 85% of the total cell RNA and it can easily be separated from the other
components [126-128].

Ribosomal RNA can be fractionated into three components by chromatography on MAK or by centrifugation in a sucrose gradient. Two of these
components, which were discovered earlier, were called "heavy" and "light"
components; the third, now known as "5S RNA," can easily be separated from
the main mass of ribosomal RNA [129-130].

The heavy component of ribosomal RNA in specimens obtained from
animal and plant tissues has a sedimentation constant of 28-32S; the sedimentation constant of RNA obtained from bacterial ribosomes is characteristically 23S. This corresponds to a molecular weight of about $1.6 \cdot 10^6$ for
RNA from animals and plants and $1.1 \cdot 10^6$ for RNA from bacteria.

The light component of ribosomal RNA has a sedimentation constant
of between 16S and 18S, corresponding to a molecular weight of about $5 \cdot 10^5$.
The 5S component has a molecular weight of approximately 40,000, corresponding to a polynucleotide chain of 120 nucleotide residues.

It was shown not long ago that the heavy component of ribosomal RNA
from mammalian cells is in fact a complex of two RNA molecules [485]. By
treatment with agents rupturing hydrogen bonds, besides 28S RNA, a new
component with molecular weight corresponding to a chain of 130 nucleotide
residues can also be isolated. This component of ribosomal RNA is known
as 7S RNA or 28SA RNA.

Transfer RNA (tRNA) [131-133]. This type of RNA is found in the cytoplasm of cells in a free state; quantitatively it may amount to 10-15% of the
total cell RNA. Transfer RNA can interact, in the presence of appropriate
enzymes, with aminoacyladenylates to form the corresponding aminoacyl-
tRNAs, which serve as the activated form of amino acids in protein synthesis.
Each type of tRNA can combine with only one amino acid; accordingly, the
number of types of individual tRNAs in the cell cannot be less than the number of amino acids. In fact, however, the number of types of individual
tRNAs is larger still, because many amino acids have been shown to combine
with two, and sometimes even with five specific tRNAs. The ability of a
tRNA to attach itself to an amino acid can easily be measured quantitatively;
enzymic aminoacylation is used in the purification of tRNA.

Transfer RNA can be separated from other types of RNA by fractional
precipitation, gel-filtration, and other chromatographic methods. Because
of the relatively low molecular weight of tRNA (about $3 \cdot 10^4$), effective fractionation of a mixture of tRNAs is possible [6, 8, 131, 134]. Countercurrent
distribution [135-138] and various chromatographic methods [139-145] are
used for this purpose. Other methods based on differences in the properties
of terminal groups of tRNA and aminoacyl-tRNA have also been developed;

the terminal nucleoside residue in tRNA can be oxidized by $NaIO_4$ [146-149], or the polypeptide chain can be built up from amino acid residues obtained from aminoacyl-tRNAs [150-151].   By a combination of different methods of purification, some types of tRNA have been obtained in an individual state*.

RNA of the cell nucleus [153-156, 499].   A considerable amount of RNA is present in the cell nucleus, where it is concentrated mainly in the nucleolus.   Special methods are needed to isolate RNA from the cell nucleus, because with ordinary extraction methods this RNA fraction is not solubilized. The most widely used method of obtaining RNA from cell nuclei is by extraction with hot phenol.   By varying the extraction temperature, different fractions of nuclear RNA can be obtained.   The pattern observed is complex, but it can be reliably concluded that besides ribosomal RNA and transfer RNA, there are at least another three types of RNA in the cell nucleus, differing in molecular weight and nucleotide composition from the types described above.

The first type is an RNA similar to ribosomal RNA in composition, but with a much higher molecular weight.   This type of RNA has been shown to be the precursor of ribosomal RNA [157]†.   The presence of RNAs with sedimentation constants of 45S and 32S is generally accepted; investigation of nuclear RNA by electrophoresis in polyacrylamide gel has demonstrated the existence of several additional components [158].

The second type of RNA differs from all other components of cellular RNA in its characteristic nucleotide composition, which is close to that of the cell DNA [159, 160].   Much of the RNA of this type has a sedimentation constant close to that of the light component of ribosomal RNA.   However, the existence of RNA fractions of similar composition, but much more highly polymerized (with a sedimentation constant of 40-60S or more), has been demonstrated [161-163].

Finally, an RNA of low molecular weight (corresponding to a content of about 40 nucleotide residues) has been obtained from chromosomes.   Its nucleotide composition is unusual [164-166].   Further details regarding different nuclear RNAs of low molecular weight are given in [486-491].

Cytoplasmic messenger RNA [154, 156, 167-169].   According to the generally accepted scheme of protein biosynthesis, the cytoplasm of living cells contains a special type of RNA, the nucleotide sequence of which determines the amino acid sequence of the synthesized protein.   The validity of this scheme has been confirmed by experiments in which polypeptides of a

---

*A general method of tRNA purification based on acylation of the amino group in the aminoacyl-tRNA by the derivative of an aromatic acid with strongly hydrophobic properties, and subsequent chromatographic separation of these acylaminoacyl-tRNAs from other tRNAs, has recently been suggested [152].

†An RNA of low molecular weight, identified as the precursor of tRNA, has recently been discovered in the cytoplasm of HeLa cells [493].

definite amino acid sequence were obtained in a cell-free system of protein synthesis in the presence of synthetic polynucleotides (see page 64) and specific virus proteins were obtained in the presence of virus RNAs. As yet, however, it has not been possible to obtain an RNA, in an individual state, which could induce the synthesis of a particular protein by interaction with ribosomes and aminoacyl-tRNA.

The discovery of RNA fractions distinguished by a high rate of biosynthesis and by a nucleotide composition similar to that of DNA of the corresponding cells, has often been described. These fractions can induce the incorporation of radioactive amino acids into protein in the presence of a cell-free system of protein biosynthesis. Taking all these facts into consideration, it seems likely that these are forms of messenger RNA.

Isolation of messenger RNAs is difficult because of their heterogeneity (because of the many different types of proteins synthesized in the cell) and the relatively high rate of their breakdown compared with other types of RNA. During isolation of RNA, marked degradation of the messenger RNAs frequently takes place. This is reflected in the contradictory results given for the molecular weight of this type of RNA. Sedimentation constants of 6-14S, 18S, and as much as 35-40S have been given by different authors for messenger RNAs from similar natural sources.

The difficulties encountered in the isolation of messenger RNAs can largely be overcome by the use of animal cells specializing in the synthesis of one protein. For example, an RNA fraction with sedimentation constant 9S, corresponding to messenger RNA for globin synthesis in its molecular weight [170], has been isolated from the reticulocytes of rabbits, but no direct experimental confirmation of its biological role has yet been obtained.

Other types of RNA. RNA has been detected in all subcellular particles investigated.

Intracellular membranes contain an RNA which differs in its composition and sedimentation constant from ribosomal RNA [171-174].

As was mentioned above, mitochondria and chloroplasts contain their own DNA and they are largely independent of the cell nucleus as regards the transfer of genetic information. Protein synthesis takes place inside these subcellular particles; it is natural to suppose that they must contain ribosomal, transfer, and messenger RNA. This hypothesis has been confirmed, and it has further been demonstrated that ribosomal RNA from the mitochondria of Neurospora crassa [175-177], and also tRNA from the same source [178, 179], and tRNA from rat liver [180] differ appreciably from the corresponding cytoplasmic RNAs. A similar situation is found with ribosomal RNA from the chloroplasts of Euglena gracilis [181], which is indistinguishable in its sedimentation constant from bacterial ribosomal RNAs, but which differs appreciably from the ribosomal RNA found in the cytoplasm of these algae.

According to a recently published report, RNA has been found in the cell wall of higher plants [182].

IV. Structure of the polynucleotide chain [183]

The material in the two preceding sections shows that the isolation of individual nucleic acids in an intact state is a complex problem which has so far been solved satisfactorily only in the case of RNAs of low molecular weight (such as tRNA, 5S RNA and, perhaps, other components of ribosomal RNA), and also for RNA and DNA from viruses. Naturally the question of determination of the complete chemical structure can be considered only in the case of these compounds.

However, a correct understanding of the basic principle governing the structure of nucleic acids and the method by which the monomer units are linked together (Formula V) was obtained in the early 1950s, when only highly degraded specimens of nucleic acids were available for investigation.

The presence of a phosphodiester bond in nucleic acids as the main type of link between the monomer units was demonstrated by the results of potentiometric titration [184]. Structure V (R = H) for DNA [185] is the only possible type of structure which agrees with the isolation of deoxynucleoside-5'-phosphates VII by enzymic hydrolysis of DNA (with intestinal phosphodiesterase [186]) and of 3',5'-diphosphates of pyrimidine deoxynucleosides VIII by acid hydrolysis of DNA [187].

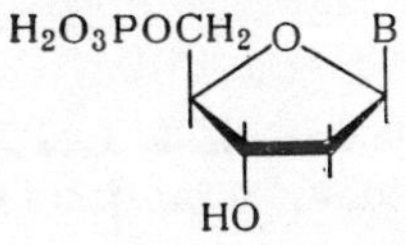

$$H_2O_3POCH_2 \quad O \quad B \qquad\qquad H_2O_3POCH_2 \quad O \quad B \qquad\qquad H_2O_3POCH_2 \quad O \quad B$$

HO H_2O_3PO HO   OH

VII 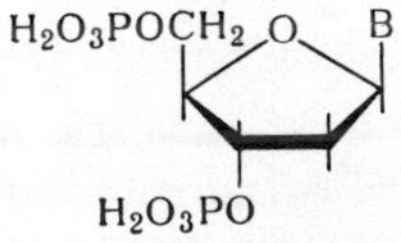   VIII 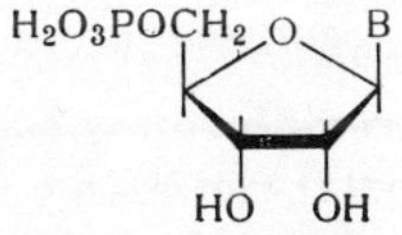   IX

B denotes base residue

In the case of RNA, participation of the 5'-hydroxyl group of the ribose residue in the formation of the phosphodiester bond was demonstrated by the formation of ribonucleoside-5'-phosphates (IX) with an almost quantitative yield by the hydrolysis of RNA with phosphodiesterase from snake venom [188]. During alkaline hydrolysis of RNA a mixture of nucleoside-2'-phosphates (X) and nucleoside-3'-phosphates (XI)* is formed. This is in agreement both with structure V (R = OH) for RNA, and with an isomeric structure for it containing 2'-5'-phosphodiester bonds, and with any structure in which 2'-5'- and 3'-5'-phosphodiester bonds alternate in random order. Investigation of the mechanism of alkaline hydrolysis of RNA has shown [185] that this result is attributable to the participation of the hydroxyl group at the C2' position of the ribose residue in the process of rupture of the phosphodiester bond and in the intermediate formation of nucleoside-2',3'-cyclic phosphates (XII), which are

---

* This mixture is frequently designated "nucleoside-2'(3')-phosphate".

subsequently split with the formation of a mixture of X and XI (further details are given in Chapter 10).

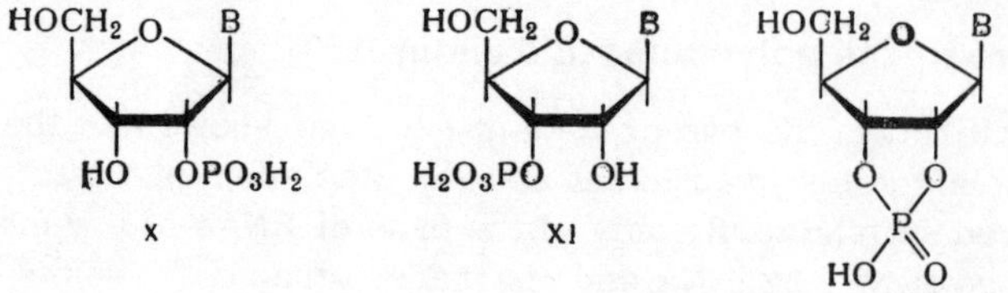

B denotes base residue

The choice between the two types $(2'-5'-$ or $3'-5'-)$ of phosphodiester bonds and determination of the true nature of the internucleotide bond in RNA were based on a study of the results of enzymic hydrolysis. The action of pancreatic ribonuclease on RNA leads to the formation of pyrimidine nucleoside-3'-phosphates and of oligonucleotides terminating in a nucleoside-3'-phosphate residue. Investigation of the action of this enzyme on synthetic model compounds demonstrated that alkyl esters of pyrimidine nucleoside-3'-phosphates are easily hydrolysed to nucleoside-3'-phosphates XI, while the corresponding derivatives of the nucleoside-2'-phosphates are not attacked by the enzyme. Consequently, a 3'-5'-phosphodiester bond also exists in RNA, the natural substrate of pancreatic ribonuclease [189]. This conclusion was later confirmed by the isolation of nucleoside-3'-phosphates during hydrolysis of RNA by the action of splenic phosphodiesterase. In this reaction, moreover, no intermediate formation of cyclic phosphate XII takes place, and the possibility of migration of the phosphoryl group is ruled out [190, 191].

Structure V for DNA and RNA was first suggested by Todd and Brown [185], and it has been confirmed by more recent findings, notably by the identity of synthetic oligonucleotides with 3'-5'-phosphodiester bonds and the hydrolysis products of nucleic acids.

Since the monomer units in the polynucleotide chain are all linked together in the same way, abbreviated formulae are often used to designate the structure of the polynucleotide chain, as follows:

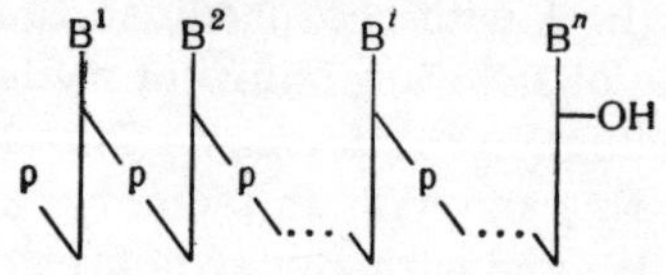

B^1....B^i....B^n– base residues

In these formulae the sugar residue is represented as a vertical straight line, and the letter p, as usual, denotes a phosphomonoester or phosphodiester group. Abbreviated formulae of this type are sometimes used also for the hydrolysis products of nucleic acids; abbreviated Formulae VIIa-XIIa, for example, corrspond to compounds with the VII-XII type of structure.

VIIa　　VIIIa　　IXa　　Xa　　XIa　　XIIa

B denotes base residue

Abbreviations for polynucleotides [192] recommended by the Commission on Nomenclature of the International Union of Pure and Applied Chemistry and the International Biochemical Union are based on the same principles.

The evidence for this basic principle of nucleic acid structure was obtained by the use of highly degraded forms of RNA and DNA, and strictly speaking there are no grounds for the assertion that native nucleic acids consist of a single polynucleotide chain, rather than of a series of long polynucleotide segments linked together by bonds of different types.

This second hypothesis is improbable, however, on biogenetic grounds. Enzymes catalysing the formation of both polyribo- and polydeoxyribonucleotide chains, in which the nucleotide residues are linked by 3'-5'-phosphodiester bonds, from nucleoside-5'-triphosphates have now been purifed. Biologically active RNA of phage Q_β [193] and DNA of bacteriophage ϕX174 [194] have now been obtained in vitro with the aid of these enzymes. This result shows that even polynucleotides with a molecular weight of several millions do not contain structural bonds of any other type than phosphodiester bonds. Autoradiographic studies of replication of DNA from Escherichia coli show that this huge molecule is built in accordance with a single plan and contains no abnormal bonds [44].

Native DNA from the cells of higher plants and animals is a more complex problem. The suggestion has been made that in this case the native DNA molecule is built of subunits, consisting of polynucleotide chains connected together by segments of polypeptides or amino acids. Among the possible types of linkages, an ester bond between the carboxyl group of an amino acid and the 3'-hydroxyl group of a nucleoside residue, or alternatively a phosphoester bond between a nucleoside-5'-phosphate residue and a hydroxyl group of a hydroxyamino acid residue [195], or a phosphoramido bond with the amino group of an amino acid residue [196], has been suggested. However, there is no direct experimental evidence in support of these hypotheses.

V. Terminal groups of the polynucleotide chain

The terminal nucleoside residues in the polynucleotide chain* can have free or phosphorylated 3'- or 5'-hydroxyl groups; there are accordingly four

---

*The terminal residue of a nucleoside (or nucleotide), linked to the polymer chain through a hydroxyl group only at the C5' position is customarily called the 3'-terminal residue of the polynucleotide chain. In the same way, the residue of a nucleoside (or nucleotide) linked to the polymer chain only through the hydroxyl group in the C3' position is called the 5'-terminal residue. In the usual method of writing abbreviated formulae of polynucleotides (for single-chain polymers), the 3'-terminal residue is on the right of the formula and the 5'-terminal residue on its left.

possible types of structure of nucleic acids, as shown in the Formulae XIII-XVI:

$$pNpNp\cdots NpN \qquad NpNp\cdots NpNp$$
$$\text{XIII} \qquad\qquad \text{XIV}$$
$$pNpNp\cdots NpNp \qquad NpNp\cdots NpN$$
$$\text{XV} \qquad\qquad \text{XVI}$$

Determination of the nature of the terminal groups of the polynucleotide and identification of the terminal nucleoside residues are two of the first problems to be solved before the structure of nucleic acids and oligonucleotides can be established. The way to their solution lies in the identification of characteristic fragments in the total hydrolysis products of the polynucleotide or in specific labelling of the terminal nucleoside residues.

By splitting the polynucleotide chain through rupture of $P-O$ bonds (the O and C5' position) (this cleavage can be achieved by enzymic hydrolysis with the phosphodiesterase of Lactobacillus acidophilus or of calf spleen, and in the case of RNA through the action of alkali also) the following characteristic fragments are formed: a nucleoside, if the 3'-terminal residue of the polynucleotide chain is free, and a nucleoside-3',5'-diphosphate if the 5'-terminal residue is phosphorylated (in the case of alkaline hydrolysis, the product is a nucleoside-2'(3'),5'-diphosphate).

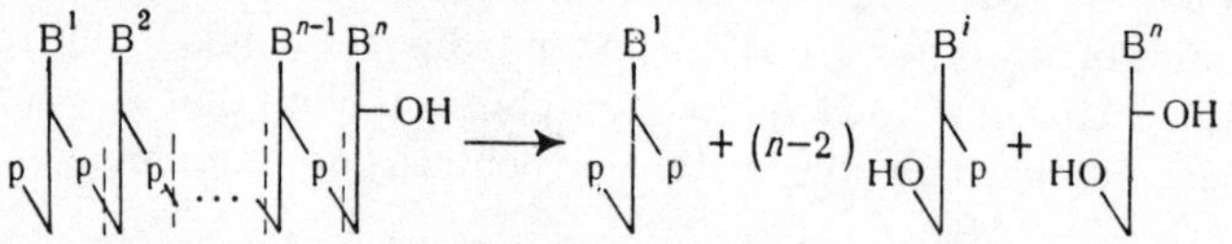

$B^1\ldots B^i\ldots B^n$ — base residues

On the other hand, by hydrolysis with rupture of the $P-O$ bonds (the O in the C3' position) (by means of snake venom phosphodiesterase), similar fragments are formed from the 3'-terminal phosphorylated residue of the nucleotide and the 5'-terminal residue of the nucleoside with a free hydroxyl group:

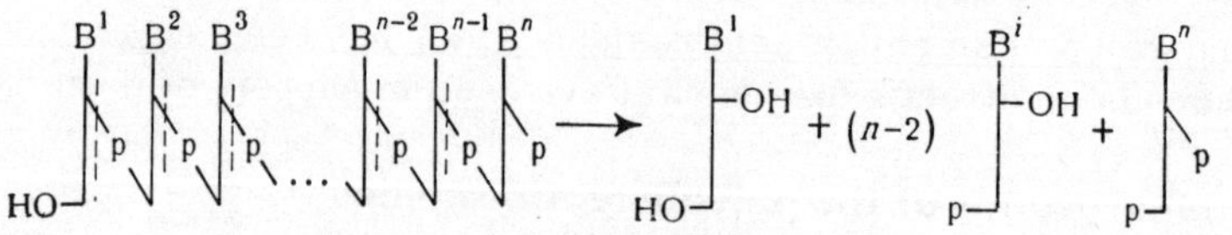

$B^1\ldots B^n$ — base residues

It would seem, therefore, that by using one of the cleavages examined above, a structure of type XIII or XIV can be reliably identified and both terminal nucleoside residues of the polymer chain can be determined: in the case of polymers of the XV or XVI type, on the other hand, this problem can be solved by a combination of two types of cleavage.

In practice, however, the matter is not so straightforward. Hydrolysis of polynucleotides with a terminal phosphate group takes place smoothly if only chemical methods of degradation are used, but in the case of hydrolysis by enzyme action, an essential condition for the rapid progress of the reaction is absence of a phosphate group at the 3'-end of the polynucleotide chain in the case of snake venon phosphodiesterase and at the 5'-end in the case of spleen phosphodiesterase (see page 51). Before enzymic hydrolysis, the terminal phosphate groups must therefore first be removed by the action of phosphomonoesterase, which causes disappearance of the specific fragment formed from the phosphorylated end of the chain.

Substantial technical difficulties arise during the determination of specific hydrolysis products formed from terminal groups of the polynucleotide chain in the presence of a large excess of products arising from the central components of the chain. This is particularly important in the case of nucleic acids with high molecular weight. A substantial increase in the sensitivity of determination of terminal groups can be obtained by radioactive labelling. Chemical or enzymic methods can be used for this purpose.

The terminal phosphate group of the polynucleotide chain can be converted into a phosphoanilide by interaction with ^{14}C-aniline and dicyclohexylcarbodiimide [197]; a more general procedure is conversion of the terminal nucleotide residue to a diester of pyrophosphoric acid [198] by reaction with ^{14}C-methyl phosphomorpholidate.

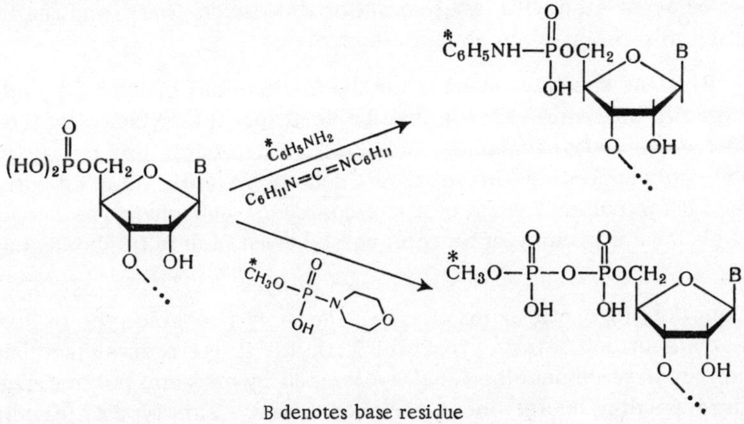

B denotes base residue

Derivatives of this type can be obtained from the 5'-end of the nucleotide residue in RNA and from the 3'- and 5'-terminal residues of the nucleotide in DNA (see Chapter 10); after degradation of the polymer by the action of alkali or enzymes, the corresponding derivatives can easily be identified.

To introduce a radioactive label into the 3'-terminal residue of the nucleoside in RNA, advantage is taken of the ability of a vicinal diol group to be oxidized by periodate and the interaction between the resulting dialdehyde and reagents such as ^{35}S-thiosemicarbazide [199], ^{14}C-semicarbazide [200], ^{3}H-isonicotinic acid hydrazide [201], or ^{3}H-sodium borohydride [202].

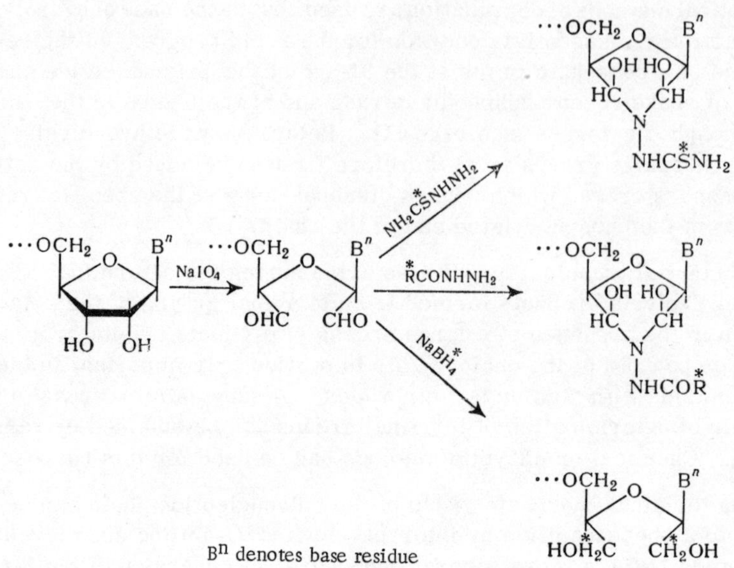

B^n denotes base residue

Free hydroxyl groups at the 3'- or 5'-end of a polydeoxyribonucleotide chain can be acetylated with labelled acetic anhydride [203], and the resulting acetates are resistant to enzymic hydrolysis.

Finally, for specific labelling of the 5'-terminal hydroxyl group in DNA, a specific enzymic reaction can be used: phosphorylation by the action of ^{32}P-adenosine-5'-triphosphate [204-206]. This reaction is catalysed by the enzyme polynucleotide kinase from Escherichia coli, infected with phages T2 or T4. If the polymer ends in a nucleoside-5'-phosphate residue, the terminal phosphate group can be removed by treatment with phosphomonoesterase.

From the biogenetic point of view, the most probable type of structure of the polynucleotide chain is structure XIII, for it has been shown that both ribo- and polydeoxyribonucleotides are formed by enzymic polymerization of the corresponding nucleoside-5'-triphosphates. This type of structure has in fact been found in many cases: in transfer RNAs, 5S RNA, in other components of ribosomal RNA, and in some virus RNAs. The RNAs of some viruses have a nucleoside-5'-triphosphate [207-210] or a nucleoside-5'-diphosphate [504] residue at the 5'-end of the chain.

Many DNAs of relatively low molecular weight are characterized by a circular structure, represented schematically in Formula XVII:

$$\boxed{\text{—NpNp}\cdots\cdots\text{NpN—}}$$

XVII

Circular DNA * structures of this type linked by covalent bonds, have been demonstrated for DNA of phage ϕX174 and the corresponding replicative form of DNA, the DNAs of polyoma, papilloma, herpes simplex and SV40 viruses, the replicative forms of DNA of several other viruses, and mitochondrial DNA [16, 26, 78]. Electron-microscopic studies of mitochondrial DNA have shown that circular molecules can join together like links in a chain to form dimers, trimers, and tetramers, analogous to catenanes [211-213].

The existence of DNA molecules in a circular form was first demonstrated by the electron microscope. The covalent character of the bond in these compounds is proved by the impossibility of their hydrolysis by exonucleases (see page 51) even in the presence of phosphomonoesterase. The appearance of single breaks in the polynucleotide chain of circular DNAs as a result of the action of endodeoxyribonucleases leads to characteristic changes in the conformation of the molecules (see Chapter 4), which can be detected by changes in the sedimentation coefficient and viscosity and also by observation under the electron microscope, although the measured length of the molecules and, consequently, the molecular weight are unchanged under these circumstances.

The circular structure of the DNA of phage ϕX174 was confirmed by its complete enzymic synthesis. A biologically active DNA could be obtained only after discovery of the enzyme catalysing the cyclization of linear polynucleotide chains.

Circular DNA structures are possibly even more widespread. Autoradiographic and genetic data show, for example, that intact bacterial DNA has a circular structure [44]. The presence of circular DNAs has been demonstrated by electron microscopy in specimens of DNA from higher animals and plants [214], but the covalent character of the cyclic linkage in them has not yet been proved.

VI. Structure of nucleoside components of nucleic acids

Nucleic acids differ from other biopolymers in the relatively small variety of monomer units composing them. The monomer units of nucleic acids are customarily divided into major and minor components. The term major components of nucleic acids covers monomer units of universal distribution found in large quantities (not less than 5%) in the polymers. The

---

*A clear distinction is not always drawn in the biological literature between circular DNAs linked by covalent bonds (XVII) and linear DNA molecules which may assume a circular conformation through noncovalent interaction between the ends of the polynucleotide chain (see, for example, LIII on page 47).

content of minor components is considerably smaller (as a rule not more than 2%); they are found in by no means all nucleic acids. Usually the minor components can be regarded as derivatives of the major components; formed from them by simple chemical reactions (such as alkylation, hydrogenation, and so on).

The classification of the components of nucleic acids in accordance with their distribution also rests on a biogenetic basis. Whereas the major components of nucleic acids are synthesized as mononucleotides, which then undergo conversion into nucleoside-5'-triphosphates and polymerization with the formation of nucleic acids, the minor components are usually formed from the major components already present in the polymer. The nucleoside residue at a certain point of the polynucleotide chain in such cases undergoes specific methylation, hydrogenation, and so on, with the resulting formation of a residue of a minor component.

1. Major components of RNA

As was mentioned above (see page 15), residues of four nucleosides are found in RNA: adenosine (Ia), guanosine (IIa), cytidine (IIIa), and uridine (IVa). These compounds were first isolated from yeast RNA at the beginning of this century; it was later shown that they are present in all types of RNA so far investigated.

It has been shown by the classical methods of organic chemistry that adenosine is 9-(β-D-ribofuranosyl)-6-aminopurine, and that guanosine is 9-(β-D-ribofuranosyl)-2-aminodihydropurinone-6; the pyrimidine nucleosides, cytidine and uridine, are 1-(β-D-ribofuranosyl)-4-aminodihydropyrimidinone-2 and 1-(β-D-ribofuranosyl)-tetrahydropyrimidinedione-2,4 * respectively. The structure of nucleosides has been confirmed by chemical synthesis and by the results of x-ray structural analysis†.

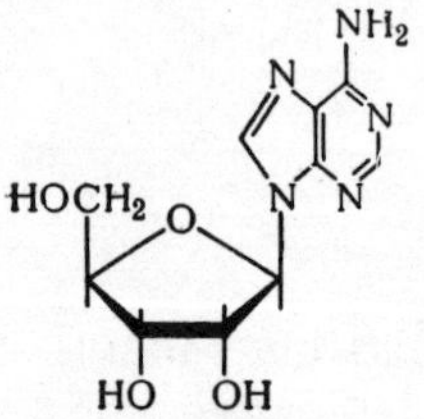

Ia 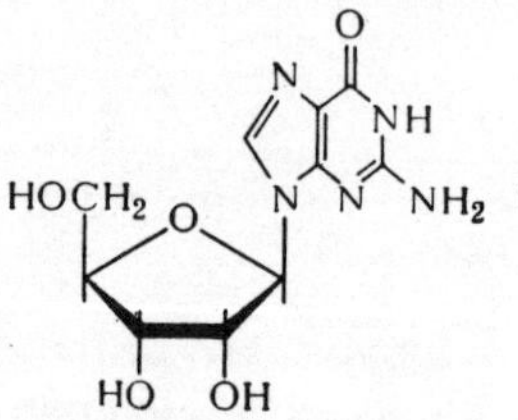   IIa 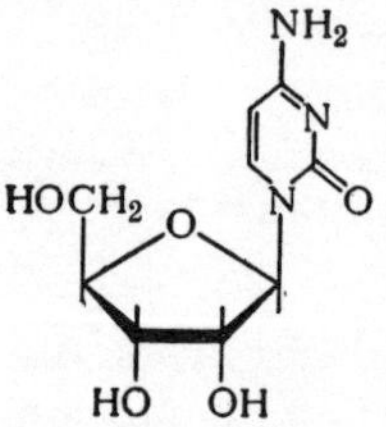   IIIa 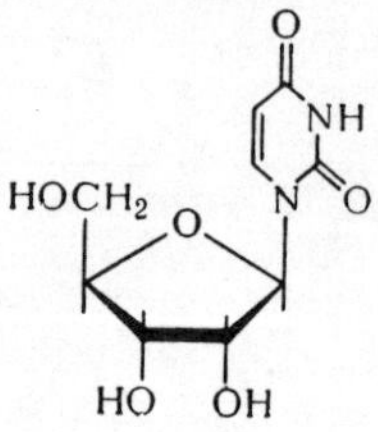   IVa

---

*Besides the system of numeration of atoms of the pyrimidine ring in nucleosides adopted in this book (taken from "Chemical Abstracts"), another system is found in the literature, in which uridine and cytidine are designated as 3-(β-D-ribofuranosyl)-tetrahydropyrimidinedione-2,6 and 3-(β-D-ribofuranosyl)-6-aminodihydropyrimidinone-2 respectively.

†Details of how the structures of the ordinary nucleosides and nucleotides were established are summarized in numerous monographs, e. g. [215-217].

2. Minor components of RNA

Progress in chromatographic analysis of the nucleotide composition of RNA has shown that many RNA specimens contain a wide assortment of different nucleosides: minor components derived from the major components. Over 20 compounds of this type are now known [218]. Transfer RNA is particularly rich in minor components, but they are also found in the heavy and light components of ribosomal RNA and also in chromosomal RNA.

The commonest method used to isolate minor components of RNA is that suggested by Hall [219]. This consists of the enzymic hydrolysis of RNA down to nucleosides by the action of a mixture of snake venom phosphodiesterase and alkaline phosphomonoesterase from Escherichia coli. The mixture of nucleosides is further fractionated by distributive chromatography on Celite. Mild conditions of hydrolysis of the polymer are combined in this case with high efficiency of separation of the monomers, and by using it many hitherto unknown minor components of RNA have been found.

Ten minor components of RNA which can be regarded as analogues or derivatives of uridine are now known. These include 5-methyluridine (ribothymidine, XVIII) [220], the ribonucleoside analogue of thymidine, 3-methyluridine (XIX) [221], another product of methylation of uridine, 4-thiouridine (XX) [222], the thio-analogue of uridine, 5,6-dihydrouridine (XXI) [223], and 5-hydroxyuridine (XXII) [224].

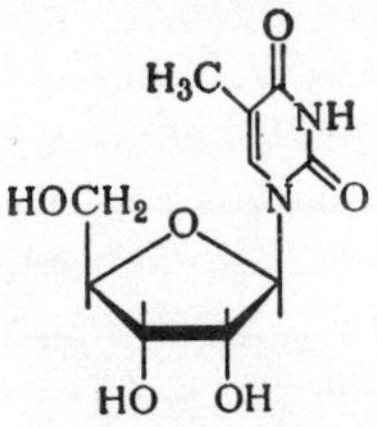

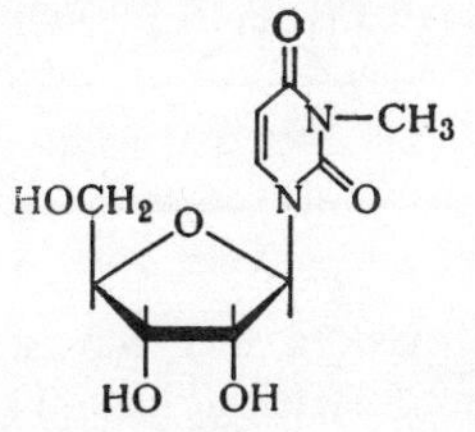

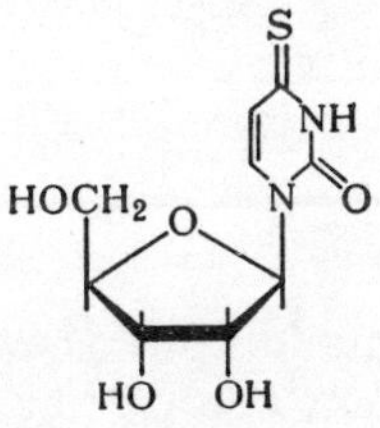

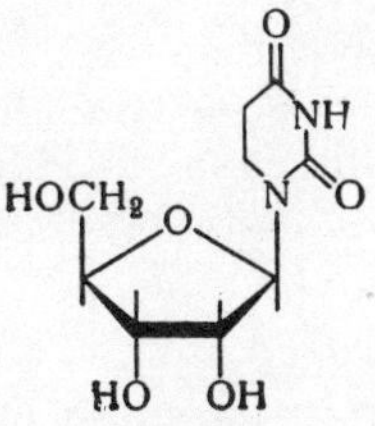

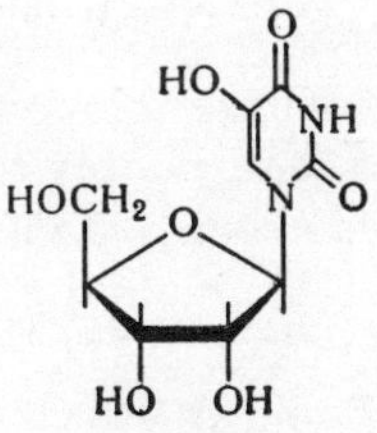

5-Methyluridine and 5-hydroxyuridine were first discovered during analysis of total RNA preparations from rat liver and yeast, 3-methyluridine and 5,6-dihydrouridine were isolated from yeast tRNA. Chromosomal RNA from pea shoots contains large quantities (up to 25%) of 5,6-dihydrouridine [164, 166]. Chromosomal RNA from ascites tumours of rats contains 5-methyl-5,6-dihydrouridine [492]. 4-Thiouridine is found in tRNA from

E. coli. Reports were recently published of the discovery of the methyl ester
of 5-carboxymethyluridine (XXIII; R = CH_3OCOCH_2) [225] in wheat germ tRNA,
and of 5-hydroxymethyluridine (XXIII; R = $HOCH_2$) [226] in tRNA from L tissue
culture cells, and of the isolation of the methyl ester of 5-carboxymethyl-2-
thiouridine (XXIV; R = CH_3OCOCH_2) [227] from yeast tRNA. The compound
5-methylaminomethyl-2-thiouridine (XXIV; R = CH_3NHCH_2) has been isolated
from tRNA from E. coli [228, 229].

XXIII   XXIV

Besides these components with an N-glycoside bond, tRNA also includes
quite considerable amounts of an unusual nucleoside with a C-glycoside bond.
This is the compound known as pseudouridine, or 5-(β-D-ribofuranosyl)-uracil
(XXV) [230]; it is also found in ribosomal RNA.

XXV

This formula shows the structure of the natural isomer of pseudouridine
("pseudouridine C"); by treatment with acids and alkalies a mixture of isomers
is formed; these have been identified as 5-(α-D-ribofuranosyl)-, 5-(β-D-ribo-
pyranosyl)-, and 5-(α-D-ribopyranosyl)-uracils (see Chapter 11).

The structures of the minor components of RNA which are cytidine derivatives exhibit much less variety than in the case of uridine derivatives. Most
characteristic of this group are the various methylated cytidines isolated from
yeast tRNA, in which the methyl group may be attached to a carbon atom, as
in 5-methylcytidine (XXVI) [231], to a nitrogen atom of the heterocyclic ring,
as in 3-methylcytidine (XXVII) [221], or to the nitrogen atom of the exocyclic amino group, as in 4-exo-N- methylcytidine* (XXVIII) [232]. The compound 2-thiocytidine is found in tRNA isolated from E. coli [229]. The isolation of 5-hydroxymethylcytidine (XXIX) [231], and also of a 3-alkylcytidine
in which the alkyl group differs from methyl [233], from tRNA has been
described. During determination of the structure of serine tRNAs isolated
from yeast, 4-exo-N-acetylcytidine (XXX) was found as a minor component [234].

---

*Although the nomenclature used in the book is not that usually adopted in the West to identify the
position of exocyclic substituents, because of its unambiguity and convenience it was decided to re-
tain it in the translated version.

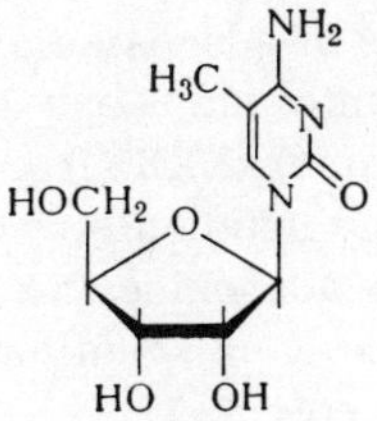

XXVI 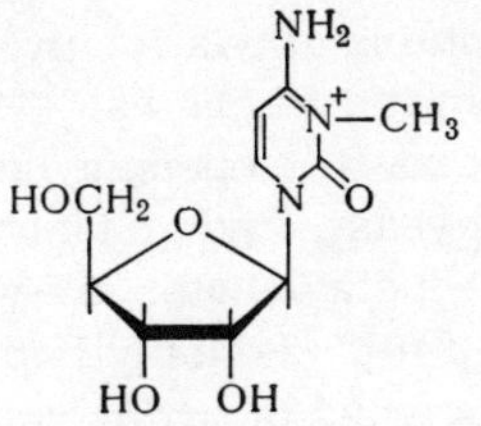   XXVII 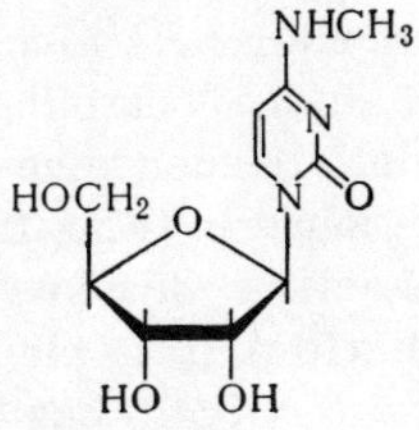   XXVIII

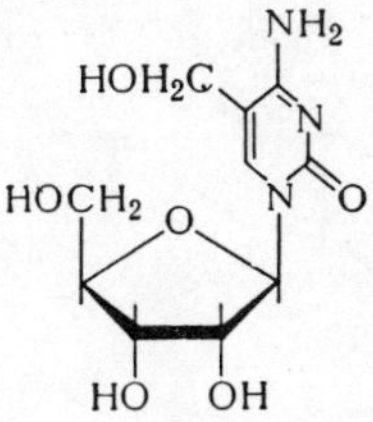

XXIX 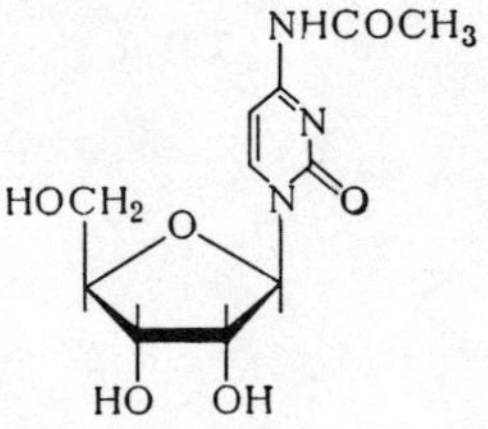   XXX

After alkaline hydrolysis of total RNA from yeast, a small quantity of nucleotides which are derivatives of 1-(α-D-ribofuranosyl)-cytosine (α-cytidine) can be isolated [235]. However, it is not yet clear whether this nucleoside is a natural component of RNA.

Minor components of RNA which are adenosine derivatives are mainly methylated derivatives: 1-methyladenosine (XXXI) [236], 2-methyladenosine (XXXII) [220], 6-exo-N-methyl- and 6-exo-N,N-dimethyladenosines (XXXIII; R = H or CH₃) [220].

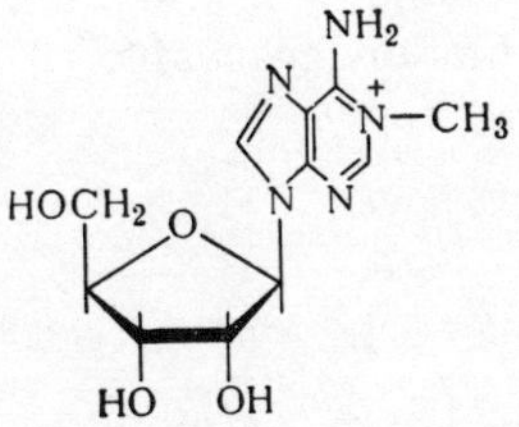

XXXI 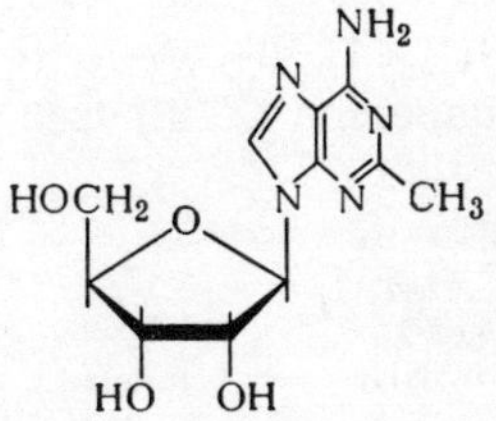   XXXII 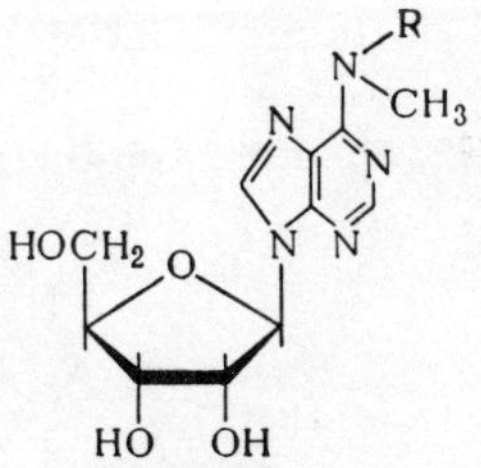   XXXIII

Transfer RNA also includes a residue of a deamination product of adenosine: inosine (XXXIV) and its 1-methyl derivative (XXXV) [237].

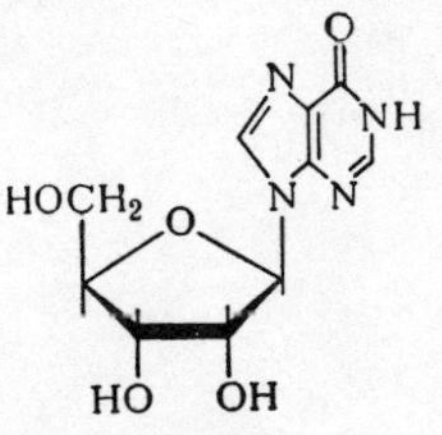

XXXIV 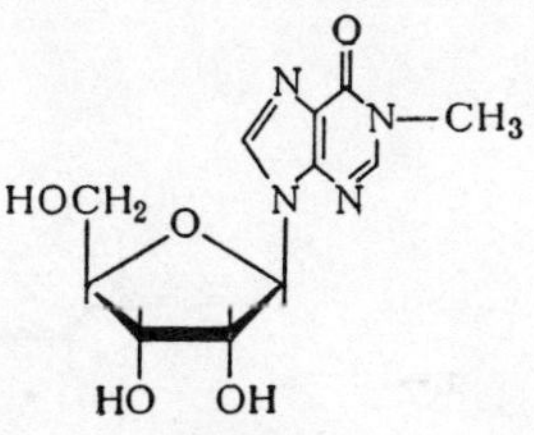   XXXV

Transfer RNA also contains alkylation products of adenosine at the 6-exo-N atom, containing isoprenoid chains. The compound 6-exo-N-(γ,γ-dimethylallyl)-adenosine (6-exo-N-isopentenyladenosine) (XXXVI) [238, 239] has been found in tRNA from yeast, higher plants, and mammals, while 6-exo-N-(cis-4"-hydroxy-3"-methyl-buten-2"-yl-1")-adenosine (XXXVII) occurs in tRNA from plants [240]. 2-Methylthio-6-exo-N-(γ,γ-dimethylallyl)-adenosine (XXXVIII) has been found in tRNA from E. coli [241].

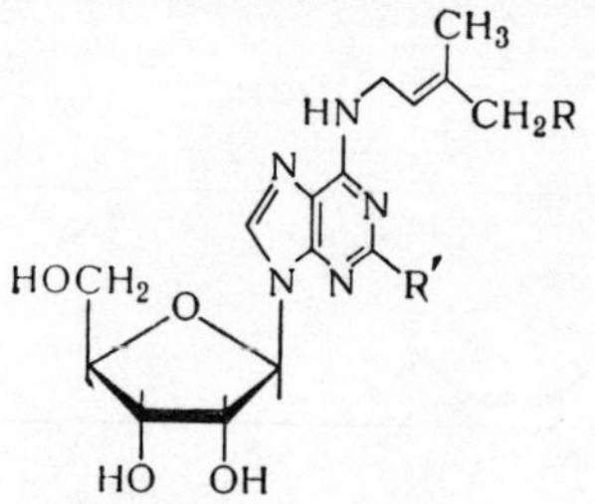

XXXVI (R=R'=H)

XXXVII (R=OH, R'= H)

XXXVIII (R=H, R'=SCH$_3$)

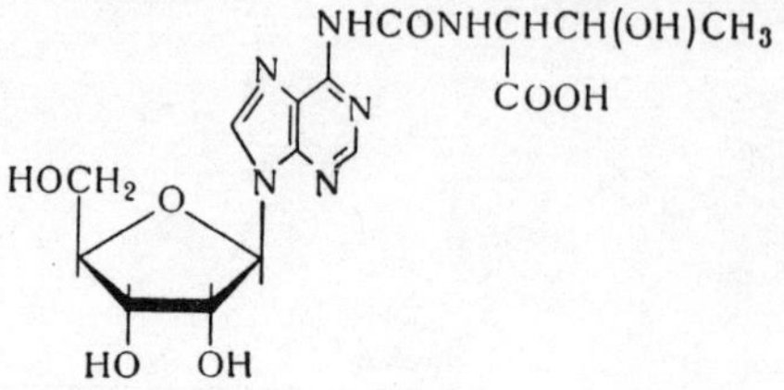

The isolation of various 6-exo-N-aminoacyladenosines has also been reported [242, 243]. Recently, however, the structure of these minor components has been re-examined. According to new information [500], the adenosine derivative isolated previously is N-[9-(β-D-ribofuranosyl)-purinyl-6-carbamoyl]-threonine (XXXIX).

The minor components of RNA which are guanosine derivatives include several of its methylation products: 1-methylguanosine (XL) [244], 2-exo-N-methylguanosine (XLI; R = H) [244], 2-exo-N,N-dimethylguanosine (XLI; R = CH$_3$) [249], and 7-methylguanosine (XLII) [245].

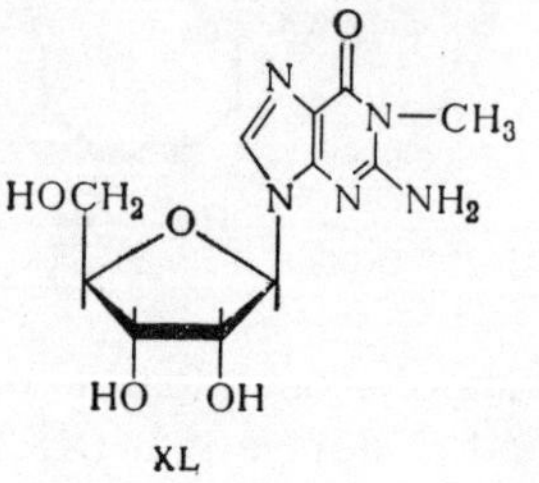

XL

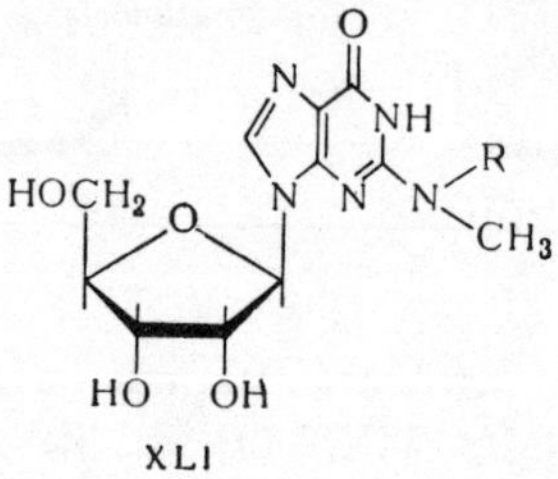

XLI

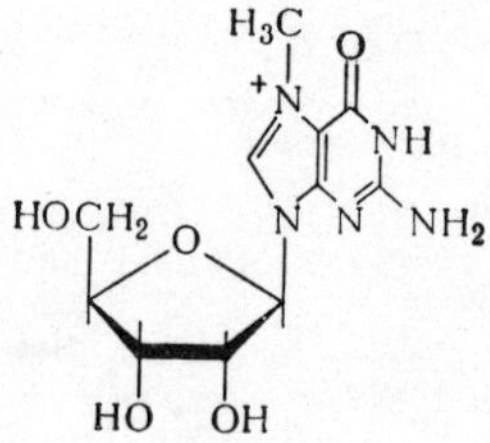

XLII

Many minor components of RNA, as this list shows, are products of biochemical methylation of major components at different atoms of the hetero-cyclic ring and its substituents. Methylation can also take place at the hydro-xyl group of the ribose residue, as is shown by the isolation of 2'-O-methyl-ribonucleosides (XLIII) from the products of enzymic hydrolysis of several RNAs, and also from nucleotides resistant to the action of alkali. Derivatives of adenosine [246, 247], guanosine [247], cytidine [247], uridine [247], pseudo-uridine [247], and 4-exo-N-methylcytidine [248], methylated in the ribose residue, have been isolated.

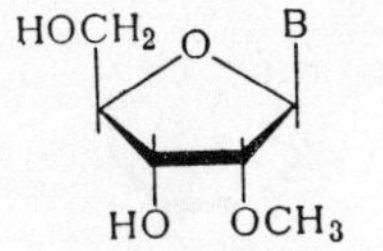

XLIII

B denotes base residue

3. Major components of DNA

It was stated above that three of the four major components of DNA, namely 2'-deoxyadenosine (Ib), 2'-deoxyguanosine (IIb), and 2'-deoxycytidine (IIIb), are analogous in structure to the corresponding ribonucleotides, while the fourth major component, thymidine (IVb), differs in containing an addi-tional methyl group at the C5 position of the pyrimidine ring.

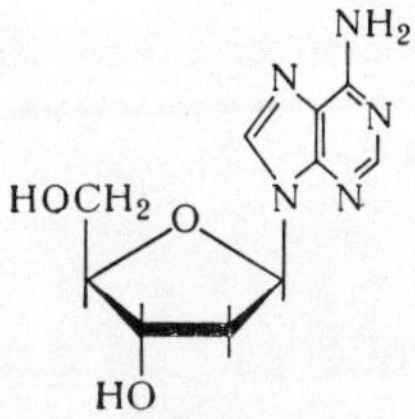

Ib

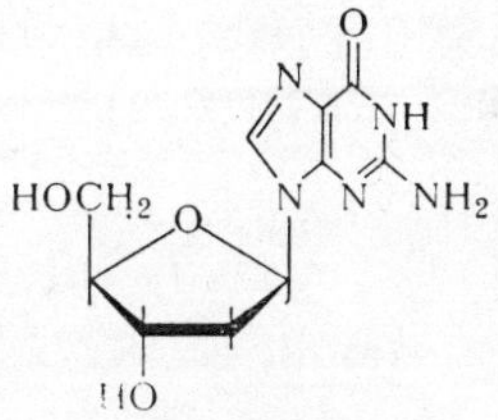

IIb

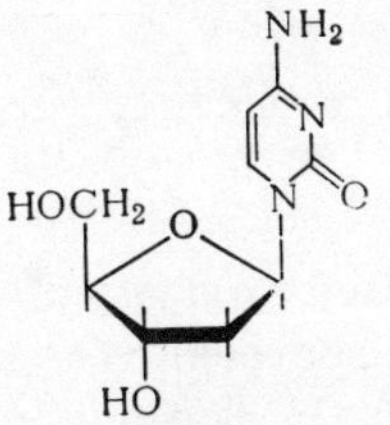

IIIb

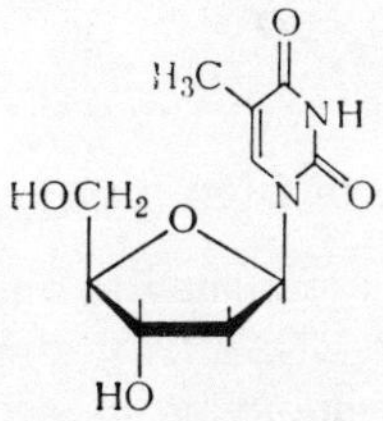

IVb

These nucleosides, first identified in hydrolysis products of calf thy-mus DNA, have subsequently been isolated from DNAs from most other biological sources. The exceptions are the DNAs of certain phages, which contain un-usual pyrimidine components.

In the DNAs of the T-even phages (T2, T4, and T6), for example, 2'-deoxycytidine is completely replaced by 5-hydroxymethyl-2'-deoxycytidine (XLIVa) [249], which is also found as a component of DNA in the form of its glycosides at the hydroxymethyl group of the heterocyclic ring [250, 251]. The α-D-glucopyranosyl- (XLIVb), β-D-glucopyranosyl- (XLIVc), and α-(6-β-D-glucopyranosyl)-D-glucopyranosyl-5-hydroxymethyl-2'-deoxycytidine (XLIVd) (α-gentiobiosyl-5-hydroxymethyl-2'-deoxycytidine) derivatives have been isolated.

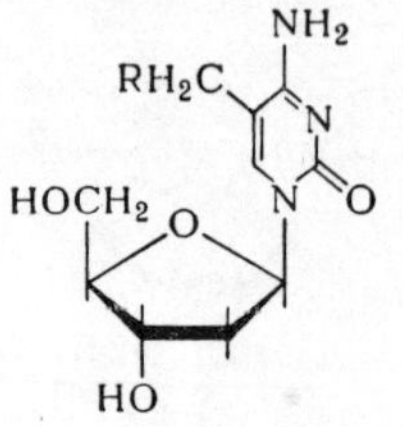

XLIV

where R denotes the following
residues:

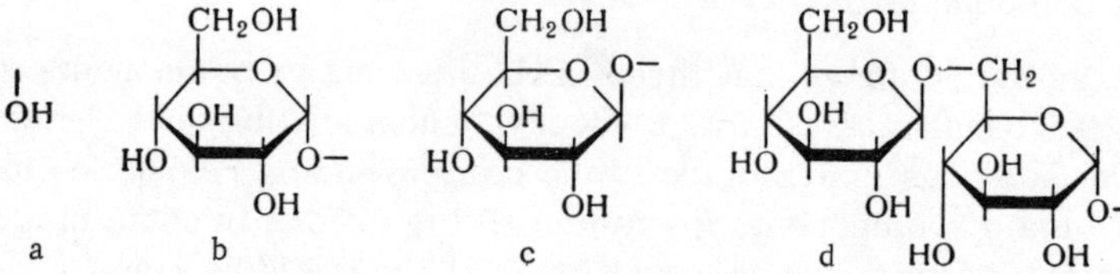

a   b   c   d

In the DNA of phage SP8, thymidine is replaced by 5-hydroxymethyl-2'-deoxyuridine (XLV) [252], while in the DNA of phage PBS-1 it is replaced by 2'-deoxyuridine (XLVI) [253].

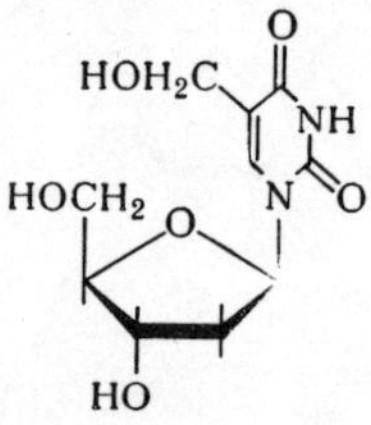

XLV 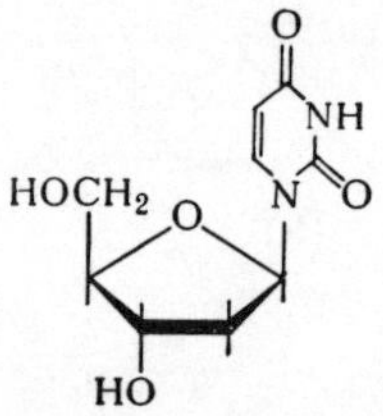   XLVI

4. Minor components of DNA [254]

Several minor components, methylation products of normal DNA components, are found in DNAs from different sources. The most widespread of these is 5-methyl-2'-deoxycytidine (XLVII) [255], which is present in large amounts in the DNA of plants and in small amounts in the DNA of mammals, fishes, and insects. The presence of residues of 6-exo-N-methyl-2'-deoxyadenosine (XLVIII) has been demonstrated [256] in DNAs isolated from several bacteria and viruses. Methylated derivatives of 2'-deoxyguanosine have also been demonstrated recently in DNA: 1-methyl- (XLIX), 2-exo-N,N-dimethyl- (L), and 7-methyldeoxyguanosine (LI).

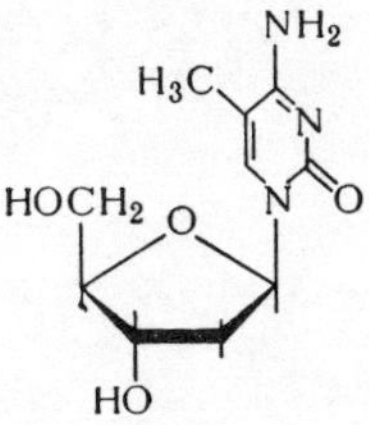

XLVII 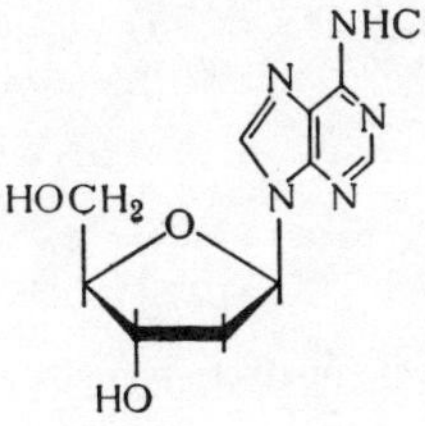   XLVIII 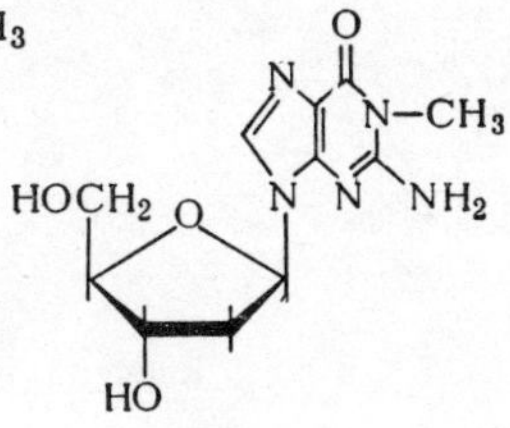   XLIX

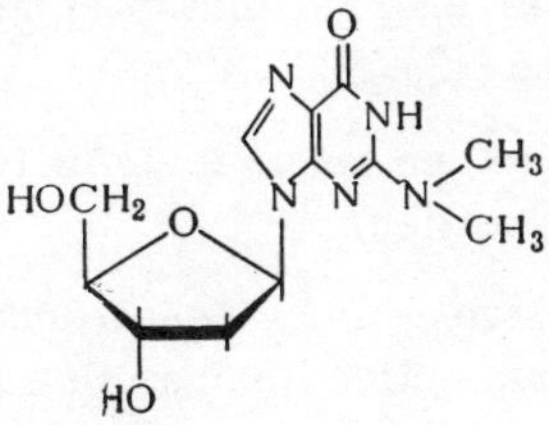

L 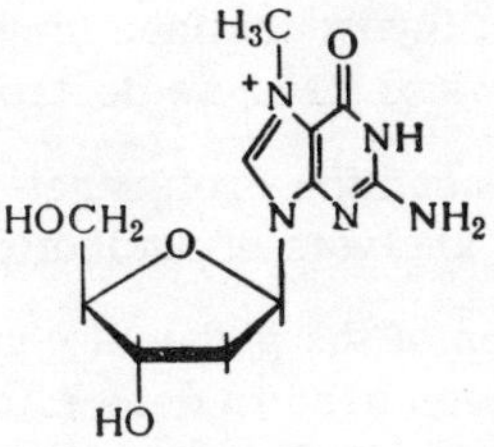   LI

VII. Nucleotide composition and the determination of identical nucleotide sequences in polynucleotides

An important characteristic of the different types of nucleic acids is the composition of their constituents. The results of analyses of the composition of nucleic acids, conducted in the late 1940s and early 1950s, provided a decisive argument for rejecting the older view of nucleic acids as tetranucleotides or as polymers containing repeating tetranucleotide sequences. These results established the foundations of modern concepts of the macromolecular structure of DNA.

Earlier studies of nucleic acid composition used methods based on the analysis of the products formed by acid hydrolysis of biopolymers. By vigorous acid hydrolysis (72% perchloric acid, 100°C, or 85% formic acid, 175°C) a mixture of purine and pyrimidine bases is formed through rupture of the N-glycoside bonds; under milder conditions (1 N hydrochloric acid, 100°C) a mixture of purine bases and pyrimidine nucleoside-2'(3')-phosphates is formed from RNA. These mixtures can be fractionated by paper chromatography and their components can be determined spectrophotometrically [258]. Alkaline hydrolysis of RNA (0.5 N KOH, 37°C) is also widely used. This leads to the formation of a mixture of nucleoside-2'(3')-phosphates which can be fractionated by ion-exchange chromatography [259], and by paper electrophoresis [260] and chromatography [261]. More recently, the method of thin-layer chromatography [262] has been extensively used for the analysis of the hydrolysis products of nucleic acids. The fairly vigorous conditions used in hydrolysis of the polynucleotides may cause some degree of degradation of their components. Mistakes associated with partial deamination of cytidine derivatives to uridine derivatives, taking place in a strongly acid or alkaline medium are of the greatest importance [263]:

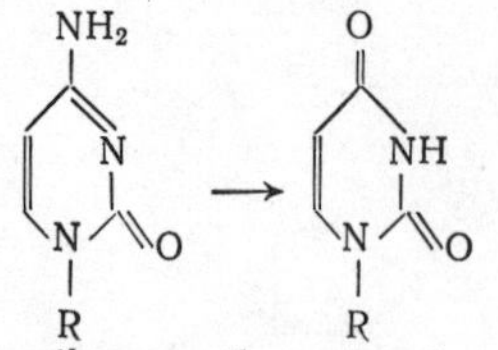

R denotes ribose or ribose phosphate residue

Some minor components of nucleic acids undergo a far greater degree of degradation. Dihydrouridine, for example, under ordinary conditions of alkaline hydrolysis of RNA, is destroyed (see page 397).

For this reason the preference now is to use enzymic hydrolysis of polymers for the analysis of nucleotide composition.

Investigation of the physical characteristics of double-stranded complexes has also been used to determine the nucleotide composition of DNA. Close correlation exists between the melting temperature [27] or buoyant density [28] of these complexes and their nucleotide composition.

Analysis shows that, with rare exceptions, all four major components characteristic of the particular type of polynucleotide are present in nucleic acids, while the minor components may vary widely. Chargaff's rule [264], the equivalence of the content of adenine and thymine, on the one hand, and of guanine and cytosine (or cytosine + 5-methylcytosine), on the other hand, holds good for most specimens of DNA. Other relationships, known generally as Chargaff's rules, can be deduced from this first rule purely arithmetically: equivalence of the content of purines and pyrimidines and equivalence of the content of bases with a keto group (guanine + thymine) and bases with an amino group (adenine + cytosine). Conversely, the ratio between the sum of guanine and cytosine and the sum of adenine and thymine differs from unity, and usually cannot be expressed by a ratio of whole numbers. This ratio characterizes the nucleotide composition of a given polymer, and it is known as the base ratio or coefficient of specificity. The sum of the guanine and cytosine contents in a given DNA, expressed in per cent, is also frequently used for this purpose.

The meaning of Chargaff's rule for DNA came to be understood after Watson and Crick had suggested their model of DNA structure: equivalence of the content of those pairs of bases which are complementary during formation of the double-stranded complex. The composition of single-stranded DNAs, such as the DNA of phage ϕ X174, does not obey Chargaff's rules.

Analysis of the nucleotide composition of DNAs from different sources [265-267] shows that whereas in bacteria the sum of the guanine and cytosine, expressed in per cent, may vary considerably (for example, 37% for Bacillus cereus, 50% for Escherichia coli, 73% for Mycobacterium phlei), in the case of higher animals and plants the figure is 40-49%. Extranuclear DNA and satellite DNA may have larger or smaller contents of G + C than the main

component of DNA from the cell nucleus of a given species, but the differences
are usually small. An important exception in this respect is the satellite DNA
from the tissues of some species of crabs, which contains less than 3% of
G + C and is virtually an almost pure copolymer of deoxyadenylic and thymidyl-
ic acids.

The composition of double-stranded complexes of virus RNA also obeys
Chargaff's rule for DNA (with replacement of thymine by uracil), while the
RNAs of reovirus and wound tumour virus of plants have 44 and 38% G + C
respectively. Equivalence between the content of complementary bases is
not usually observed with the other types of RNA.

Analysis of total RNA preparations (which evidently reflects the composition of the components of the ribosomal RNAs fairly closely) shows that,
as a rule, the base ratio is greater than unity and changes very little from
one species to another even in bacteria [265]; the composition of ribosomal
RNA usually differs substantially from that of DNA. Conversely, the composition of some fractions of nuclear RNA and of the cytoplasmic messenger RNA
of higher animals is characterized by base ratios less than unity in value,
and it thus approaches the composition of total DNA. The nucleotide composition of specimens of total RNA (and also the composition of the heavy and
light components of ribosomal RNA) obeys Chargaff's rule for RNA closely
[268]: the ratio between the content of bases with a keto group and bases with
an amino group is close to unity. The significance of this relationship for the
macromolecular structure of ribosomal RNA is not yet clear. The nucleotide
composition of different RNAs within cells of the same type has recently been
analysed. The results showed small yet definite differences between the components of ribosomal and nuclear RNA, as are shown in Table 1.1 for polyribonucleotides from cells of a HeLa culture [499].

The fact that two polynucleotides have the same composition does not
mean that they are identical in the chemical sense even if their molecular
weights are equal: they can differ in the sequence of their nucleotides. Since
the determination of nucleotide sequence (see the next section) is a complex
problem which has not yet been completely solved, approximate methods have
been developed for estimating agreement between the nucleotide sequences of
different polynucleotides. These methods are widely used in biochemical
research.

The first method, usually called "hybridization" [269], is based on the
following principle. If a double-helical DNA complex is heated above its
melting temperature and the mixture of single-stranded polymers obtained
is then slowly cooled in the presence of another single-stranded polynucleotide, besides the restoration of the original complex, a certain quantity of a
"hybrid" double-stranded complex is also formed. This hybrid contains polynucleotide chains previously belonging to different macromolecules. The
more nucleotide sequences complementary to the corresponding segment of
one strand of the original DNA are present in the chain of the added polymer,

TABLE 1.1. Nucleotide Composition of Various Types of RNA from HeLa Cells [499]

Type of RNA	Content of particular type of RNA in cell, %	Content of bases of RNA, %				Base ratio
		cytosine	adenine	guanine	uracil	
Cytoplasmic RNA:						
28S rRNA...........	53	32	16	36	16	2.1
18S rRNA	24	27	21	30	22	1.3
7S rRNA	1	28	21	28	23	1.3
5S rRNA	1	26	18	34	22	1.5
tRNA	12	27	22	27	24	1.2
mRNA	3	24	26	21	28	0.83
Other types of RNA	2	-	-	-	-	-
RNA of cell nucleus:						
45S RNA...........	1	33	13	37	17	2.3
32S RNA...........	3	33	14	37	16	2.3
DNA-like RNA	-	22	26	21	31	0.75

the more the hybrid complex is formed. Hybrid complexes can be obtained between two types of DNA or between DNA and RNA. Their formation can be demonstrated and their content estimated by equilibrium ultracentrifugation in a density gradient or by the ability of double-stranded complexes to be held up by nitrocellulose filters. The most convenient method for this purpose, however, is to use columns containing DNA "entrapped" in agar [270, 271]. Polynucleotides with a molecular weight of $(0.5-1.0) \cdot 10^6$ are held up in such a column only through the formation of complexes with the DNA. The quantity of polynucleotide held up in the column is easily determined. The ability of different polynucleotides to compete with each other for binding with the same DNA can be investigated by this method; it can also be used to determine the presence of common nucleotide sequences in different RNA specimens.

Essentially the same principle (cyclization) is used to detect what are called "sticky ends" of double-stranded DNAs: mutually complementary single-stranded terminal segments of the polynucleotide chain (shown schematically in Formula LII). A DNA with this structure is found in certain bacteriophages, for example in phage λ [272]. In other cases, notably the DNA of phage T3, such polynucleotides can be obtained by brief enzymic hydrolysis of the native DNA [273], for instance, by treatment with the exonuclease III from Escherichia coli (see page 51).

Heating polynucleotide LII at a temperature a little below its melting temperature, followed by slow cooling leads to the formation of the circular structure LIII on account of noncovalent interaction between the complementary "sticky ends." This type of structure can be found with the electron microscope. This method has been used to demonstrate that the terminal segments

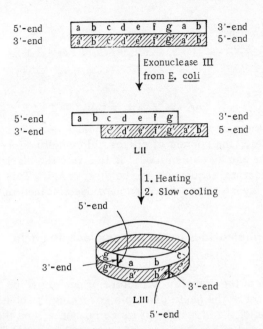

of the polynucleotide chains of DNAs of phages T3 and T7 have the same
sequence. By a combination of hybridization and cyclization a distinction
can be drawn between virus DNAs containing a single nucleotide sequence
and those consisting of an assortment of circularly transposed fragments
[274] (see page 21).

Another method of polymer identification used to study polymers with
a similar nucleotide sequence is the method of "nearest neighbour analysis"
[275-277]. This method can be applied when the polynucleotide which is to
be studied can be obtained from radioactive precursors*.

A mixture of nucleoside-5'-triphosphates (see page 77), one component
of which contains radioactive phosphorus, is subjected to enzymic polymerization. The resulting polymer is then hydrolysed to mononucleotides in
such a way that the P – O bond (the O in the C5' position) is ruptured; in the
case of DNA this is done by hydrolysis with a mixture of deoxyribonuclease
from micrococci and phosphodiesterase from snake venom, and in the case
of RNA by alkaline hydrolysis. After hydrolysis, the radioactive phosphate
is incorporated into the nucleoside-3'-phosphate whose residue in the polymer chain was next to the residue of the introduced labelled precursor. For
example, the relative content of the dinucleotide sequences ApA, GpA, UpA,
and CpA can be determined by subjecting α-^{32}P-adenosine-5'-triphosphate
to polymerization and measuring the radioactivity of the corresponding nucleo-
side-3'-phosphates after hydrolysis:

---

*Nearest neighbour analysis can also be undertaken with the acid of chemical cleavage of poly-
nucleotides by a number of methods. (Fuller details are given on page 52.)

$$
\begin{bmatrix} \text{ppp}\overset{*}{\text{A}} \\ \text{pppG} \\ \text{pppU} \\ \text{pppC} \end{bmatrix} \xrightarrow[\text{ization}]{\text{polymer-}} \cdots \text{Up}^*|\text{Ap}^*|\text{Ap}|\text{Up}|\text{Gp}^*|\text{Ap}|\text{Cp}^*|\text{Ap}|\text{Up}^*|\text{Ap}\cdots
$$

$$\downarrow \text{hydrolysis}$$

$$\text{Ap}^* + \text{Up}^* + \text{Cp}^* + \text{Gp}^* + \text{Ap} + \text{Up} + \text{Cp} + \text{Gp}$$

By repeating the analysis for each nucleoside-5'-triphosphate used in preparing the polymer, the content of all possible combinations of dinucleotides in that polymer can be determined. In this way the identity or non-identity of two nucleotide sequences can be discovered. This method is widely used for the investigation of polynucleotides obtained in vitro with the aid of enzymes.

VIII. Sequence of nucleotides in the polynucleotide chain

[278-281, 501, 502]

Determination of the complete structure of nucleic acids requires not only the establishment of the basic principle governing the construction of the polymer, elucidation of the structure of its components, and their quantitative determination in the polymer; the order of the monomer units in the polymer chain must also be established. Recent work has shown that the polynucleotide chain of nucleic acids, except in the case of satellite DNA from crabs [mostly poly-d(ApTp)] and, possibly, certain other satellite DNAs, is not built of repeating units. On the other hand, as the results of nearest neighbour analysis show, the distribution of nucleotide units in the chain differs considerably from the mean statistical distribution. According to modern biochemical concepts, the specific nucleotide sequence in the DNA chain contains all information concerning the amino acid sequence in the proteins of a particular organism.

Establishment of nucleotide sequence is a highly complex task; the problem has been solved in practice at present only for RNAs of relatively low molecular weight, such as tRNA and 5S RNA. Methods of determination of nucleotide sequence in the polynucleotide chain of practical value at the present time are based on partial hydrolysis of the polynucleotides. Before examining the basic principles underlying the determination of polynucleotide structure and the application of these principles to natural nucleic acids of various types, we shall be well advised to discuss methods of selective cleavage of the polynucleotide chain now in use.

1. Partial cleavage of polynucleotides

There are two possible approaches to this problem: successive detachment of monomer residues from one end of the polynucleotide chain and cleavage of the polymer into units by rupture of the internal phosphodiester bonds. In either approach both chemical and enzymic methods can be used; at present the latter are more popular.

The chemical method of successive detachment of nucleoside residues from the 3'-end of the polyribonucleotide chain [282, 283] is based on the following scheme. Polynucleotide LIV, containing the only diol group of the whole polymer at the 3'-end of the chain, is treated with periodate; the resulting dialdehyde LV is selectively split by β-elimination with the formation of polynucleotide LVI, which is one unit shorter. After enzymic dephosphorylation of LVI, polynucleotide LVII, identical in its terminal carbohydrate group with the original LIV, is formed; this new, shortened nucleotide can then be resubjected to the same stepwise degradation. Several convenient methods of stepwise degradation of nucleosides by detachment of terminal nucleoside residues on this basis have been developed [284-288]. The problem of obtaining mild conditions for the degradation has been successfully solved and convenient techniques have been developed for identification of the bases formed.

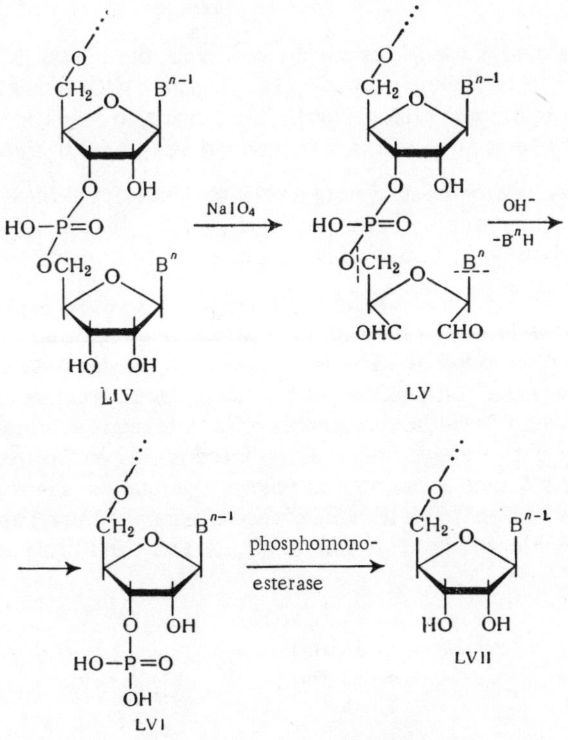

B^{n-1}, B^n – base residues

To detach the 5'-terminal residue of DNA, a method based on catalytic oxidation of the primary hydroxyl group and subsequent β-elimination has been suggested [289]:

c

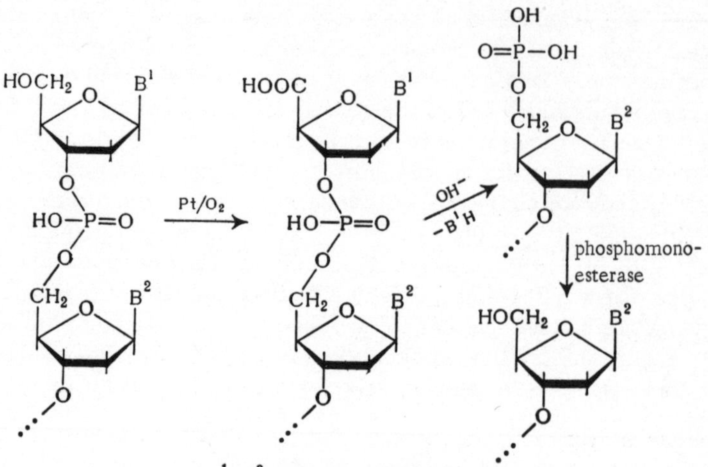

B^1, B^2 denotes base residues

Detachment of nucleotides one by one from the 3'-end of a polydeoxy-ribonucleotide can be done by treating the polymer with a mixture of dimethyl sulphoxide and acetic anhydride [290]. This leads to oxidation of the 3'-terminal hydroxyl group to a carbonyl compound and to β-elimination.

Successive detachment of nucleotide residues from the ends of a poly-nucleotide chain can also be achieved by the action of exonucleases [4], as mentioned previously*. Some information about the most widely used enzymes of this class is given in Table 1.2.

The process of enzymic degradation can be controlled by analysing the detached mononucleotides at various times: in many cases it is more con-venient to analyse the composition of the undegraded fragments [291, 292]. This last approach can be demonstrated with reference to establishment of the structure of octanucleotide LVIII, isolated after partial hydrolysis of yeast alanine tRNA (see Scheme 1 on page 57, sequence 41-47). After treat-ment of this octanucleotide with snake venom phosphodiesterase, heptanucleo-tide LIX, hexanucleotide LX, pentanulceotide LXI, and tetranucleotide LXII were isolated:

GpGpGpApGpApGpU	LVIII
GpGpGpApGpApG	LIX
GpGpGpApGpA	LX
GpGpGpApG	LXI
GpGpGpA	LXII

Comparison of their nucleotide composition with the composition of the original oligonucleotide shows beyond all doubt that the 3'-end of the LVIII chain has the structure –GpApGpU. In this case the structure of the tetranucleotide

---

*Successive detachment of nucleotides can also be carried out by means of polynucleotide phos-phorylase (see page 77). The products of the reaction between an oligoribonucleotide and phos-phoric acid, catalysed by this enzyme, are nucleoside-5'-diphosphates.

TABLE 1.2. Action of Various Exonucleases on Polynucleotides [4]

Enzyme	Substrate	Reaction product	End of polynucleotide chain from which mononucleotides detached
Phosphodiesterase from snake venom	Single-stranded polynucleotide with free 3'-OH end group	pN	3'-OH
Phosphodiesterases from bovine spleen and <u>Lactobacillus</u> <u>acidophilus</u>	Single-stranded polynucleotide with free 5'-OH end group	Np	5'-OH
Exonuclease I from <u>Escherichia</u> <u>coli</u>	Single-stranded polydeoxyribonucleotide with free 5'-OH end group	dNp	5'-OH
Exonuclease III from <u>Escherichia</u> <u>coli</u>	Double-stranded complex of polydeoxyribonucleotide; both chains have 3'-OH end groups	pdN	3'-OH (both chains)

LXII can be written equally certainly on the basis of the results of determination of the 3'-terminal group (see page 31), because the only nucleoside formed by alkaline hydrolysis is adenosine:

$$GpGpGpA \xrightarrow{\text{OH}^-} 3Gp + A$$
LXII

Both the chemical and enzymic methods of determination of nucleotide sequence by successive detachment of the terminal monomers of the polynucleotide chain have their limitations: those of the chemical method are due to the nonquantitative course of the process and to partial degradation of pseudouridine by the action of periodate; those of the enzymic method are due to occasional ruptures of phosphodiester bonds in the middle of the polynucleotide chain (through contamination with other enzymes). It is therefore rarely possible to determine the sequence of more than 5–6 nucleotide residues by this method. Nevertheless, it is widely used for the analysis of oligonucleotides formed as the result of partial degradation of the RNA chain.

The other principle on which monomer sequence in the polynucleotide chain can be established is based on its breakdown into smaller units; methods of cleavage of the chain at the site of specific monomers or monomer combinations are essential for this purpose. This can be done chemically by increasing the lability of the phosphoester bond in derivatives of ribose- or 2-deoxyribose-3-phosphate by conversion of the sugar residue into its aldehyde form and subsequent β-elimination [293]:

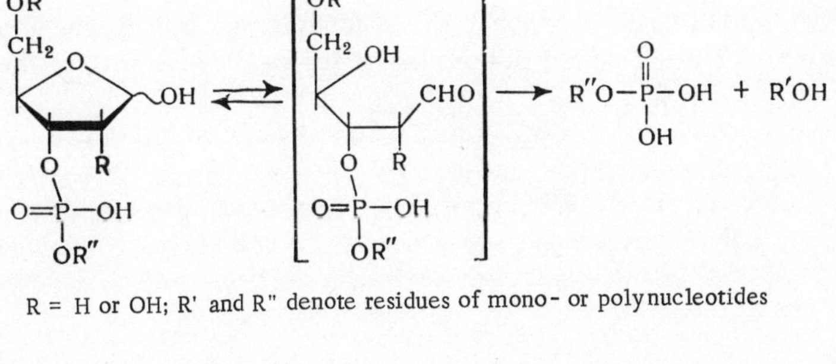

R = H or OH; R' and R" denote residues of mono- or polynucleotides

In practice, certain types of bases (designated X) must be selectively removed from the polynucleotide LXIII. The polynucleotide thus obtained, containing residues of pentose-3',5'-diphosphate (LXIV) with a free glycoside centre in the ribose moiety, can be broken down under relatively mild conditions with rupture of the P−O (the O in the C3' position) bond, with the formation of oligonucleotides not containing nucleosides with X bases*.

Conversion of DNA into polynucleotides of the LXIV type has been extensively studied: by acid hydrolysis, apurinic acid can be obtained [294, 295], by treatment with hydrazine, apyrimidinic acid is formed [296], while by the action of nitrous acid and hydroxylamine, acytidylic deaminated DNA is produced [297]. Similar conversions have recently been carried out for the more labile RNA: treatment with hydroxylamine yields deuridylic RNA [298, 299].

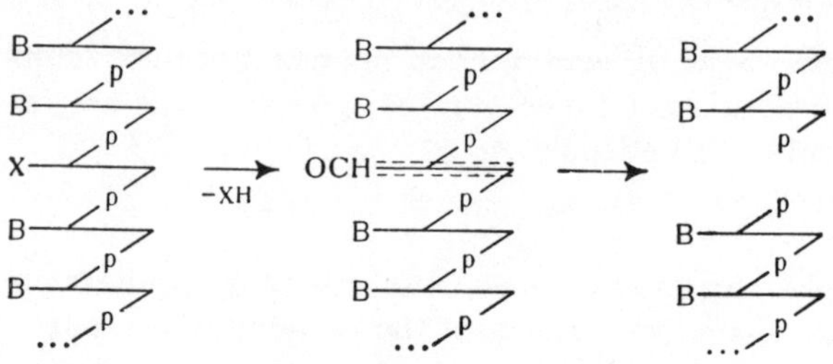

LXIII   LXIV

B denotes base residue

The preparation of polynucleotides of the LXIV type and their degradation are discussed in greater detail in Chapters 7 and 10.

For specific enzymic cleavage of the polynucleotide chain, the action of endonucleases is used. This method is most widely employed for RNA, because of the ready availability of two enzymes with well-marked and thoroughly investigated specificity: pyrimidyl ribonuclease (pyrimidyl-RNase; pancreatic RNase) and guanyl-RNase (RNase T_1 from takadiastase; RNase from actinomycetes). Pyrimidyl-RNase catalyses the hydrolysis of phosphodiesters of nucleosides which are derivatives of 2-ketopyrimidines with no substituent in the N3 position. As a result the corresponding nucleoside-3'-phosphates or their derivatives are obtained; nucleoside-2',3'-cyclic phosphates are intermediate products.

Pyrimidyl-RNase ruptures phosphodiester bonds formed by derivatives of uridine and cytidine and also breaks down the phosphodiesters of some minor components of RNA, especially pseudouridine, 5,6-dihydrouridine, and ribothymidine†. The effect of the nature of the pyrimidyl ring of the nucleo-

---

* As the result of vigorous alkaline hydrolysis of polymer LXIV, the nucleoside residue next to the residue of the split nucleoside is converted into nucleoside-3',5'-diphosphate (pNp). This can be used for nearest neighbour analysis (see page 47).

† Careful investigations of model compounds have shown that ordinary specimens of pyrimidyl-RNase can also split derivatives of purine nucleosides at a very low reaction velocity [300]. The reason evidently is that these specimens contain other specific RNases as impurities [301]. In structural investigations, however, this side reaction can be disregarded.

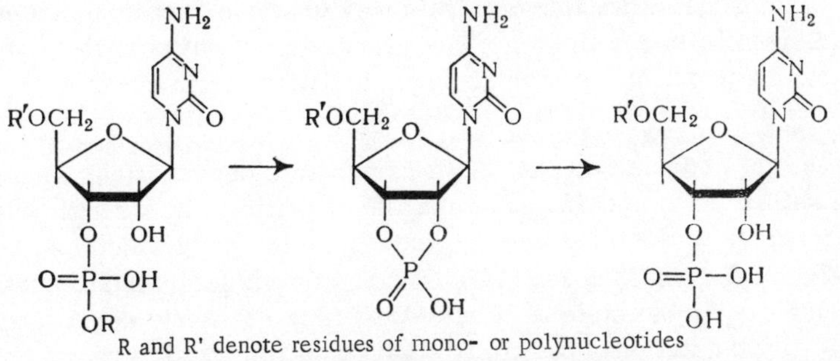

R and R' denote residues of mono- or polynucleotides

side on the velocity of the reaction catalysed by pyrimidyl-RNase has been examined by Witzel [302] (see also [303]). Hydrolysis of RNA by pyrimidyl-RNase yields a mixture of 3'-phosphates of pyrimidine nucleosides and oligo-nucleotides terminating in a pyrimidine-3'-phosphate residue. The velocity of the enzyme reaction depends strongly on the nature of the detached residue and the conformation of the polynucleotide (see below); specific partial cleavage of polynucleotides can take place if the velocity of the reaction is reduced.

Guanyl-ribonuclease (RNase T_1 from takadiastase [304]) catalyses the hydrolysis of phosphodiesters of nucleoside-3'-phosphates which are 6-keto-purine derivatives unsubstituted in the N1 and N7 positions (see below).

Of this group of nucleosides, guanosine is a major component of RNA, while inosine, 2-exo-N-methylguanosine and 2-exo-N,N-dimethylguanosine are minor components. The reaction products are the corresponding nucleo-side-3'-phosphates or oligonucleotides with a terminal guanosine-3'-phos-phate. Just as with pyrimidyl-RNase, specific cleavage of polynucleotides can be carried out with guanyl-RNase.

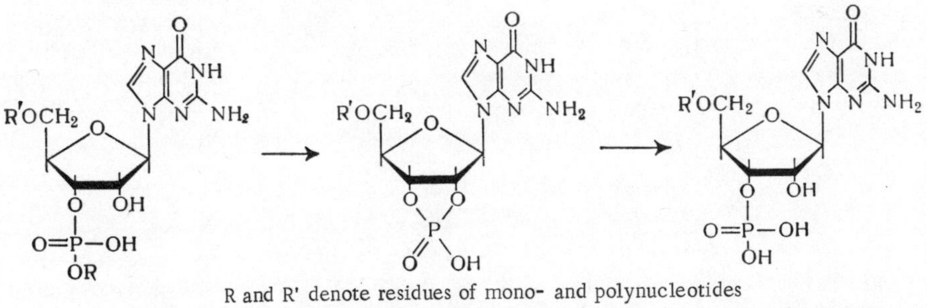

R and R' denote residues of mono- and polynucleotides

The RNase isolated from actinomycetes possesses closely similar specificity [305]; in addition to the types of nucleotides listed above it also catalyses the hydrolysis of derivatives of 1-methylguanosine-3'-phosphate.

Intermediate products of the rupture of phosphodiester bonds by guanyl-RNase, as by the action of pyrimidyl-RNase, are cyclic nucleoside-2',3'-phosphates, which are then converted into nucleoside-3'-phosphates. In some cases this second stage of the reaction takes place very slowly and cyclic phosphates accumulate in large quantities.

Besides the endonucleases referred to above, other enzymes with less marked specificity are known. The properties of some of the endonucleases which can be used for structural analysis are given in Table 1.3.

Finally, mixed chemical and enzymic methods can be used for the hydrolysis of polynucleotides. Before enzymic degradation, a nucleotide can be subjected to specific chemical modification, as a result of which one of the types of nucleotides with a phosphodiester bond which can be split by the action of the enzyme used is converted into a derivative resistant to enzyme action. An example of this method is the specific degradation of modified RNA by the action of pyrimidyl-RNase on its cytidine residues after preliminary alkylation of the uridine residues in the RNA chain at the N3 position by treatment with carbodiimide derivatives [306, 307] or after splitting of the uridine residues by hydroxylamine [308]. Another example of this

TABLE 1.3. Some Endonucleases Hydrolysing Polynucleotides [4]

Enzyme	Substrate	Types of bonds preferentially ruptured	End products of hydrolysis of polynucleotide
RNase T$_2$ from takadiastase	RNA	. . . Ap Np. . .	Nucleoside-3'-phosphates
RNase from Bacillus subtilis	RNA	. . . Gp Ap. . . . . . Gp Gp. . .	Nucleoside-3'-phosphates and dinucleotides
RNase from Escherichia coli	RNA	. . . Ap Np. . . . . . Up Np. . .	Nucleoside-3'-phosphates
Acid RNase from spleen	RNA	. . . Ap Np. . .	Nucleoside-3'-phosphates and short oligonucleotides
Nuclease from micrococci	RNA, single-stranded DNA	. . . Ap Np. . . Up Np. . . similarly for the deoxy series	Nucleoside-3'-phosphates and dinucleotides
Pancreatic DNase I	Double-stranded DNA complexes	d(. . . pA pT. . .) d(. . . pA pC. . .) d(. . . pG pT. . .) d(. . . pG pC. . .)	Nucleoside-5'-phosphates and dinucleotides (65%) + trinucleotides (25%)
DNase II from pig's spleen	The same	d(. . . Gp Gp. . .) d(. . . Ap Cp. . .)	Large oligonucleotides
DNase I from Escherichia coli	"	-	Oligonucleotides with terminal 5'-phosphate residue and mean chain length ~7
Streptodornase	"	d(. . . pN pG. . .) d(. . . pN pA. . .)	Oligonucleotides

type is the specific degradation of RNA by guanyl-RNase at the inosine and 2-exo-N,N-dimethylaminoguanosine residues after the preliminary action of glyoxal on guanosine and 2-exo-N-methylguanosine residues [309]. These reactions will be examined more fully below.

In another possible combination of chemical and enzymic methods, the enzyme itself is modified chemically, thereby changing its specificity. For example, the breakdown of 5S RNA by partially alkylated pyrimidyl-RNase yields larger oligonucleotide fragments than the action of the untreated enzyme [310].

2. Principles of the unit method

The only method which has so far enabled the complete structure of several RNAs to be established is the unit method, which is based on reconstruction of the structure of a polynucleotide from data concerning the structure of fragments obtained by its partial degradation. The degradation must be carried out by at least two independent methods, differing in their specificity. For RNA, the usual procedure is hydrolysis by pyrimidyl- and guanyl-RNases. After separation of the resulting short oligonucleotide fragments [311-317], their structure is determined by the methods reviewed below.

The structure of trinucleotides follows unambiguously from determination of the terminal nucleotide residues (see page 31). In some cases the structure of longer polynucleotides can also be established unambiguously by cross-cleavage by an endonuclease of different specificity, as is shown below for one of the tetranucleotides isolated after hydrolysis of alanine tRNA by pyrimidyl-RNase

$$\text{GpGpApCp} \xrightarrow{\text{guanyl-RNase}} \text{2Gp} + \text{ApCp}$$

In other cases, successive detachment of nucleotides from the end of the chain or partial hydrolysis of an oligonucleotide by one or other method may prove useful.

The known specificity of enzymes makes it possible to recreate some partial sequences of the polynucleotide chain. The original polynucleotide is then split up into larger units (partial hydrolysis with guanyl-RNase is usually used for this purpose), and segments of nucleotide sequences which will allow the fragments obtained previously to be unambiguously fitted into the polynucleotide chain are sought in them, and in this way the complete structure of the polynucleotide can be recreated. The success of this reconstruction is entirely dependent on the presence of nonrepeating segments in the nucleotide sequence which can be used as reference points. End groups and also minor components or oligonucleotides occurring only once in the products of enzymic hydrolysis of the polynucleotide can be used for this purpose.

3. Investigation of the primary structure of polynucleotides

tRNA [521, 522]. The first example of determination of the complete structure of a nucleic acid was published in 1965 by Holley and his collaborators [318], who described their elucidation of the structure of alanine tRNA from yeast. We shall examine this work in more detail because it illustrates a concrete application of the principles described above †.

Partial hydrolysis of alanine tRNA by the action of guanyl-RNase yielded two large fragments (Scheme 1), each of which corresponded to about half of the molecule [319]. Fragment (a), containing the 3'-terminal group of tRNA, incorporates the minor components 1-methylinosine (position 39) and ribothymidine (55), which occur only once in this RNA; they can be used as reference points for recreating the nucleotide sequence in the molecule. The specimen of alanine tRNA studied was a mixture of two closely related compounds, for in some molecules the uridine residue in position 48 was replaced by a 5,6-dihydrouridine residue. On hydrolysis, the corresponding mixture of closely related oligonucleotides was formed. These usually could not be separated and spectrophotometric determination showed that they contained a fractional number of moles of uridine. Because of this characteristic property, the nucleotides containing such a fragment can be unequivocally identified, and it can be used as a reference point like the minor components; subsequently it will be referred to as component $\overset{*}{U}$.

The 3'-terminal nucleotide sequence of alanine tRNA is found in the composition of the hexanucleotide UpCpCpApCpCp (B6, see Scheme 1) formed by complete hydrolysis of tRNA by guanyl-RNase. After partial hydrolysis by this same enzyme, two fragments containing this sequence are isolated. One of them (A4) consists of 11 nucleotide residues, and after complete hydrolysis with guanyl-RNase it gives B6 and also the pentanucleotide ApCpUpCpGp (B5), which unequivocally solves the structure of A4. Since, after complete hydrolysis of A5 by pyrimidyl-RNase, the tetranucleotide GpGpApCp (C3) is isolated, the structure of A5 is also unequivocally solved, and the sequence 59–76 of the original tRNA is reconstructed.

The 1-methylinosine residue is a component of the tetranucleotide B1 obtained by complete hydrolysis of tRNA by guanyl-RNase, and of the decanucleotide A1 isolated after partial hydrolysis by this same enzyme. The nucleotide sequence of A1 is largely covered by the nucleotide sequence of the octanucleotide C1 found after hydrolysis of tRNA by pyrimidyl-RNase, which ends in component $\overset{*}{U}$. On the other hand, octanucleotide C1 is a

---

† Here and later in the book for the sake of convenience of description certain inessential details will be omitted. The results are not considered in their historical order, and only the minimum of information necessary for establishment of the structure is given. Facts of a confirmatory nature are usually omitted.

(a) 3'-terminal fragment

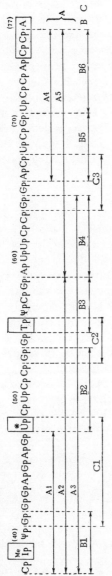

(b) 5'-terminal fragment

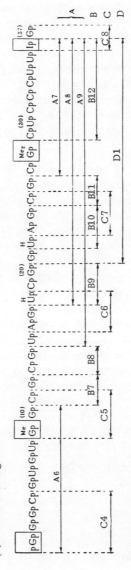

Reconstruction of molecule

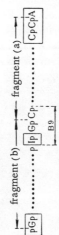

Scheme 1. Determination of the primary structure of alanine tRNA from yeast; A1...A9) products of partial hydrolysis by guanyl-RNase; B1...B12) products of complete hydrolysis by guanyl-RNase; C1...C9) products of complete hydrolysis by pyrimidyl-RNase; D1) product of partial hydrolysis of products A by guanyl-RNase. Besides the standard abbreviations (see page 8), the following symbols are used: G^Me for 1-methylguanosine; G^{Me_2} for 2-exo-N,N-dimethylguanosine; I for 1-methylinosine; U for 5,6-dihydrouridine; U for a mixture of uridine and 5,6-dihydrouridine. End groups and minor nucleotides used as reference points during reconstruction of the sequence are distinguished.

component of fragment B2, so that the nucleotide sequence 38–53* can ultimately be reconstructed. The whole of this sequence is found in oligonucleotide A2 isolated after partial hydrolysis with guanyl-RNase. After complete
hydrolysis of A2 by the same enzyme, as well as products expected on the
basis of existing data, one mole of guanosine-3'-phosphate and the nonrepeating tetranucleotide TpΨpCpGp (B3) also are formed. Since ribothymidine
occurs only once in the molecule and since the trinucleotide GpGpTp (C2) is
formed by hydrolysis of tRNA with pyrimidyl-RNase, the nucleotide sequence
in A2 (38–58) is unequivocally established. Finally, after partial hydrolysis
of tRNA by guanyl-RNase, oligonucleotide A3 is found. After complete hydrolysis with guanyl-RNAase, this oligonucleotide yields the products formed
by hydrolysis of A2, and also the hexanucleotide B4 with which we are familiar.
In this way, the complete nucleotide sequence of the 3'-terminal fragment can
be reconstructed.

   The nucleotide sequence of the 5'-terminal fragment of alanine tRNA
[fragment (b) in Scheme 1] has been established in the same way. The terminal guanosine-5'-phosphate residue and the minor components 1-methylguano-
sine (9), 2-exo-N,N-dimethylguanosine (28), and inosine (36) can be used as
reference points. Sequence 1–11 is obtained by comparing the structure of
fragments A6, C4, and C5, while sequence 27–36 is obtained from the structure of fragments A7 and B12 respectively. Some difficulties arose during
the reconstruction of segment 18–26. Complete hydrolysis of fragment A8,
containing this segment, by guanyl-RNase yielded not only hydrolysis products of A7, but also two trinuleotides containing dihydrouridylic acid (B9 and
B10), the dinucleotide CpGp (B11), and guanosine-3'-phosphate. The presence
of the trinucleotide ApGpCp (C7) in the products of hydrolysis of tRNA by
pyrimidyl-RNase suggests that fragments B10 and B11 are linked together.
By partial hydrolysis with guanyl-RNase, oligonucleotide A8 was split into
trinucleotide B9 and a fragment consisting of 16 nucleotide residues (D1); consequently, B9 is the 5'-end of the A8 sequence. The 5'-end of oligonucleotide
D1 may have the structure LXV or LXVa.

$$\overset{H}{Gp}Up \mid ApGpCp \mid GpCp \mid \overset{Me_2}{Gp}Cp \mid \cdots \qquad \text{LXV}$$

$$Cp \mid \overset{H}{Gp}Gp\overset{}{Up} \mid ApGpCp \mid \overset{Me_2}{Gp}Cp \mid \cdots \qquad \text{LXVa}$$

   The choice between them was made on the basis of the fact that the only
trinucleotide product of hydrolysis of D1 by pyrimidyl-RNase is ApGpCp,
$$\overset{H}{}$$
whereas in the case of structure LXVa, GpGpUp would also have been present.
Comparison of the products formed by hydrolysis of fragments A8 and A9 by

---

*It has recently been shown [505] that fragment C1 is in fact a nonanucleotide and not an octa-
nucleotide, and that the sequence suggested by Holley and collaborators for alanine tRNA must be
altered: it includes an additional guanosine residue between positions 47 and 48.

guanyl-RNase enables the chain to be lengthened by another three nucleotide units, and in this way the sequence 15-35 is obtained.

The products of hydrolysis of the 5'-fragment (b) of alanine tRNA by the action of pyrimidyl-RNase include the nonrepeating nucleotide IpGp (C8); this unequivocally demonstrates the structure of the 3'-terminal segment of this fragment. The closeness of the minor component to the end of the chain of this fragment demonstrates unequivocally that the two CpGp dinucleotides (fragments B7 and B8) must lie on segment 11-14 of the sequence; since the structures of the two dinucleotides are identical, the sequence of the whole fragment (1-36) is unequivocally established.

Naturally both investigated fragments (a) and (b) have terminal groups characteristic of the complete tRNA molecule, and they can therefore be joined together in only one way [part (c) of Scheme 1]; that this reconstruction is correct is confirmed by isolation of the nonrepeating trinucleotide IpGpCp (C9) after hydrolysis of tRNA with pyrimidyl-RNase. The nucleotide sequence along the whole chain of the tRNA molecule is thus fully established.

Similar methods have been used to determine the structure of other transfer RNAs. The results of determination of the structure of tRNAs from yeast specific for the transfer of serine [320], tyrosine [321], valine [322], and phenylalanine [323] were published in 1966-1967; reports of the determination of the structure of other forms of tRNA have appeared subsequently [324-327, 506-512]. The primary structure of different tRNAs shows certain general features. These will be examined in more detail in Chapter 4, which deals with the macrostructure of nucleic acids.

5S RNA. The task of determining the structure of 5S RNA is more complex than in the case of tRNA. To begin with, this polynucleotide has a somewhat longer chain (120 nucleotide residues), and secondly, it contains no minor components, thus reducing the number of reference points and making it essential to obtain more extensive overlapping of the sequences of the isolated fragments. However, even in this case the problem can be solved by analysis of the structures of the products of partial hydrolysis of the polynucleotide by guanyl-RNase. It simply means that many more such fragments must be isolated and identified. An example of how the problem has been solved is the determination of the structure of 5S RNA from tissue cultures of KB cells [328].

A more interesting approach is that due to Brownlee and collaborators, which they used to investigate the structure of 5S RNA from Escherichia coli [329]. This was the first time that the sequence in polynucleotides of this type was completely established. This approach is based on the use of a much greater variety of methods for partial hydrolysis of the polynucleotides. In addition to partial hydrolysis by guanyl-RNase, as was used by Holley and collaborators, in this case two other enzymic methods of partial degradation were used: partial hydrolysis by pyrimidyl-RNase and partial hydrolysis by splenic acid RNase, which does not possess absolute specificity (see page 54), and also two combined chemical and enzymic methods. In the first of these, the polynucleotide was treated with N-cyclohexyl-N'-(β-methylmorpholinylethyl)-carbodiimide tosylate which reacts with the uridine residue to give derivative LXVI resistant to the action of pyrimidyl-RNase. On subsequent treatment with this enzyme only phosphodiester bonds formed by cytidine-3'-phosphate are ruptured (see page 54).

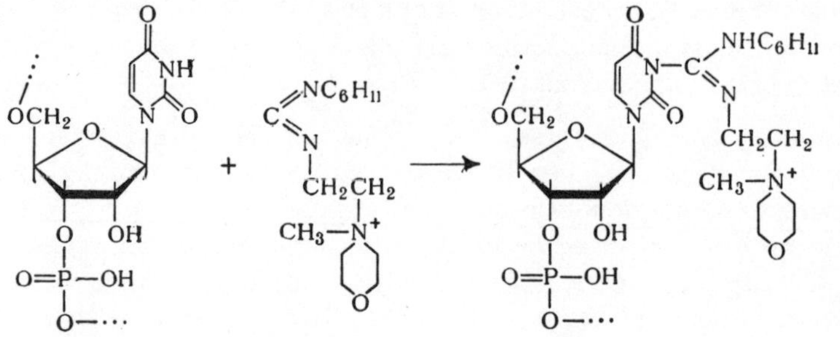

LXVI

The second method consists of partial methylation of the polynucleotide by dimethyl sulphate followed by treatment with guanyl-RNase. The phosphodiester bonds of the guanosine residues methylated at N7 are not ruptured.

An illustration of the effectiveness of this approach is given by determination of the 5'-terminal sequence of 5S RNA consisting of 16 nucleotides [Scheme 2, part (a)]. The reference point here is the terminal uridine-5'-phosphate residue: it is a component of dinucleotide F1 obtained by complete hydrolysis with guanyl-RNase, and of the trinucleotide E1, obtained by combined chemical and enzymic cleavage at the cytidine residues. During hydrolysis of the polynucleotide by guanyl-RNase after partial methylation, the hexanucleotide D1 is isolated; its structure can be unequivocally deduced from the fact that tetranucleotide CpCpUpGp (F2) is present in the products of complete hydrolysis of RNA by guanyl-RNase, while the isomeric product CpUpCpGp is absent. The fragment B1 can be isolated from the products of partial hydrolysis with pyrimidyl-RNase, while after complete hydrolysis by this enzyme, besides other products, two moles of the trinucleotide GpGpCp (G1 and G2) are obtained; its structure is therefore unequivocally reconstituted. Furthermore, comparison of the hydrolysis products of B1 and of the fragment C1 formed by the action of splenic acid RNase enables the Cp residue to be added to the 3'-end of the reconstituted sequence. Comparison of C1 with fragment A1 obtained by partial hydrolysis by guanyl-RNase enables a further Gp residue to be added to the known sequence; finally, fragment B2 contains yet another Up residue. The hexanucleotide D2 isolated from the products of cleavage of the polynucleotide by guanyl-RNase after its partial methylation (before the action of the enzyme) has a tetranucleotide sequence which overlaps that of fragment B2; in this way two further residues of the nucleotide can be added to the 3'-end of the chain, so that the reconstruction of fragment 1 is completed.

In the same way fragments 2, 3, 4, and 5 have been reconstructed [see part (b) of Scheme 2]; as initial reference points the 3'-terminal residue of the RNA and the nonrepeating decanucleotides found in the complete hydrolysis products of the polymer after treatment with guanyl-RNase were used (sequences 24-33 and 87-96).

The final reconstruction of the molecule presented considerable difficulty, because the same sequence GpUpApGp occurs both at the 3'-end of fragments 1-4 and at the 5'-end of fragments 2-5. The position of fragments 1 and 5 is unequivocally determined by their end groups; finally, oligonucleotides enabling fragments 2, 3, and 4 to be linked were isolated from the products of hydrolysis of 5S RNA by guanyl-RNase.

<u>Other components of ribosomal RNA</u>. The molecular weight of the 23S and 16S components of ribosomal RNA is much higher than that of 5S RNA, and the task of determining the complete nucleotide sequence of these polymers has not yet been accomplished.

The structure at the 5'-terminal residue of ribosomal RNA from several sources has been successfully established [330, 331]; the 5'-terminal residue of the nucleoside in the polymer was labelled with radioactive phosphate with the aid of polynucleotide kinase (see page 34). The 16S component

(a) 5'-terminal fragment (fragment 1)

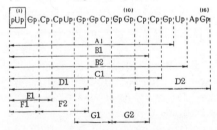

Scheme 2. Determination of primary structure of 5S RNA from <u>Escherichia coli</u>: A1) product of partial hydrolysis by guanyl-RNase; B1, B2) products of partial hydrolysis by pyrimidyl-RNase; C1) product of partial hydrolysis by splenic acid RNase; D1, D2) products of hydrolysis by guanyl-RNase after preliminary methylation; E1) product of hydrolysis by pyrimidyl-RNase after treatment with carbodiimide; F1, F2) products of complete hydrolysis by guanyl-RNase; G1, G2) products of complete hydrolysis by pyrimidyl-RNase.

(b) Reconstruction of sequence of molecule

fragment 1

fragment 2

fragment 3

fragment 4

fragment 5

pUp ···GpUpApGp Cp Gp Cp Gp Gp Up Gp GpUp Cp Cp Cp Ap Cp CpUp Gp ApCp Cp Cp CpApUp GpCp Cp Gp ApApCpUp Cp

ApGp ApApGpUp Gp ApApApCpGpCpCp GpUp Ap Gp Cp Gp Cp Cp Gp ApUp Gp GpUp ApGp Up GpUp Gp Gp Gp GpUp CpUp Cp

CpCpCp ApUp Gp Cp Gp Ap GpAp GpUp ApGp Gp Gp ApAp CpUp Gp Cp CpApGp Gp Cp Ap U

of ribosomal RNA from E'scherichia coli, for example, has been found to have
the pentanucleotide sequence pApApApUpGp. . . , while the 23S component has
the trinucleotide sequence pGpGpUp. . . .

The analysis of products formed by complete hydrolysis of high-molecular weight components of ribosomal RNAs by the action of pyrimidyl- and
guanyl-RNases has been successfully accomplished [332, 513]; partial
hydrolysis of these nucleotides by the action of these enzymes has also been
described [333, 334], and in this way fragments comparable in molecular
weight with tRNA or 5S RNA were obtained. By using methylated nucleoside
components of ribosomal RNA as the label, the nucleotide sequence in segments adjacent to these components in the composition of 16S and 23S RNAs
from Escherichia coli was successfully determined [514].

Virus RNAs. Less progress has been made with the study of the
structure of virus RNAs.

By means of the chemical method of successive detachment of mononucleotides, the existence of a GpCpCpCpA sequence has been demonstrated
at the 3'-end of tobacco mosaic virus RNA [287]; this result has also been
confirmed by another method [335, 515]. By a combination of different methods, an identical undecanucleotide sequence has been established at the 3'-end
of RNAs of phages S2 [288, 515, 516], MS2 [336, 515], and R17 [516], and a
sequence of 16 nucleotides has been determined at the 3'-end of the RNA of
phage Q_β [516]. It is interesting to note that the 3'-terminal tetranucleotide
sequence in all cases so far investigated is identical.

An original method was used to analyse the 5'-terminal sequence of the
RNA of phage Q_β [337]. This RNA can be obtained by biosynthesis from
labelled precursors, and it is thus possible for the "nearest neighbour" method to be used (see page 47). On the basis of the distribution of radioactive
label in the 5'-terminal oligonucleotide obtained by hydrolysis of the polymer
with pyrimidyl-RNase, and by the use of α-^{32}P-guanosine-, α-^{32}P-adeno-
sine-, and α-^{32}P-cytidine-5'-triphosphates in turn, the terminal sequence
pppGpGpGpGpApApCpC. . . can be unequivocally reconstructed. Another investigation [517] has shown that the RNA isolated from phage Q_β is nonhomo-
geneous. Besides molecules containing the above-mentioned 5'-terminal
sequence, other molecules are present containing an additional guanosine
residue at the 5'-end. For information on the 5'-terminal sequence of RNA
from phage MS2, see [518, 519].

Hydrolysis of tobacco mosaic virus RNA by means of guanyl-RNase
yielded three long oligonucleotides of nonrepeating sequence [338]; to localize
their position in the RNA molecule the method of partial deproteinization of
the virus RNA was used. Deproteinization of virus RNA by the action of
dodecyl sulphate is known to take place successively, starting from the 5'-end
of the RNA chain [339]. By interrupting the deproteinization process at different times and by cleavage of the liberated RNA, the relative position of the
different fragments can be determined.

Several long oligonucleotides have been isolated from the products of partial hydrolysis of RNA from phage R17 by the action of guanyl-RNase T_1, and the sequence of one of them has been established [520]. It contains 57 mononucleotide residues; its nucleotide sequence corresponds to the sequence of amino acids 81-99 in the protein of the phage membrane.

DNA. Determination of the complete base sequence in natural DNAs is a task for the future. This is because of the extremely high molecular weight of DNA as well as the absence of specific methods of cleavage of its molecule which would yield oligonucleotide fragments of sufficient length and, at the same time, allow their sequence to be unequivocally determined.

Methods of chemical hydrolysis of DNA available at present (see page 51) yield polypyrimidine or polypurine fragments usually containing six or seven mononucleotide residues [340-342]. Such fragments could probably be useful for determination of the structure of longer DNA fragments, but specific methods of obtaining such fragments have not yet been developed. The method of chemical cleavage has been used to solve some special problems of DNA structure, notably the relative position of the glycosylated and nonglycosylated residues of 5-hydroxymethylcytidine in DNA of the T-even phages [343].

On the other hand, methods of DNA cleavage (for example, by means of hydrodynamic shearing forces) exist which produce fragments too large for structural investigations; the nucleotide sequence of these fragments can be estimated approximately by the hybridization method (see page 45). For example, by controlled cleavage of DNA by means of ultrasound, followed by separation of the fragments by centrifugation in a caesium chloride density gradient, and studying their hybridization with different types of RNA, the approximate length of DNA segments responsible for the biosynthesis of certain components of ribosomal RNA has been determined, and their mutual position established [344].

Some very promising methods, based on the use of electron microscopy, have been used to determine the nucleotide sequence in very long polynucleotides resembling DNA. The distance between the bases in the polynucleotide chain of denatured DNA is about 7 Å. It can be hoped that by the use of electron microscopes of sufficiently high resolving power it will be possible to see individual bases in the polynucleotide chain directly and in this way to establish the sequence of the monomer components. Before this method can be put into practice, two problems will have to be solved. First, substantial improvements will be required in the technique of electron microscopy and the method of preparation of specimens. Second, methods of chemical modification of DNA enabling the different types of bases to be selectively labelled must be developed. It is important that the label introduced must be clearly visible in the electron microscope, i.e., the modifying agent must contain atoms of the heavy metals (lead, mercury, copper, or uranium) or functional groups capable of forming stable complexes with the cations of these metals.

Reagents which satisfy at least some of these requirements already exist;
they will be mentioned below. However, no practical attempt has yet been
made to prove the value of this interesting approach to determination of the
nucleotide sequence in polynucleotide chains.

IX. Synthetic polynucleotides

It will be clear from the last section that the problem of establishing
the primary structure of nucleic acids is highly complex and has so far been
solved only in a few of the simplest cases. For this reason, model oligonucleotides and polynucleotides of known structure are of great importance
to the investigation of the link between the structure and the reactivity and
biological activity of these compounds. Some polynucleotides of this type
(see page 48) have been obtained from natural sources, but the systematic
use of model polynucleotides for the solution of physicochemical and biological problems has only become possible after preparative methods have
become available for obtaining such compounds by chemical or enzymic synthesis. Chemical methods of synthesis are important for the preparation of
oligonucleotides; for the preparation of polynucleotides, on the other hand,
enzymic or combined chemical and enzymic methods are used.

1. Chemical methods of synthesis [345–348]

The key stage in oligonucleotide synthesis is the formation of the $3'$–$5'$-
phosphodiester bond between nucleoside residues composing the polymer.
Two principal methods are now used for this purpose.

In the first method (the "phosphodiester" method) the internucleotide
bond is formed by interaction between an activated derivative of a monoester
of phosphoric acid (the nucleotide component) and the hydroxyl group of the
second (nucleoside) component:

$$\underset{\substack{|\\ \text{OH}}}{\overset{\substack{\text{O}\\ \|}}{\text{RO}-\text{P}-\text{OH}}} \longrightarrow \underset{\substack{|\\ \text{OH}}}{\overset{\substack{\text{O}\\ \|}}{\text{RO}-\text{P}-\text{X}}} \xrightarrow{\text{R'OH}} \underset{\substack{|\\ \text{OH}}}{\overset{\substack{\text{O}\\ \|}}{\text{RO}-\text{P}-\text{OR'}}} + \text{HX}$$

LXVII

R and R' represent different radicals.

As a rule the activated nucleotide derivative LXVII is obtained from the
nucleotide component directly in the reaction mixture by treatment with a
suitable condensing agent (for a comparative study of these agents, see [349,
350]). The substances most widely used for this purpose are dicyclohexylcarbodiimide (LXVIII) (the first reagent used for nucleotide polymerization)
[351] and chloroanhydrides of aromatic sulphonic acids, primarily mesitylene
sulphonylchloride (LXIX, R = CH$_3$) and 2,4,6-tri-isopropylbenzene sulphonylchloride [LXIX, R = (CH$_3$)$_2$CH] [352].

$$C_6H_{11}N=C=NC_6H_{11}$$

LXVIII

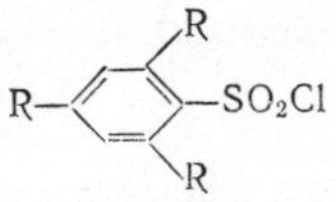

LXIX

The mechanism of formation of the phosphodiester bond by the action of dicyclohexylcarbodiimide and the other similar reagents is a complex one [353, 523]. Compounds probably having the structure of an imidoylphosphate (LXX) or mixed anhydride (LXXI) are formed first; these are subsequently converted through a series of stages into the trimetaphosphate derivative LXXII, which is evidently the true activated derivative of the nucleotide component. No such derivative can be formed from a diester of phosphoric acid.

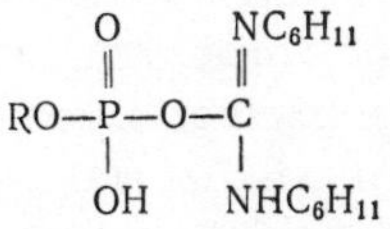

LXX 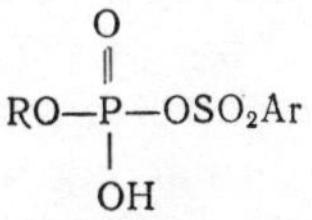   LXXI 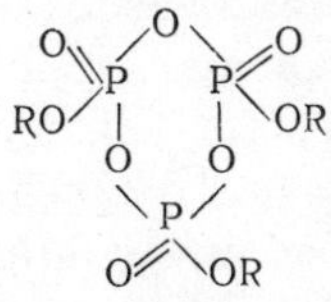   LXXII

R represents different radicals

For this reason no further conversion of the diesters takes place, even when there is an appreciable excess of the nucleoside component; because of this fact, not only derivatives of nucleosides, but also protected derivatives of nucleotides (see below) also can be used as the nucleoside component.

Activation of phosphodiesters is, however, possible as the result of treatment with chloroanhydrides of aromatic sulphonic acids. This reaction is the principle of another approach to oligonucleotide synthesis (the "triester" method of synthesis) [354]:

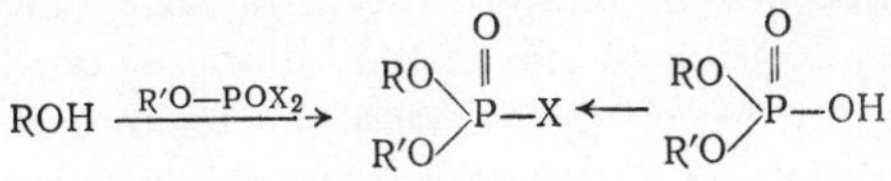

$$ROH \xrightarrow{R'O-POX_2}$$

LXXIII

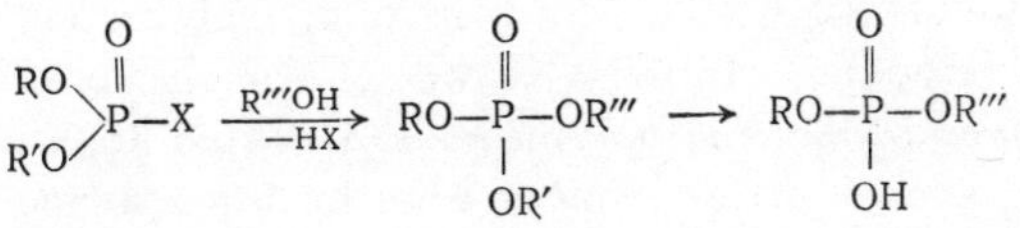

LXXIII   LXXIV

R and R''' denote nucleoside residues, R' protecting group

The intermediate activated compounds of the phosphodiesters evidently have the structure LXXIII (X = SO_2Ar); similar compounds can also be obtained from the corresponding nucleosides. For example, by phosphorylation of nucleosides by means of β,β,β-trichloroethylphosphodichloride $CCl_3CH_2OP(O)Cl_2$ the activated diester LXXIII (X = Cl, R''' = CCl_3CH_2) is formed [355]. The R' group in the original diester or phosphorylating agent is so chosen that it can be removed selectively from the intermediate triester LXXIV without rupturing the internucleotide bond, so that the conversion to the desired phosphodiester can be made.

Polynucleotides can be regarded as polymers of nucleoside-3'-phosphates or of nucleoside-5'-phosphates; correspondingly, the nucleotide component taking part in the reaction may be a derivative of nucleoside-3'- or of nucleoside-5'-phosphate. Among the deoxyribonucleotides, the second pathway is that most commonly adopted, while the first is usual with the ribonucleotides.

The activated derivative of the nucleotide component can react not only with the hydroxyl group of the nucleoside component, but also with other nucleophilic functional groups of compounds present in the reaction mixture (amino groups and hydroxyl groups of bases, phosphate groups, hydroxyl groups of the carbohydrate residue at C2'). To eliminate these side reactions, the functional groups in the nucleotide and nucleoside components must be protected. Examples of the use of various protecting groups will be examined below; the essential feature is that only protective groups which can be detached under mild conditions not involving isomerization or rupture of the internucleotide bond may be used (see Chapter 10).

The possibility of interaction between the activated derivative of the nucleotide and water means that oligonucleotide synthesis must be carried out under strictly anhydrous conditions; in some cases this causes difficulty because of the sparingly soluble nature of the nucleotide derivatives in organic solvents.

Notwithstanding the use of protected derivatives of nucleotides and nucleosides, some side reactions (for example, the formation of pyrophosphates during synthesis from phosphomonoesters) still take place; for this reason, careful chromatographic purification of the reaction products is necessary. A method which considerably simplifies the purification of the reaction products is fixation of one of the components of the reaction mixture on a polymer carrier [356-360]. This polymer can easily be separated from the other components of the reaction mixture. The reaction product fixed to the polymer can undergo further conversions, which considerably simplifies the multistage syntheses. Finally, after all stages have been completed the product can be freed from the polymer and isolated in a pure state. Great interest is being shown at the present time in this approach to oligonucleotide synthesis*. The nonspecificity of chemical methods of formation of an internucleotide bond, which are a disadvantage of the chemical approach to oligonucleotide synthesis (preparation of the protected derivatives of nucleosides and nucleotides requires multistage syntheses) in this case confers several advantages, because a very wide variety of nucleoside derivatives, including synthetic analogues of the components of nucleic acids, can be

---

* An essential condition of success in multistage synthesis on polymer carriers is the quantitative course of individual stages of the process: it is not yet possible to achieve this by existing methods of formation of the phosphodiester bond.

introduced into oligonucleotide synthesis. This provides many new and extensive opportunities for investigation of the link between structure and function of natural polynucleotides.

Oligodeoxyribonucleotides. The simplest approach to the preparation of homogeneous oligodeoxyribonucleotides is by polymerization (or, more accurately, polycondensation) of mononucleotides. For example, by polymerization of thymidine-5'-phosphate by the action of dicyclohexylcarbodiimide, oligonucleotides with the general formula $(pdT)_n$ and with a degree of polymerization up to 15 were obtained [361-363]. An important side reaction taking place during polymerization is the formation of cyclic oligonucleotides; to suppress this reaction, a "polymerization terminator" (3'-O-acetylthymidine-5'-phosphate) is added to the reaction mixture; in this case oligonucleotides in which the 3'-terminal hydroxyl group is blocked are formed, so that the possibility of cyclic transformation is ruled out. A similar method has been used for the polymerization of other deoxynucleoside-5'-phosphates. Derivatives with a protected heterocyclic ring were used in the reaction: 4-exo-N-anisoyldeoxycytidine-5'-phosphate (LXXV; R = H), 6-exo-N-benzoyl-deoxyadenosine-5'-phosphate (LXXVI; R = H), or 2-exo-N-acetyldeoxyguanosine-5'-phosphate (LXXVII; R = H) as the polymerized component and their 3'-O-acetates (LXXV-LXXVII; R = CH_3CO) as polymerization terminators. After detachment of the protecting groups by treatment with ammonia, oligocytidylic [364], oligoadenylic [365], and oligoguanylic [366] acids containing a 5'-terminal phosphate group are obtained.

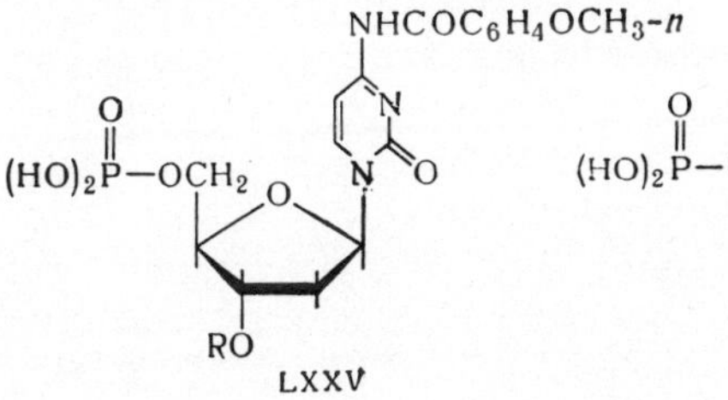

LXXV 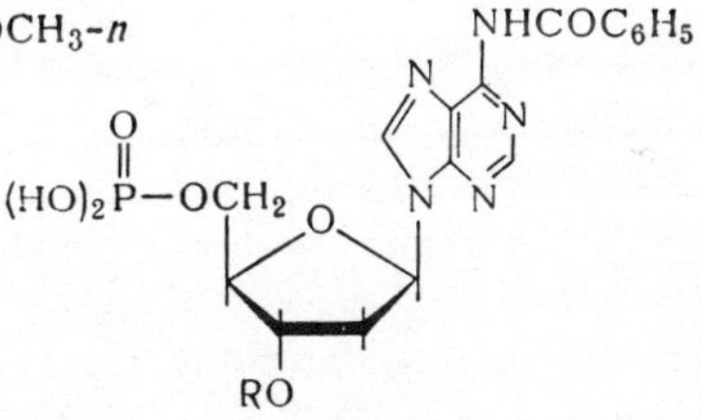   LXXVI

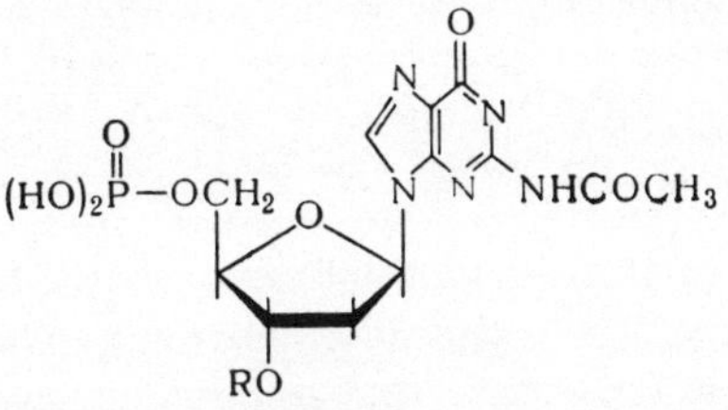

LXXVII

If the compound used as polymerization terminator is the acetyl derivative of a nucleoside-5'-phosphate, which differs in the structure of its heterocyclic ring from the polymerized component, oligonucleotides of the general structure $(pN)_m pX$ can be obtained. Such compounds have been obtained, for example, by polymerization of a mixture of thymidine-5'-phosphate and 4-exo-N,3'-O-diacetyldeoxycytidine-5'-phosphate [362]:

$$n\text{pdT} + \text{pdC}\overset{\text{Ac}}{-}\text{Ac} \xrightarrow[\text{2. NH}_4\text{OH}]{\begin{array}{c}\text{1. Dicyclohexyl-}\\\text{carbodiimide}\end{array}} (\text{pdT})_n\text{pdC}$$

An approach based on the polymerization of protected deoxynucleoside-3'-phosphates has also been investigated [367]; in this case the results have been less satisfactory.

Dinucleotides with a 5'-terminal phosphate group can also be used as the polymerized component (details concerning the preparation of these compounds are given below); as a result of this unit polymerization, oligonucleotides containing a repeating dinucleotide sequence are formed [368, 369]. In the case of 5'-phosphothymidylyl-(3' → 5')-2-exo-N-acetyldeoxyguanosine, for example:

$$n\text{pdTpdG}\overset{\text{Ac}}{} \xrightarrow[\text{2. NH}_4\text{OH}]{\begin{array}{c}\text{1. Dicyclohexyl-}\\\text{carbodiimide}\end{array}} (\text{pdTpdG})_n \quad n = 2 - 6$$

A similar approach, the polymerization of a tri- or tetranucleotide unit, has also been used to obtain oligonucleotides with a repeating trinucleotide [370, 371] or tetranucleotide [372] sequence. Another approach, based on the stepwise lengthening of the oligonucleotide chain, is also frequently used for the synthesis of compounds of this type.

In this case, oligonucleotides with any predetermined sequence of monomer components can be obtained. In the simplest case of synthesis of a dinucleoside monophosphate, both components of the reaction, nucleotide and nucleoside, are monomers. The first synthesis of a natural dinucleoside monophosphate [373], namely thymidylyl-(3' → 5')-thymidine, was successfully accomplished by the phosphotriester method. The reaction between 5'-O-acetylthymidine-3'-benzylphosphorochloridate (LXXVIII; R' = Ac; R" = PhCH$_2$; X = Cl) and 3'-O-acetylthymidine (LXXIX; R''' = Ac) yielded an intermediate triester LXXX (R' = R''' = Ac; R" = PhCH$_2$), which, after hydrogenolysis to remove a benzyl group, and deacetylation by treatment with alkali, was converted into the dinucleoside phosphate LXXXI (see page 69).

Similar syntheses by the phosphotriester method have been undertaken more recently [354, 355]. The nucleotide component used in this case was 5'-O-tritylthymidine-3'-(β,β,β-trichloroethyl)-phosphate (LXXVIII; R' = Tr; R" = CCl$_3$CH$_2$; X = Cl) or 5'-O-tritylthymidine-3'-(β-cyanoethyl)-phosphate (LXXVIII; R' = Tr; R" = CNCH$_2$CH$_2$; X = OH), activated by 2,4,6-triisopropyl-benzene sulphonyl chloride. Detachment of the β,β,β-trichloroethyl group from the intermediate triester can be achieved by reduction with zinc − copper couple, while the β-cyanoethyl group can be removed by mild alkaline treatment.

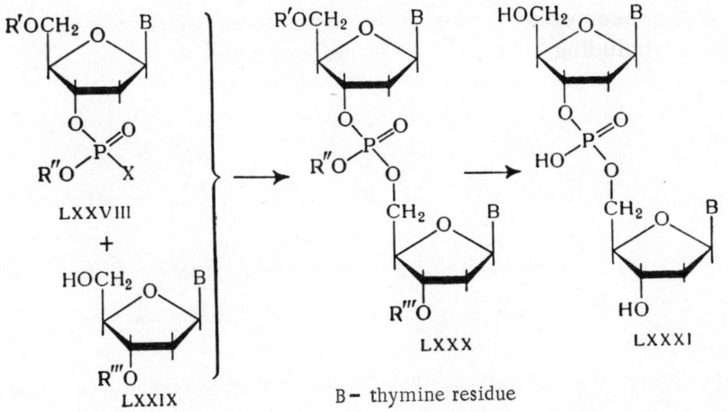

LXXVIII
+
HOCH₂

LXXIX

LXXX

LXXXI

B— thymine residue

For stepwise oligonucleotide synthesis by the phosphodiester method, derivatives of nucleoside-5'-phosphates are usually used as the nucleotide component. For example, thymidylyl-(3' → 5')-thymidine has been obtained by the reaction between 3'-O-acetylthymidine-5'-phosphate (LXXXII) and 5'-O-tritylthymidine in the presence of dicyclohexylcarbodiimide [351] (see page below).

After treatment with dilute acid to remove the triphenylmethyl group, and with ammonia to remove the acetyl group, the resulting protected dinucleoside monophosphate LXXXIIIa can be converted into dinucleoside monophosphate LXXXI. The intermediate product LXXXIIIb formed after detachment of only one of the protecting groups, can be reused as the nucleoside component for the formation of a phosphodiester bond. Its condensation with the corresponding protected nucleoside-5'-phosphate for example, LXXV-LXXVII; (R = COCH₃) yields the protected trinucleoside diphosphate; this process of lengthening of the nucleotide chain at the 3'-end can be repeated further.

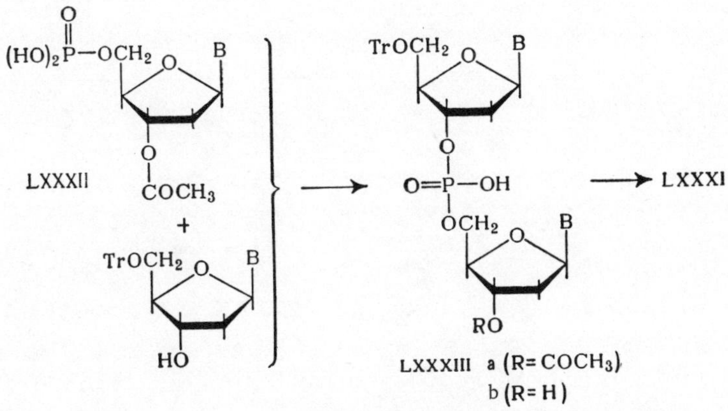

LXXXII

LXXXI

LXXXIII a (R= COCH₃)
b (R= H)

B— thymine residue

A similar series of reactions has been used for the synthesis of many oligodeoxynucleotides [369, 374-378], including a dodecanucleotide containing thymidine and cytidine residues:

$$
d(Tr—T) \xrightarrow[\text{2. OH}^-]{\substack{\text{1. pdT-Ac dicyclo-}\\ \text{hexylcarbodiimide}}} d(Tr—TpT) \xrightarrow[\text{4. OH}^-]{\substack{\text{3. pdC-Ac; dicyclo-}\\ \text{hexylcarbodiimide}}}
$$

$$
\longrightarrow d(Tr—TpTp\check{C}) \xrightarrow[\substack{\text{9-10. Reactions 3 and 4}\\ \text{Repetition of reac-}}]{\substack{\text{5-8. Reactions 1 and 2}\\ \text{repeated twice}}}
$$

$$
\longrightarrow d(Tr—TpTp\check{C}pTpTp\check{C}) \xrightarrow{\text{tions 5-10 twice}}
$$

$$
\longrightarrow d[Tr—TpTp\check{C}p(TpTp\check{C}p)_3] \xrightarrow[\text{12. OH}^-]{\text{11. H}^+} d(TpTpCp)_4
$$

where Č represents a 4-exo-N-anisoyldeoxycytidine residue

A similar method, namely the use of an oligonucleotide as the nucleoside component and lengthening of the polymer chain from the 3'-end, has also been used for the synthesis of oligonucleotides on a polymer carrier [357-360]. In this case, another alternative form was successfully put into effect: the polymer chain was lengthened from the 5'-end by the use of a 3'-protected oligonucleotide as the nucleoside component (a protected nucleoside in the first stage) [379].

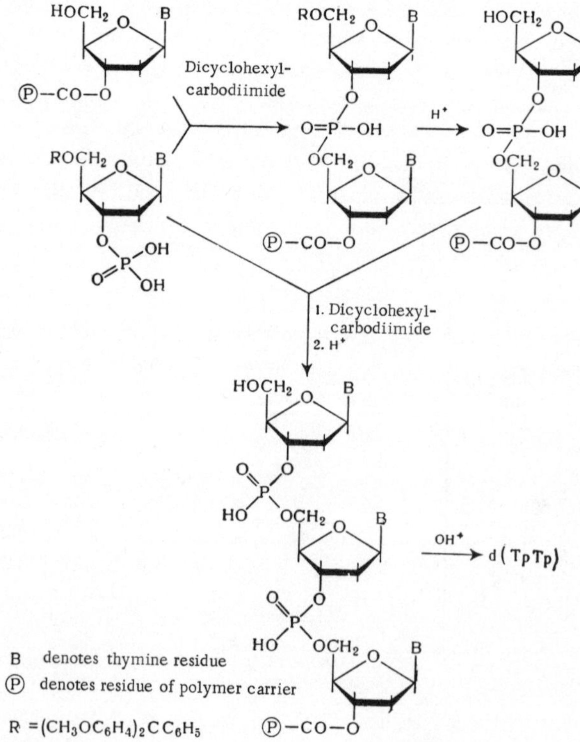

B   denotes thymine residue
Ⓟ   denotes residue of polymer carrier

R = (CH₃OC₆H₄)₂CC₆H₅   Ⓟ—CO—O

In this case, however, it is more convenient to use an oligonucleotide with a 3'-terminal phosphate group as the nucleotide component [356, 380], as illustrated in the following scheme:

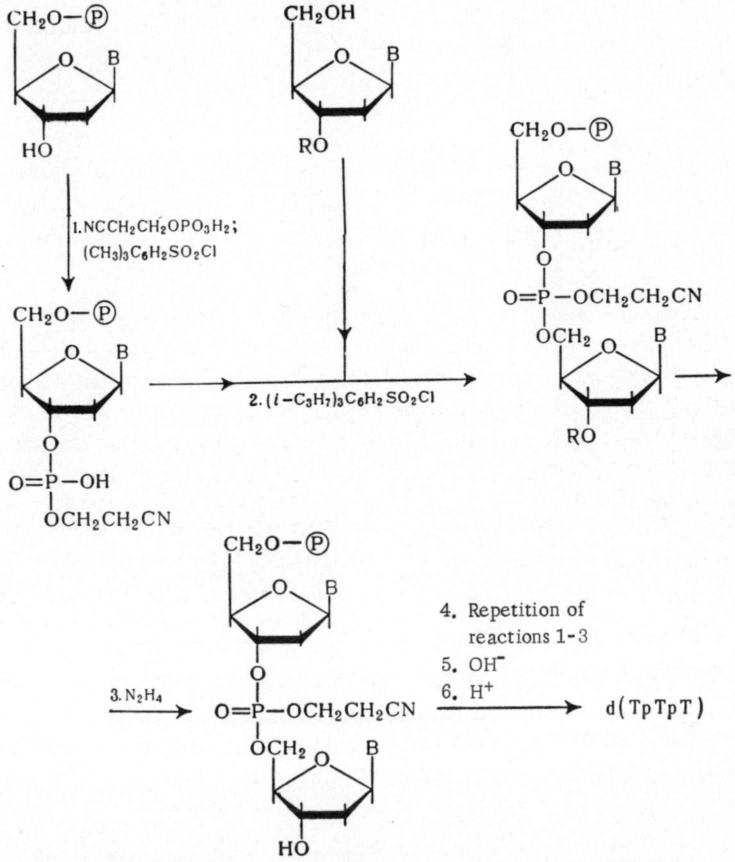

B represents thymine residue; P residue of polymer carrier; R = $C_6H_5COCH_2CH_2CO-$

A similar approach to the lengthening of the oligonucleotide chain has also been successfully used for the reaction without a polymer carrier [354, 355].

An oligonucleotide with 5'-terminal phosphate group can also be used as the nucleotide component [381, 382]. Oligonucleotides of this type can be obtained by stepwise lengthening of the oligonucleotide chain from the 3'-end by using as the nucleoside component derivatives of nucleoside-5'-phosphates with a protected phosphate group of the LXXIV type, such as nucleoside-5'-(β-cyanoethyl)-phosphates (LXXXIVa) [383-385], nucleoside-5'-(β,β,β-trichloroethyl)-phosphates (LXXXIVb) [386, 387], or condensation products of deoxynucleoside-5'-phosphates with 2',3'-(dimethoxybenzylidene)-uridine (LXXXIVc) [388, 389].

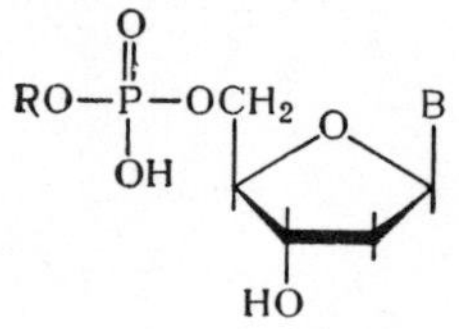

a) $R = NCCH_2CH_2-$

b) $R = CCl_3CH_2-$

c) $R = -OCH_2$

LXXXIV

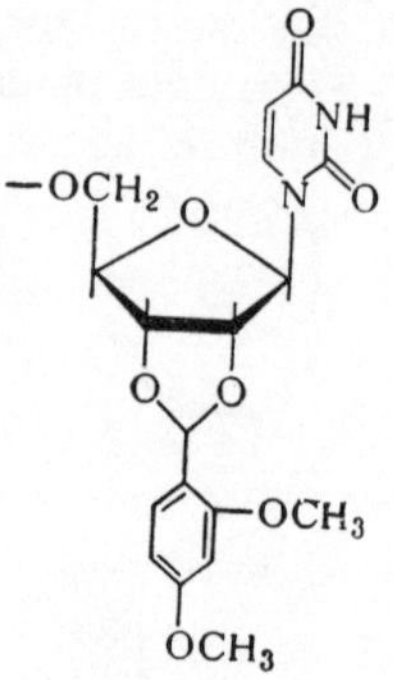

B denotes base residue

By condensation of the oligomer nucleoside component with an oligomer nucleotide component, the polymer chain can be lengthened by several components at once, which greatly facilitates the preparation of long oligonucleotides. Examples of the effectiveness of this method are the synthesis of an oligonucleotide containing 24 thymidine residues by condensation of two dodecanucleotides [390], and the synthesis of a dodecanucleotide with different nucleoside residues:

$$d[NCCH_2CH_2-pT(pT)_{11}] + d[(pT)_{11}pT-Ac] \xrightarrow[\text{2. OH}^-]{\substack{\text{1. Mesitylene sul-}\\ \text{phonyl chloride}}} d(pT)_{24}$$

$$d(NCCH_2CH_2-pTpTp\breve{A}p\breve{A}pTpT) + d[p\breve{A}p\breve{C}p\breve{A}p\breve{A}pTp\breve{A}-Ac] \xrightarrow[\text{2. OH}^-]{\substack{\text{1. Mesitylene sul-}\\ \text{phonyl chloride}}}$$

$$\longrightarrow d(pTpTp\breve{A}p\breve{A}pTpTp\breve{A}p\breve{C}p\breve{A}p\breve{A}pTp\breve{A})$$

where $\breve{A}$ represents a 6-exo-N-benzoyldeoxyadenosine residue and $\breve{C}$ a 4-exo-N-anisoyldeoxycytidine residue.

Oligoribonucleotides. Methods of chemical synthesis are much less highly developed in the case of this series of compounds than with the oligodeoxynucleotides. Only the trinucleoside diphosphates are relatively accessible [391, 392], and there have been few investigations of the synthesis of oligonucleotides with a higher degree of polymerization [393-395]. This is due principally to the high lability of the internucleotide bond of ribonucleotides in an alkaline medium and to its ability to undergo isomerization in an acid medium (see Chapter 10), which complicates the choice of protecting groups; difficulties also arise because of the need to protect the 2'-hydroxyl group of the nucleosides and nucleotides used in the reaction.

The usual approach to oligoribonucleotide synthesis is by the stepwise increase in the length of the oligonucleotide chain from the 5'-end. To synthesize dinucleoside phosphates, 2',5'-protected ribonucleoside-3'- phosphates (LXXXV) are usually used as the nucleotide component, and 2,3'-protected ribonucleosides (LXXXVI) as the nucleoside component.

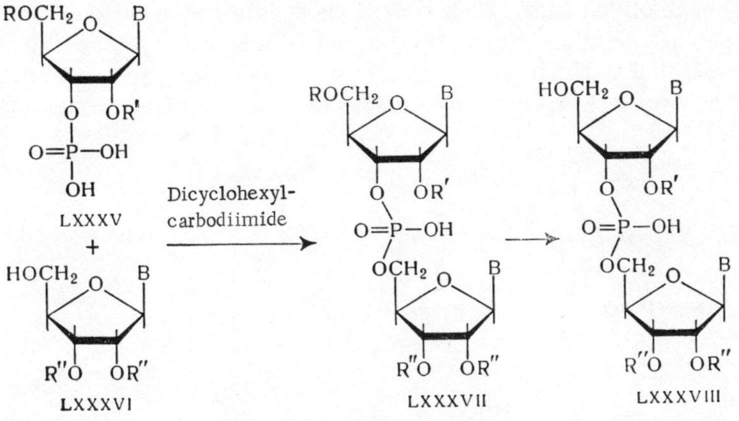

B denotes base residue

For instance, 2',5'-di-O-acetylnucleoside-3'-phosphates and 2',3'-O-(p-methoxybenzylidene)- or 2',3'-O-(p-dimethylaminobenzylidene)-nucleosides can conveniently be used for the reaction [396, 397]. After mild alkaline and acid hydrolysis, the protected dinucleoside LXXXVII (R = R' = Ac; R" + R" = >CHAr) is converted into free dinucleoside monophosphate LXXXVIII (R' = R" = H).

To obtain a further increase in length of the chain, it is essential that the 5'-hydroxyl group of the intermediate protected dinucleoside can be liberated under conditions when the remaining protected groups remain unattacked, i.e., that the conversion from the protected LXXXVII to the partially protected dinucleoside monophosphate LXXXVIII (where R' and R" represent protected groups) is possible. This partially protected dinucleoside phosphate can be used as the nucleoside component for further synthesis: its condensation with nucleotide LXXXV yields a protected trinucleoside diphosphate.

The most successful approach has been that based on the use of 5'-O-mono- (or di-) methoxytrityl-2'-O-acetylnucleoside-3'-phosphates as the nucleotide component and 2',3'-di-O-acetyl- (or benzoyl-) nucleosides as the nucleoside component (the amino groups of the heterocyclic rings of both components also must be protected). In this way all 64 possible trinucleoside diphosphates composed of the ordinary nucleotides found in RNA have been obtained [39].

Another approach, based on the use of an acetyl (alkaline-labile) group to protect the 5'-hydroxyl group of the nucleotide component, has proved to be rather more difficult to put into practice [398-400], but a number of trinucleoside diphosphates have been successfully obtained from 5'-O-acetyl-2'-O-(α-ethoxyethyl)-nucleoside-3'-phosphates and 2',3'-O-ethoxymethylidene-nucleosides (dimethylaminomethylidene derivatives were used to protect the amino groups of the ring, see page 366) [392, 401, 402].

Dinucleotides with a 3'-terminal phosphate group can be obtained by the direct phosphorylation of dinucleoside monophosphates [403] or by synthesis using protected derivatives of nucleoside-3'-phosphate [404–407] or of nucleoside-2',3'-cyclic phosphates [407–409] as the nucleoside component, as for example:

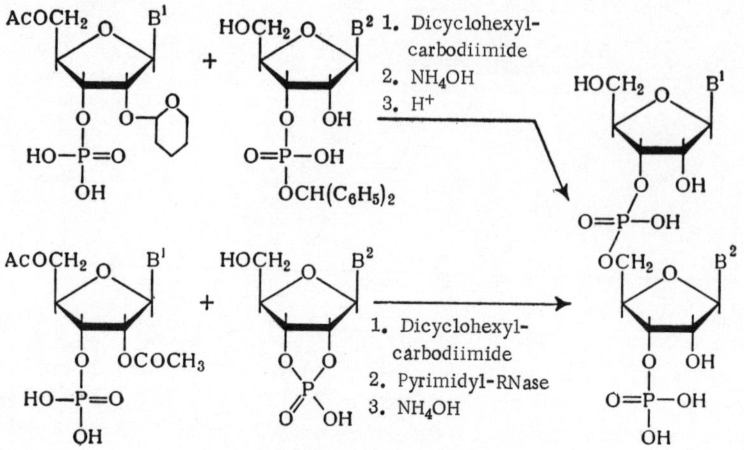

B^1, B^2 represent base residues

Another approach to the preparation of dinucleoside monophosphates has also been suggested. This is based on the condensation of a 2',3'-disubstituted nucleoside-5'-phosphate and a 2',5'-disubstituted nucleoside [410, 411], as for example:

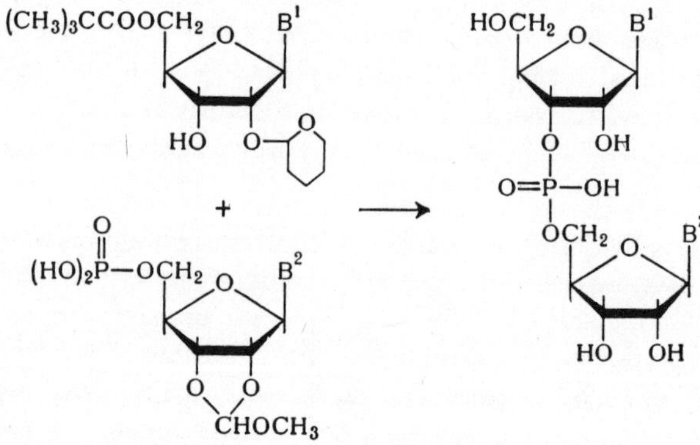

B^1, B^2 represent base residues

2. Principles of enzymic synthesis of oligonucleotides and polynucleotides

Three different groups of enzymes are used for the enzymic synthesis of the phosphodiester bond in oligonucleotides and polynucleotides.

The cleavage of poly- and oligoribonucleotides by the action of ribonucleases has already been examined (see page 69). The first stage of the reaction, the conversion of the phosphodiester of the ribonucleoside into a cyclic phosphate, is largely reversible, and under certain conditions by treatment of the cyclic phosphate of a mono- or oligonucleotide (the nucleotide component) with an excess of nucleoside or of nucleoside-3'-phosphate (the nucleoside component), a 3' → 5'-phosphodiester bond can be formed:

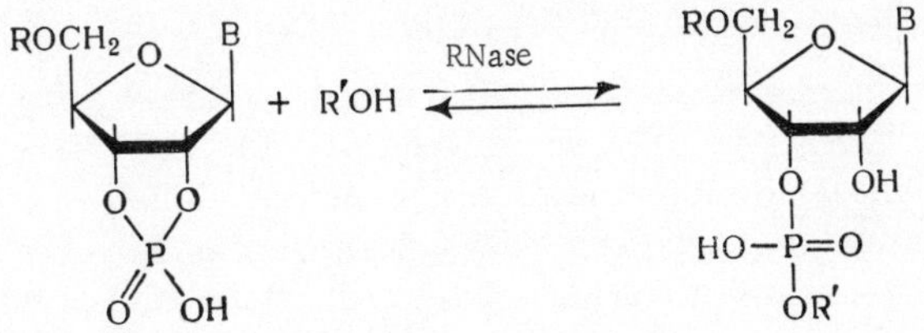

In this case B represents the base residue, R a hydrogen atom or residue of a mono- or oligonucleotide, and R' the residue of a nucleoside or nucleotide.

The possibility of using ribonucleases for the synthesis of an internucleotide bond was first discovered in the case of pyrimidyl-RNase (pancreatic RNase) [412–415]. Similar reactions have been carried out with guanyl-RNase T_1 [416–419] and guanyl-RNase from actinomycetes [420], and also with some other RNases possessing less marked specificity relative to the base structure of the nucleotide component [421].

Another group of enzymes catalyses the formation of oligonucleotides or polynucleotides with 3' → 5'-phosphodiester bonds from ribo- or deoxyribonucleoside-5'-triphosphates or ribonucleoside-5'-diphosphates. The reactions taking place can be represented schematically by the equations:

$$n\text{pppN} \longrightarrow (\text{pN})_n + n\text{H}_4\text{P}_2\text{O}_7$$
$$n\text{ppN} \rightleftarrows (\text{pN})_n + n\text{H}_3\text{PO}_4$$

Some of the properties of the enzymes of this group used to prepare polynucleotides are summarized in Table 1.4.

The enzymes listed in this table possess narrow specificity relative to the structure of the sugar moiety of the nucleoside diphosphate or triphosphate; at the same time, their specificity relative to the nature of the heterocyclic base is comparatively broad. Besides diphosphates and triphosphates of the nucleosides composing nucleic acids, in some cases analogues of natural nucleotides can be incorporated into a polymer.

Only derivatives of mononucleotides can undergo polymerization catalysed by the enzymes of this group; attempts to achieve the corresponding polymerization of derivatives of dinucleotides and oligonucleotides have proved unsuccessful. The molecular weight of the resulting polymer largely depends on the conditions of polymerization; often both oligonucleotides (for example, trinucleoside diphosphates if the reaction is catalysed by polynucleo-tide phosphorylase) and polynucleotides with a molecular weight of several millions can be obtained.

As a rule the presence of a certain quantity of oligonucleotide or poly-nucleotide, which may have various functions to perform, is essential in the incubation mixture together with the enzyme and mononucleotide precursors for the enzymic polymerization of nucleoside-5'-diphosphates and triphos-phates to take place. These functions may be as follows.

1) An oligonucleotide with a 3'-terminal hydroxyl group may act as a polymerization "primer," i.e., as an acceptor to which the mononucleotide residues undergoing polymerization are attached. The sequence of the mono-mers in the growing polymer chain is independent of the monomer sequence in the added oligonucleotide or polynucleotide; the added primer is incor-porated into the reaction product, of which it constitutes the 5'-terminal sequence. This function of the oligonucleotide or polynucleotide is observed for reactions catalysed by terminal deoxynucleotidyl transferase and poly-nucleotide phosphorylase.

2) An oligonucleotide with a 3'-terminal phosphate group can sharply accelerate enzymic polymerization without itself being incorporated into the reaction product and without affecting the monomer sequence in it. In this case it is said that the oligonucleotide acts as an "initiator." The mechanism of action of initiators is not understood; the only reliably authenticated case of such a function of an oligonucleotide is the reaction catalysed by polynucleo-tide phosphorylase.

3) Finally, an oligonucleotide or polynucleotide may act as a "template" determining the nucleotide sequence in the resulting polymer, without itself being incorporated in the reaction product. In the case of template synthesis the reaction product is a complementary copy of the polynucleotide chain of the template. This case is observed in reactions catalysed by DNA-polymerase, RNA-polymerase, and RNA-synthetase.

Sometimes the function of the oligonucleotide or polynucleotide in enzymic polymerization is more complex. For example, if double-stranded complexes of oligodeoxynucleotides with chains of different lengths are added to DNA-polymerase, they are made up into complete double-stranded com-plexes. One of the chains of the complex in this case acts as the polymeriza-tion primer and the other as the template determining the nucleotide sequence of the synthesized segment of polymer.

If only one nucleoside-5'-triphosphate (of the four possible) is introduced into a reaction catalysed by RNA-polymerase, a homopolymer is synthesized, but in this case its structure is not a complementary copy of the added poly-nucleotide. This is an example of synthesis by reiteration: polymerization begins on the segment of the polynucleotide template containing a sequence of at least three nucleotide residues complementary to the single added

TABLE 1.4. Enzymes Catalysing the Polymerization of Nucleoside-5'- di- and triphosphates

Enzyme	Usual source for isolation	Structure of substrate	Possibility of polymerization without addition of oligo- or poly-nucleotide	Possibility of functioning of oligo- or poly-nucleotide as			Literature cited
				initiator	primer	template	
DNA-polymerase (duplicase)	Escherichia coli, calf thymus	Deoxyribonucleoside-5'-triphosphate	+	-	+	+	Surveys: 41, 422-424
Terminal deoxy-nucleotidyl trans-ferase (addase)	Calf thymus	Ditto	-	-	+	-	425-427
Polynucleotide phos-phorylase	Azotobacter vinelandii, Micrococcus lysodeik-ticus	Ribonucleoside-5'-di-phosphate	+(?)	+	+	-	Surveys: 428, 429
RNA-polymerase (transcriptase)	Rat liver, Azotobacter vinelandii, Escherichia coli	Ribonucleoside-5'-tri-phosphate	+	-	-	+	Surveys: 424, 428, 430-432
RNA-synthetase (RNA-replicase)	Escherichia coli, infected with phages f2, MS2, Q_β, etc.	Ribonucleoside-5'-tri-phosphate	-	-	-	+	433-435

nucleoside triphosphate. The product formed as a result of this partial copying of the template sequence acts as the primer for subsequent enzymic polymerization, leading to the formation of a homopolynucleotide.

In some cases enzymic polymerization can also take place in the absence of added oligonucleotide or polynucleotide. Ordinary specimens of polynucleotide phosphorylase will catalyse the polymerization of nucleoside diphosphates without the addition of primer or initiator. However, on further purification, the enzyme loses this property, which is evidently due to the presence of oligonucleotides in the upurified enzyme preparations. Under certain conditions polynucleotides can be synthesized without a template by means of DNA- and RNA-polymerases; these cases will be examined below.

The specificity of the enzymes now being discussed relative to the structure of the added oligonucleotide or polynucleotide as a rule is very broad. The only exceptions are the RNA-synthetases, which utilize RNAs of the corresponding phages much more efficiently as the template. Both polydeoxynucleotides and oligodeoxynucleotides, starting with a trinucleoside diphosphate, can act as primers for terminal deoxynucleotidyl transferase. However, polymerization also takes place in the presence of short oligodeoxynucleotides. Very slow polymerization is observed with trinucleoside diphosphates, but the reactions with hepta- to decadeoxynucleotides take place at appreciably higher velocities. RNA-polymerases from microorganisms can use single- and double-stranded polydeoxyribonucleotides and polyribonucleotides, and also oligodeoxyribonucleotides with a degree of polymerization of 5 and over and oligoribonucleotides of approximately the same size, as template.

Finally, a member of yet another group of enzymes which can be used to obtain synthetic polynucleotides is a recently discovered enzyme* which catalyses the formation of phosphodiester bonds between the 5'-terminal phosphate group and the 3'-terminal hydroxyl group of oligodeoxynucleotides incorporated in double-stranded complexes [436-438]. This reaction can be represented schematically as follows:

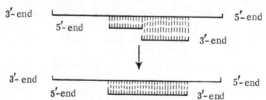

Both polydeoxyribonucleotides (double-helical DNA complexes, one chain of which contains ruptured phosphodiester bonds) and relatively short oligodeoxyribonucleotides with a degree of polymerization of 6-7 can take part in this enzymic reaction [439].

---

* This enzyme is known by the following names: polynucleotide-joining enzyme, DNA-ligase, and DNA-sealase.

Oligo- and polydeoxyribonucleotides. Enzymic methods of synthesis have very limited application in the field of oligodeoxyribonucleotides. The conditions must be chosen so that the terminal deoxynucleotidyl transferase catalyses the attachment of only a small number of mononucleotide units to the added primer. In this way, oligomer units can be obtained, such as [440]:

$$d[pT(pT)_5(pA)_{8-14}]; \qquad d[pT(pT)_5(pG)_{8-12}] \quad \text{or} \quad d[pT(pT)_3(pC)_{1-2}(pT)_{1-2}]$$

Methods of obtaining homopolymers and polymers containing a repeating oligonucleotide sequence have reached the highest level of development in the polydeoxyribonucleotide series.

Polymers of the first type can be synthesized with the aid of terminal deoxyribonucleotidyl transferase. Starting from oligodeoxyadenylic acids and 2'-deoxyadenosine-5'-triphosphate, for instance, polydeoxyadenylic acid can be obtained [441]:

$$(pdA)_3 + nppp dA \longrightarrow (pdA)_{n+1}$$

The reaction products are single-stranded polydeoxynucleotides, whose degree of polymerization in some cases may reach 600 *.

Double-stranded complexes of homogeneous polydeoxynucleotides can be obtained by means of DNA-polymerase. If the enzyme is incubated, without a template, with a mixture of deoxycytidine-5'-triphosphate and deoxyguanosine triphosphate (or deoxyinosine-5'-triphosphate), after a certain induction period (poly-pdC)·(poly-pdG) or (poly-pdC)·(poly-pdI) complexes are synthesized [442, 443]. After denaturation and centrifugation in a caesium sulphate gradient, the corresponding single-stranded polymers are obtained [444].

$$npppdG + npppdC \xrightarrow[\text{without template}]{\text{DNA-polymerase}} (pdG)_n \cdot (pdC)_n$$

In the analogous reaction with DNA-polymerase and without a template, a polymer with an alternating sequence of nucleosides is formed from a mixture of deoxyadenosine-5'-triphosphate and thymidine-5'-triphosphate [445]:

$$npppdA + npppdT \xrightarrow[\text{without template}]{\text{DNA-polymerase}} d(pApT)_n$$

To obtain homogeneous polydeoxyadenylic and polythymidylic acids, polymerization in the presence of a template must be used. Either polyadenylic or polyuridylic acid [446, 447], and also synthetic oligodeoxyribonucleotides, such as a $(pdA)_7 \cdot (pdT)_{11}$ complex [448] may be used as the template.

---

* The relative accessibility of certain homodeoxypolynucleotides makes it possible to obtain the corresponding oligodeoxynucleotides by their enzymic degradation. DNase I and nuclease from micrococci are used for this purpose (see page 54).

$$n\text{ppppd}A \xrightarrow[\quad (pdA)_7\cdot(pdT)_{11} \quad]{\text{DNA-polymerase}} (pdA)_n \quad n = 50\text{—}100$$

$$n\text{ppppd}A + n\text{ppppd}T \xrightarrow[\quad (pdA)_7\cdot(pdT)_{11} \quad]{\text{DNA-polymerase}} (pdA)_n \cdot (pdT)_n \quad n \approx 5000$$

The double-stranded complexes formed by this reaction can be separated by preparative centrifugation in a caesium chloride gradient after denaturation by treatment with alkali [449]. The products have a molecular weight of between 10^5 and $4 \cdot 10^6$.

Synthetic oligodeoxynucleotides containing an alternating sequence of nucleotide residues can be used as template for DNA-polymerase [448, 450]; the reaction products in this case are the corresponding polydeoxynucleotides, for example:

$$n\text{ppppd}A + n\text{ppppd}C + n\text{ppppd}G + n\text{ppppd}T \xrightarrow[\quad]{\text{DNA-polymerase} \atop [d(pTpC)_5]\cdot[d(pApG)_5]}$$
$$\longrightarrow [d(pTpC)_n \cdot [d(pApG)_n] \quad \text{Mol. wt.} \sim 1 \cdot 10^6$$

A similar approach has been used for the synthesis of polydeoxynucleotides with repeating trinucleotide [451] and tetranucleotide [452] sequences.

Polynucleotides containing analogues of natural nucleosides can also be synthesized with the aid of DNA-polymerase [453]. In particular, polynucleotides containing residues of 5-substituted pyrimidine deoxynucleotides [444] (for example, 5-bromodeoxycytidine, LXXXIX), 2'-deoxyuridine (XC) [454], and 4-thiothymidine (XCI) [455], have been obtained.

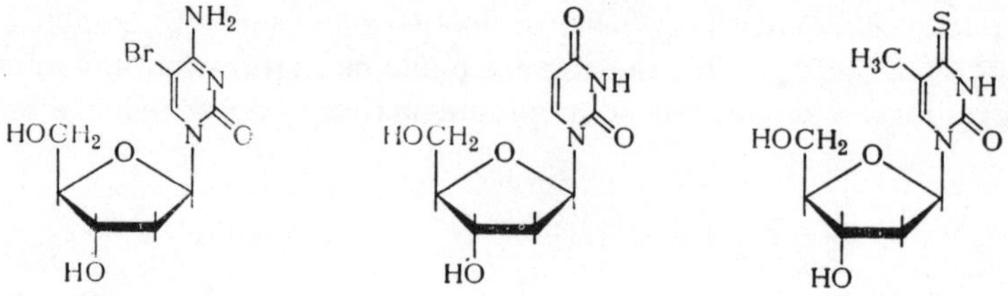

LXXXIX   XC   XCI

Heteropolydeoxynucleotides with a random arrangement of their nucleoside residues can be obtained by the reaction catalysed by terminal deoxyribonucleotidyl transferase [456-458]. This linking of synthetic oligodeoxyribonucleotides by means of polynucleotide-joining enzyme enables the synthesis of polynucleotides with a specific base sequence to be carried out. Khorana and co-workers [459], for instance, have obtained a double-stranded polynucleotide XCII corresponding in its structure to the DNA fragment responsible for synthesis of part of the sequence of alanine tRNA:

```
|———— Fragment 4 ————|—————— Fragment 1——————————|
  3'-end                                            5'-end
d(G-A-G-G-G-A-A-T-C-G-T-'A-C-C-C-T-C-T-C-A-G-A-G-G-C-C-A-A-G)
  · · · · · · · · · · · · · · · · · · · · · · · · · · ·
d( G-C-T-C-C-C-T-'T-A-G-C-A-T-G-G-G-A-G-A-G-T-C-T-C-C-G-G ) 3'-end
                  (10)                  (20)           (27)
  5'-end
|—————————Fragment 2——————————|—Fragment 3—|
```

XCII

Oligonucleotide fragments 1-4 used for the enzymic reaction were obtained by chemical synthesis. Enzymic polymerization can also be used for the partial synthesis of these polynucleotides. For instance, a polynucleotide containing fragments 1 and 2 can be built up to a 2-stranded complex of polynucleotide XCI by the action of DNA-polymerase and a mixture of deoxynucleoside-5'-triphosphates [460].

The combined use of DNA-polymerase and polynucleotide-joining enzyme has also enabled the first synthesis of a natural DNA (the DNA of phage ΦX174) to be successfully accomplished [193]. In this case, the natural single-stranded phage DNA was used as the template for synthesis. The enzymic reaction yielded a double-helical complex, the newly synthesized chain of which contained, not thymidine residues, but residues of 5-bromo-2'-deoxyuridine, since the thymidine-5'-triphosphate in the incubation mixture was replaced by 5-bromo-2'-deoxyuridine-5'-triphosphate. The complementary polynucleotides were separated by centrifugation in a density gradient, and the biosynthetic polynucleotide was then used as template in the enzymic reaction. The reaction product was a completely synthetic polynucleotide, identical in all investigated characteristics with the single-stranded DNA of phage φX174.

Oligoribonucleotides *. To obtain oligoribonucleotides, notably trinucleoside diphosphates, a number of enzymic methods have been suggested. Their undoubted advantage over chemical methods is the relative simplicity of the reaction and elimination of the need to introduce protecting groups.

Synthesis of trinucleoside diphosphates by means of ribonucleases is the most widely used method. As nucleotide component for this synthesis it is essential to use the cyclic phosphate of a mononucleotide capable of reacting with the particular enzyme (see pages 52-54 for specificity of the nucleases), or an oligonucleotide having the residue of such a nucleoside-2',3'-cyclic phosphate at its 3'-end; the nucleoside component may have any structure. For example, two types of trinucleoside diphosphates can be synthesized with the help of guanyl-RNase (X and Y denote nucleoside residues):

$$G{>}p + XpY \longrightarrow GpXpY$$
$$XpG{>}p + Y \longrightarrow XpGpY$$

If suitable conditions are chosen, relatively high yields of dinucleoside monophosphates and trinucleoside diphosphates can be obtained.

---

* Some oligonucleotides can most conveniently be obtained by cleavage of the corresponding polymers. This applies to homogeneous oligoribonucleotides readily obtained by brief alkaline hydrolysis of the corresponding homogeneous polyribonucleotides, and also to oligoribonucleotides of the $(Xp)_n Yp$ type, which are easily obtained by splitting the product of polymerization (by the action of polynucleotide phosphorylase) of ppX and ppY with a nuclease specific relative to the Yp residue. (X and Y in this case denote nucleoside residues). A number of di-, tri-, and tetranucleotides can be easily isolated from products of hydrolysis of RNA by the action of RNases.

D

Conditions have been found in which trinucleoside diphosphates are the chief product formed during the reaction catalysed by polynucleotide phosphorylase [461-463]; their structure is unequivocally determined by the structure of the added primer (dinucleoside monophosphate) and the structure of the nucleoside-5'-diphosphate undergoing polymerization, for example:

$$ApU + ppA \longrightarrow ApUpA + ApUpApA + ApUpApApA$$

Polyribonucleotides. The most convenient method of synthesis of homopolyribonucleotides is by polymerization of nucleoside-5'-diphosphates catalysed by polynucleotide phosphorylase. Depending on the reaction conditions polymers with molecular weights ranging from $3 \cdot 10^4$ to $2 \cdot 10^6$ can be obtained. Because of the broad specificity of the enzyme relative to the structure of the heterocyclic residue, not only homopolymers of nucleotides which are the major components of RNA [464, 465] can be obtained by this method, but also homopolymers which are derivatives of minor components, notably polypseudouridylic [466], polyinosinic [464], and poly-3-methyl- and poly-5-methyluridinic acids [467, 468], and also homopolymers of various methyl derivatives of cytidylic [469] and adenylic acids [470]. Other homopolymers containing residues of analogues of natural nucleosides also have been obtained, such as the 5-halogen derivatives of uridine (XCIII) [471] and cytidine (XCIV) [472], isoadenosine (XCV) [473], 2-aminoadenosine [474], and other compounds.

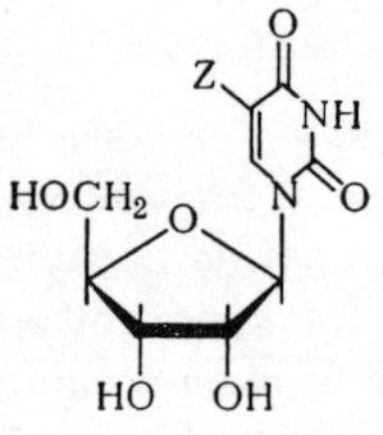

XCIII   XCIV   XCV

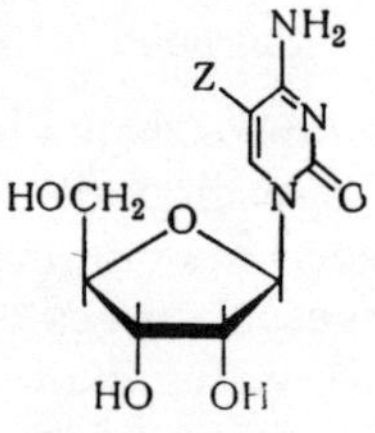

Z= F,Cl,Br,l   Z= Br,l

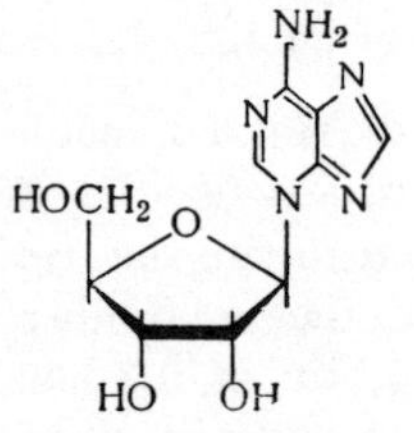

If oligonucleotides which are derivatives of another nucleotide are used as primers in the reaction with highly purified polynucleotide phosphorylase, polymer units of the type $(pX)_n (pY)_m$ are obtained. For example:

$$C(pC)_5 + nppA \longrightarrow C(pC)_5(pA)_n \quad n = 10-100$$

In particular, if dinucleoside monophosphate is the primer, a homonucleotide containing a specific nucleotide triplet at the 5'-end is formed:

$$GpU + nppA \longrightarrow GpU(pA)_n$$

To obtain homopolyribonucleotides, the reaction catalysed by RNA-polymerase also can be used. As in the case of DNA-polymerase, under certain conditions synthesis of the polymer can take place without the addition of a template. Under these circumstances, a double-stranded complex

of homopolynucleotides is formed from a mixture of adenosine-5'-triphos-
phate and uridine-5'-triphosphate [475], while a polymer with alternating
mononucleotide components is obtained from a mixture of cytidine-5'-tri-
phosphate and inosine-5'-triphosphate [476].

$$n\text{pppA} + n\text{pppU} \xrightarrow[\text{\underline{vinelandii} without template}]{\text{RNA-polymerase from \underline{Azotobacter}}} (\text{pA})_n \cdot (\text{pU})_n$$

$$n\text{pppC} + n\text{pppI} \xrightarrow[\text{\underline{vinelandii} without template}]{\text{RNA-polymerase from \underline{Azotobacter}}} (\text{pCpI})_n$$

Polyadenylic, polyuridylic, and polycytidylic acids can be obtained by
polymerization catalysed by RNA-polymerase under the conditions of synthesis by reiteration (see page   ), and different DNAs can be used as the
original template. To obtain polyguanylic acid, synthesis is carried out on
a polycytidylic acid template [477].

The RNA-polymerase reaction is widely used to prepare polyribonucleotides with an alternate sequence of monomer components. Oligodeoxyribonucleotides with a known sequence, obtained by chemical synthesis, can
be used as template [478, 479]; polymers of higher molecular weight are
formed by the use of polydeoxyribonucleotides with a repeating sequence of
di-, tri-, and tetranucleotide components as primer. For example [480]:

$$n\text{pppA} + n\text{pppG} \xrightarrow[\text{d(pTpC)}_n \cdot \text{d(pApG)}_n]{\text{RNA-polymerase}} (\text{pApG})_n$$

$$n\text{pppU} + n\text{pppC} \xrightarrow[\text{d(pTpC)}_n \cdot \text{d(pApG)}_n]{\text{RNA-polymerase}} (\text{pUpC})_n$$

Polymers containing residues of minor components of RNA and, in
particular, dihydrouridine [481] (a polymer of this compound cannot be obtained with polynucleotide phosphorylase), as well as analogues of major
components of RNA, such as 5-bromouridine [482], can be formed in the
presence of RNA-polymerase. Polymers of nucleoside antibiotics, such as
tubercidin [483] and formycin [484], analogues of adenosine, can also be obtained in this way.

By the polymerization of a mixture of nucleoside diphosphates in the
presence of polynucleotide phosphorylase, polynucleotides in which the distribution of the various monomer units is close to statistical are formed.
The problem of obtaining polyribonucleotides of known structure with a
specific nucleotide sequence has not yet been solved, although the possibility
of obtaining such compounds with the aid of the RNA-polymerase reaction,
using appropriate polydeoxyribonucleotides as template, is in no doubt.

Bibliography

1. A. G. E. Pearse, Histochemistry, Churchill, London (1953).
2. H. Swift, in: The Nucleic Acids, Vol. 1, E. Chargaff and J. N. Davidson (editors), Academic Press, New York (1955).
3. A. V. Zelenin, Luminescence Histochemistry of Nucleic Acids [in Russian], Nauka (1967).
4. V. S. Shapot, The Nucleases [in Russian], Meditsina (1968).
5. Z. Dische, in: The Nucleic Acids, Vol. 1, E. Chargaff and J. N. Davidson (editors), Academic Press, New York (1955).
6. K. S. Kirby, in: Progress in Nucleic Acid Research, Vol. 3, J. N. Davidson and W. E. Cohn (editors), Academic Press, New York (1964).
7. J. Josse and J. Eigner, Ann. Rev. Biochem., 35:789 (1966).
8. K. S. Kirby and T. L. V. Ulbricht, Ann. Repts. Progr. Chem., 63:536 (1966).
9. K. S. Kirby, Biochem. J., 66:495 (1957); 70:206 (1958).
10. E. R. M. Kay, N. S. Simmons, and A. L. Dounce, J. Am. Chem. Soc., 74:1724 (1952).
11. M. G. Sevag, D. B. Lackman, and J. Smolens, J. Biol. Chem., 124:425 (1938).
12. J. Marmur, in: Methods in Enzymology, Vol. 6, S. P. Colowick and N. O. Kaplan (editors), Academic Press, New York – London (1963), p. 726; J. Mol. Biol., 3:208 (1961).
13. H. R. Massie and B. H. Zimm, Proc. Nat. Acad. Sci. USA, 54:1636 (1965).
14. P. F. Davison, Proc. Nat. Acad. Sci. USA, 45:1560 (1959).
15. P. F. Davison, Nature, 185:918 (1960).
16. C. A. Thomas and L. A. McHattie, Ann. Rev. Biochem., 36:485 (1967).
17. P. A. Edwards and K. V. Shooter, Quarth. Rev., 19:369 (1965).
18. H. O. Smith and M. Levine, in: Methods in Enzymology, Vol. 12, Part A, L. Grossman and K. Moldave (editors), Academic Press, New York – London (1967), p. 557.
19. P. Doty, B. B. McGill, and S. A. Rice, Proc. Nat. Acad. Sci. USA, 44:432 (1958).
20. I. Rubinstein, C. A. Thomas, and A. D. Hershey, Proc. Nat. Acad. Sci. USA, 47:1113 (1961).
21. E. Burgi and A. D. Hershey, Biophys. J., 3:309 (1963).
22. F. W. Studier, J. Mol. Biol., 11:373 (1965).
23. J. Eigner and P. Doty, J. Mol. Biol., 12:549 (1965).
24. D. M. Crothers and B. H. Zimm, J. Mol. Biol., 12:525 (1965).
25. A. K. Kleinschmidt, D. Lang, D. Jacherts, and R. K. Zahn, Biochim. Biophys. Acta, 61:857 (1962).
26. A. K. Kleinschmidt, Naturwiss., 54:417 (1967).
27. J. Marmur and P. Doty, J. Mol. Biol., 5:109 (1962).
28. C. L. Schildkraut, J. Marmur, and P. Doty, J. Mol. Biol., 4:430 (1962).
29. J. Vinograd and J. E. Hearst, Fortschr. Chem. Org. Naturstoffe, 20:372 (1962).
30. N. Sueoka and T. Y. Cheng, J. Mol. Biol., 4:161 (1962).
31. R. K. Main, M. J. Wilkins, and L. J. Cole, J. Am. Chem. Soc., 81:6490 (1959); G. Bernardi, Biochim. Biophys. Acta, 174:423 (1969).
32. B. M. Albert, in: Methods in Enzymology, Vol. 12, Part A, L. Grossman and K. Moldave (editors), Academic Press, New York – London (1967), p. 566.
33. R. Sinsheimer, in: Nucleic Acids, Vol. 2, E. Chargaff and J. N. Davidson (editors), Academic Press, New York (1955).
34. Yu. Z. Gendon, in: Virology and Immunology [in Russian], L. A. Zil'ber (editor), Nauka (1964), p. 36.
35. J. A. Cohen, Science, 158:343 (1967).

36. J. A. Cohen, Bull. Soc. Chim. Biol., 50:293 (1968).
37. C. A. Thomas and J. Abelson, in: Procedures in Nucleic Acid Research, G. L. Cantoni and D. R. Davies (editors), Harper and Row, New York — London (1966), p. 553.
38. D. Freifelder, in: Methods in Enzymology, Vol. 12, Part A, L. Grossman and K. Moldave (editors), Academic Press, New York — London (1967), p. 550.
39. R. L. Sinsheimer, in: Procedures in Nucleic Acid Research, G. L. Cantoni and D. R. Davies (editors), Harper and Row, New York — London (1966), p. 569.
40. A. S. Spirin, Uspekhi Biol. Khimii, 4:93 (1962).
41. T. I. Tikhonenko, in: Biosynthesis of Protein and Nucleic Acids [in Russian], A. S. Spirin (editor), Nauka (1965), p. 193.
42. S. Kit, Ann. Rev. Biochem., 32:43 (1963).
43. L. N. Zhinkin and P. P. Rumyantsev, Textbook of Cytology [in Russian], Vol. 1, Nauka (1965), p. 355.
44. J. Cairns, J. Mol. Biol., 6:208 (1963).
45. K. I. Berns and C. A. Thomas, J. Mol. Biol., 11:476 (1955).
46. L. A. MacHattie, K. I. Berns, and C. A. Thomas, J. Mol. Biol., 11:648 (1965).
47. E. Safert and H. Venner, Z. Physiol. Chem., 344:278 (1966).
48. M. J. Pitout and I. J. Maré Nature, 215:1187 (1967).
49. I. R. Lehman, J. Biol. Chem., 235:1479 (1960).
50. K. Miura, in: Methods in Enzymology, Vol. 12, Part A, L. Grossman and K. Moldave (editors), Academic Press, New York — London (1967), p. 543.
51. S. Falkow and R. V. Citarella, J. Mol. Biol., 12:138 (1965).
52. T. F. Roth and D. R. Helinski, Proc. Nat. Acad. Sci. USA, 58:650 (1967).
53. P. T. Hikson and T. F. Roth, Proc. Nat. Acad. Sci. USA, 58:1731 (1967).
54. N. R. Cozzarelli, R. B. Kelly, and A. Kornberg, Proc. Nat. Acad. Sci. USA, 60:922 (1968).
55. D. R. Freifelder and D. Freifelder, J. Mol. Biol., 32:25 (1968).
56. S. A. Krolenok and I. A. Chernogradskaya, in: Textbook of Cytology [in Russian], Vol. 2, Yu. B. Vakhtin (editor), Nauka (1966), p. 280.
57. J. H. Taylor, in: Molecular Genetics, J. H. Taylor (editor), Academic Press, New York (1963).
58. J. A. Huberman and A. D. Riggs, Proc. Nat. Acad. Sci. USA, 55:599 (1966).
59. M. S. Sasaki and A. Norman, Exptl. Cell Res., 44:642 (1966).
60. S. Zamenhof, in: Methods in Enzymology, Vol. 3, S. P. Collowick and N. O. Kaplan (editors), Academic Press, New York — London (1957), p. 696.
61. C. Kidson, J. Mol. Biol., 17:1 (1966).
62. K. Sakabe and R. Okazaki, Biochim. Biophys. Acta, 129:651 (1966).
63. M. Oishi, Proc. Nat. Acad. Sci. USA, 60:329 (1968).
64. V. Holoubek, Anal. Biochem., 18:375 (1967).
65. A. G. Levis, V. Krsmanovic, A. Miller-Faures, and M. Errera, Europ. J. Biochem., 3:57 (1967).
66. S. Kit, J. Mol. Biol., 3:711 (1961).
67. T. Y. Cheng and N. Sueoka, Science, 141:1194 (1963).
68. P. Borst and G. J. C. M. Ruttenberg, Biochim. Biophys. Acta, 114:647 (1966).
69. W. G. Flamm, H. E. Bond, H. E. Burr, and S. B. Bond, Biochim. Biophys. Acta, 123:652 (1966).
70. H. E. Bond, W. C. Flamm, H. E. Burr, and S. B. Bond, J. Mol. Biol., 27:289 (1967).
71. N. Sueoka, J. Mol. Biol., 3:31 (1961).
72. N. Sueoka and T. Y. Cheng, Proc. Nat. Acad. Sci. USA, 48:1851 (1962); J. Mol. Biol., 4:161 (1962).

73. R. P. Klett and M. Smith, in: Methods in Enzymology, Vol. 12, Part A, L. Grossman and K. Moldave (editors), Academic Press, New York – London (1967), p. 554.

74. M. Warring and R. J. Britten, Science, 154:794 (1966).

75. Zh. G. Shmerling, Uspekhi Sovr. Biol., 59:33 (1965).

76. S. Granick and A. Gibor, Progr. Nucl. Acid Res., 6:143 (1967).

77. T. Iwamura, Progr. Nucl. Acid Res., 5:133 (1966).

78. M. Rabinowitz, Bull. Soc. Chim. Biol., 50:311 (1968).

79. H. Roels, Intern. Rev. Cytol., 19:1 (1966).

80. D. J. Luck, Proc. Nat. Acad. Sci. USA, 49:223 (1963); D. J. Luck and E. Reich, Proc. Nat. Acad. Sci. USA, 52:931 (1964).

81. M. Rabinowitz, J. H. Singlair, L. DeSalle, R. Haselkorn, and H. Swift, Proc. Nat. Acad. Sci. USA, 53:1126 (1965).

82. G. F. Kafl and M. A. Créce, in: Methods in Enzymology, Vol. 12, Part A, L. Grossman and K. Moldave (editors), Academic Press, New York – London (1967), p. 533.

83. I. B. David and D. B. Wolstenholme, J. Mol. Biol., 28:233 (1967).

84. U. Fukuhara, Proc. Nat. Acad. Sci. USA, 58:1065 (1967).

85. M. M. K. Nass, Proc. Nat. Acad. Sci. USA, 56:1215 (1966).

86. A. M. Kroon, P. Borst, E. F. J. Van Briggen, and G. J. C. M. Ruttenberg, Proc. Nat. Acad. Sci. USA, 56:1836 (1966).

87. J. H. Singlair, B. J. Stevens, N. Gross, and M. Rabinowitz, Biochim. Biophys. Acta, 145:528 (1967).

88. M. Naas and S. Naas, J. Cell. Biol., 19:593 (1963).

89. E. H. L. Chun, M. H. Voughan, and A. Rich, J. Mol. Biol., 7:130 (1963).

90. J. M. Eisenstadt and G. Brawerman, in: Methods in Enzymology, Vol. 12, Part A, L. Grossman and K. Moldave (editors), Academic Press, New York – London (1967), p. 541.

91. O. C. Richards, Proc. Nat. Acad. Sci. USA, 57:156 (1967).

92. K. S. Kirby, Biochim. Biophys. Acta, 55:545 (1962); Biochem. J., 96:266 (1965).

93. H. H. Hiatt, J. Mol. Biol., 5:217 (1962).

94. J. E. M. Midgley, Biochim. Biophys. Acta, 108:358 (1965).

95. K. Scherrer and S. E. Darnell, Biochem. Biophys. Res. Comm., 7:486 (1962).

96. R. J. Britton and R. B. Roberts, Science, 131:32 (1960).

97. E. H. McConkey, in: Methods in Enzymology, Vol. 12, Part A, L. Grossman and K. Moldave (editors), Academic Press, New York – London (1967), p. 620.

98. J. D. Mandell and A. D. Hershey, Anal. Biochem., 1:66 (1962).

99. T. Murakami, in: Methods in Enzymology, Vol. 12, Part A, L. Grossman and K. Moldave (editors), Academic Press, New York – London (1967), p. 634.

100. R. Tsanev, Biochim. Biophys. Acta, 103:374 (1965).

101. E. G. Richards, J. A. Coll, and W. B. Gratzer, Anal. Biochem., 12:452 (1965).

102. V. E. Loening, Biochem. J., 102:251 (1967).

103. D. H. L. Bishop, J. R. Claybrook, and S. Spiegelman, J. Mol. Biol., 26:373 (1967).

104. J. P. Himmel and B. S. Anderson, Arch. Biochem. Biophys., 112:443 (1965).

105. C. G. Kurland, J. Mol. Biol., 2:83 (1960).

106. A. Maeda, J. Biochem., 50:377 (1961).

107. A. S. Spirin, Biokhimiya, 26:511 (1961).

108. A. A. Hadjiolov, P. V. Venkov, and R. G. Tsanev, Anal. Biochem., 17:263 (1966).

109. H. Schuster, in: The Nucleic Acids, Vol. 3, E. Chargaff and J. N. Davidson (editors), Academic Press, New York (1960).

110. R. Markham, in: Progress in Nucleic Acid Research, Vol. 2, J. N. Davidson and W. E. Cohn (editors), Academic Press, New York (1963).

111. C. Weissman and S. Ochoa, Progr. Nucl. Acid Res., 6:353 (1967).

112. H. G. Wittmann and C. Scholtissek, Ann. Rev. Biochem., 35:299 (1966).

113. M. Girard, in: Methods in Enzymology, Vol. 12, Part A, L. Grossman and K. Moldave (editors), Academic Press, New York – London (1967), p. 581.

114. H. Fraenkel-Conrat, in: Procedures in Nucleic Acid Res., G. L. Cantoni and D. R. Davies (editors), Harper and Row, New York – London (1966), p. 480.

115. D. F. Summers, ibid., p. 488.

116. P. J. Comatos, ibid., p. 493.

117. D. Mills, D. H. L. Bishop, and S. Spiegelman, Proc. Nat. Acad. Sci. USA, 60:713 (1968).

118. P. J. Comatos and I. Tamm, Proc. Nat. Acad. Sci. USA, 49:707 (1963).

119. K. L. Tomita and A. Rich, Nature, 201:1160 (1964).

120. T. Sato, Y. Kyoguko, S. Higuchi, Y. Mitsui, Y. Itaka, M. Tsuboi, and K. Miura, J. Mol. Biol., 16:180 (1966).

121. P. H. Hofschneider, J. Ammann, and B. Francke, in: Methods in Enzymology, Vol. 12, Part A, L. Grossman and K. Moldave (editors), Academic Press, New York – London (1967), p. 613.

122. A. S. Spirin, in: Progress in Nucleic Acid Research, Vol. 1, J. N. Davidson and W. E. Cohn (editors), Academic Press, New York (1963).

123. A. A. Bogdanov and A. S. Shakulov, in: Biosynthesis of Protein and Nucleic Acids [in Russian], A. S. Spirin (editor), Nauka (1965), p. 86.

124. M. Peterman, The Physical and Chemical Properties of Ribosomes, Elsevier, Amsterdam (1964).

125. A. S. Spirin and L. I. Gavrilova, The Ribosome [in Russian], Nauka (1968).

126. K. Moldave, in: Methods in Enzymology, Vol. 12, Part A, L. Grossman and K. Moldave (editors), Academic Press, New York – London (1967), p. 607.

127. E. L. Bolton, in: Procedures in Nucleic Acid Research, G. L. Cantoni and D. R. Davies (editors), Harper and Row, New York – London (1966), p. 437.

128. J. Barlow and A. P. Mathias, in: Procedures in Nucleic Acid Research, G. L. Cantoni and D. R. Davies (editors), Harper and Row, New York – London (1966), p. 444.

129. R. Rosset, R. Montier, and J. Julien, Bull. Soc. Chim. Biol., 46:87 (1964).

130. D. G. Comb and T. Zehavi-Willner, J. Mol. Biol., 23:441 (1967).

131. G. L. Brown, Progr. Nucl. Acid Res., 2:259 (1963).

132. A. A. Baev, in: Biosynthesis of Protein and Nucleic Acids [in Russian], A. S. Spirin (editor), Nauka (1965), p. 50.

133. A. A. Baev, Uspekhi Biol. Khimii, 7:67 (1965).

134. K. Tanaka, in: Procedures in Nucleic Acid Research, G. L. Cantoni and D. R. Davies (editors), Harper and Row, New York – London (1966), p. 466.

135. J. Apgar, R. W. Holley, and S. H. Merrill, J. Biol. Chem., 237:796 (1962).

136. R. W. Holley, J. Apgar, G. A. Everett, J. T. Madison, S. H. Merrill, and A. Zamir, in: Synthesis and Structure of Nucleic Acids [in Russian], Ya. M. Varshavskii (editor), Mir (1966), p. 94.

137. M. Tada, M. Schweiger, and H. G. Zachau, Z. Physiol. Chem., 328:85 (1962).

138. B. P. Doctor, in: Methods in Enzymology, Vol. 12, Part A, L. Grossman and K. Moldave (editors), Academic Press, New York – London (1967), p. 664.

139. R. M. Bock and J. D. Cherayil, ibid., p. 638.

140. P. L. Bergquist, B. C. Bagulev, and R. K. Ralph, ibid., p. 660.

141. N. Sueoka and T. Yamane, ibid., p. 658.

142. R. Stern and U. Z. Littauer, Biochemistry, 7:3469 (1968).

143. I. Gillam, S. Millward, D. Blew, M. von Tigerstrom, E. Wimmer, and G. M. Tener, Biochemistry, 6:3043 (1967).
144. A. D. Kelmers, G. D. Novelli, and M. P. Stubberg, J. Biol. Chem., 240:3979 (1965).
145. J. F. Weiss and A. D. Kelmers, Biochemistry, 6:2507 (1967); J. F. Weiss, R. L. Pearson, and A. D. Kelmers, Biochemistry, 7:3479 (1968).
146. P. C. Zamecnick, M. L. Stephenson, and J. F. Scott, Proc. Nat. Acad. Sci. USA, 46:811 (1960).
147. M. L. Stephenson and P. C. Zamecnick, Proc. Nat. Acad. Sci. USA, 47:1627 (1961).
148. M. L. Stephenson and P. C. Zamecnick, in: Methods in Enzymology, Vol. 12, Part A, L. Grossman and K. Moldave (editors), Academic Press, New York – London (1967), p. 670.
149. M. A. Grachev, N. I. Menzorova, L. S. Sandakchiev, É. I. Budovskii, and D. G. Knorre, Biokhimiya, 31:840 (1966).
150. A. H. Mehler and A. Bank, J. Biol. Chem., 238:2888 (1963).
151. S. Simon, U. Z. Littauer, and E. Katchalski, Biochim. Biophys. Acta, 80:169 (1964).
152. I. Gillam, D. Blew, R. C. Warrington, M. von Tigerstrom, and G. M. Tener, Biochemistry, 7:3459 (1968).
153. H. Harris, in: Progress in Nucleic Acid Research, Vol. 2, J. N. Davidson and W. E. Cohn (editors), Academic Press, New York (1963).
154. L. P. Ovchinnikov, Uspekhi Biol. Khimii, 9:3 (1968).
155. G. P. Georgiev, Progr. Nucl. Acid Res., 6:259 (1967).
156. A. A. Hadjiolov, Progr. Nucl. Acid Res., 7:196 (1967).
157. R. P. Perry, Progr. Nucl. Acid Res., 6:219 (1967).
158. R. A. Weinberg, U. Loening, A. Willems, and S. Penman, Proc. Nat. Acad. Sci. USA, 58:1088 (1967).
159. K. G. Gazaryan, N. G. Shuppe, and B. D. Prokoshkin, Biokhimiya, 31:108 (1966).
160. M. I. Lerman, E. A. Vladimirtseva, V. V. Terskikh, and G. P. Georgiev, Biokhimiya, 30:375 (1965).
161. E. K. F. Bautz, J. Mol. Biol., 17:298 (1966).
162. K. Scherrer, L. Marcaud, F. Zajdela, B. Breckenridge, and F. Cros, Bull. Soc. Chim. Biol., 48:1037 (1966).
163. R. Soeiro, C. Birnboim, and J. Darnell, J. Mol. Biol., 19:362 (1966).
164. R. C. Huang and J. Bonner, Proc. Nat. Acad. Sci. USA, 54:960 (1965).
165. W. Benjamin, O. A. Levander, A. Gellhorn, and R. H. DeBellis, Proc. Nat. Acad. Sci. USA, 55:858 (1966).
166. J. Bonner and J. Widholm, Proc. Nat. Acad. Sci. USA, 57:1379 (1967).
167. F. Lipmann, in: Progress in Nucleic Acid Research, Vol. 1, J. N. Davidson and W. E. Cohn (editors), Academic Press, New York (1963).
168. I. A. Bass and V. A. Gvozdev, in: Biosynthesis of Protein and Nucleic Acids [in Russian], A. S. Spirin (editor), Nauka (1965), p. 50.
169. M. F. Singer and P. Leder, Ann. Rev. Biochem., 35:195 (1966).
170. H. Chantreune, A. Burny, and G. Marbaix, Progr. Nucl. Acid Res., 7:173 (1967).
171. V. Shapot and H. C. Pitot, Biochim. Biophys. Acta, 119:37 (1966).
172. N. P. Rodionova and V. S. Shapot, Biochim. Biophys. Acta, 129:206 (1966).
173. C. Bergeron-Bouvet and Y. Moule, Biochim. Biophys. Acta, 123:617 (1966).
174. H. W. King and W. Fitschen, Biochim. Biophys. Acta, 155:32 (1968).
175. P. J. Rogers, B. N. Preston, E. B. Titchener, and A. W. Linnam, Biochem. Biophys. Res. Comm., 27:405 (1967).

176. M. R. Rifkin, D. D. Wood, and D. J. D. Luck, Proc. Nat. Acad. Sci. USA, 58:1025 (1967).
177. H. Küntel and H. Noss, Nature, 215:1340 (1967).
178. E. Barnett and D. H. Brown, Proc. Nat. Acad. Sci. USA, 57: 452 (1967).
179. J. J. Epler and E. Barnett, Biochem. Biophys. Res. Comm., 28:328 (1967).
180. C. A. Buck and M. M. K. Nass, Proc. Nat. Acad. Sci. USA, 60:1045 (1968).
181. E. Stuts and H. Noll, Proc. Nat. Acad. Sci. USA, 57:774 (1967).
182. P. D. Phethean, L. Jervis, and M. Hallway, Biochem. J., 108:25 (1968).
183. D. M. Brown and A. R. Todd, in: Nucleic Acids, Vol. 1, E. Chargaff and J. N. Davidson (editors), Academic Press, New York – London (1955), p. 409.
184. P. A. Levene and H. S. Simms, J. Biol. Chem., 65:519 (1925); 70:327 (1927).
185. D. M. Brown and A. R. Todd, J. Chem. Soc., 52 (1952).
186. C. E. Carter, J. Am. Chem. Soc., 73:1537 (1951).
187. C. A. Dekker, A. M. Michelson, and A. R. Todd, J. Chem. Soc., 947 (1953).
188. W. E. Cohn and E. Volkin, J. Biol. Chem., 203:319 (1953).
189. D. M. Brown and A. R. Todd, J. Chem. Soc., 2040 (1953).
190. L. A. Heppel and P. R. Whitfield, Biochem. J., 60:1 (1955).
191. P. R. Whitfield, L. A. Heppel, and R. Markham, Biochem. J., 60:15 (1955).
192. IUPAC-IUB Combined Commision on Biochemical Nomenclature, Biochemistry, 5:1445 (1966).
193. M. Goulian, A. Kornberg, and R. L. Sinsheimer, Proc. Nat. Acad. Sci. USA, 58:2321 (1967).
194. S. Spiegelman, I. Haruna, I. B. Holland, G. Beaudreau, and D. Mills, Proc. Nat. Acad. Sci. USA, 54:919 (1965).
195. A. Bendich and H. S. Rosenkranz, in: Progress in Nucleic Acid Research, Vol. 1, J. N. Davidson and W. E. Cohn (editors), Academic Press, New York (1963).
196. Yu. F. Drygin, A. A. Bogdanov, and M. A. Prokof'ev, Chemistry of Natural Compounds [in Russian] (1966), p. 218.
197. R. K. Ralph, R. J. Young, and H. G. Khorana, J. Am. Chem. Soc., 84:1490 (1962).
198. U. L. RajBhandary, R. J. Young, and H. G. Khorana, J. Biol. Chem., 239:3875 (1964).
199. R. Dulbecco and J. D. Smith, Biochim. Biophys. Acta, 39:358 (1960).
200. A. Steinschneider and H. Fraenkel-Conrat, Biochemistry, 5:2729 (1966).
201. J. A. Hunt, Biochem. J., 95:541 (1965).
202. U. L. RajBhandary, J. Biol. Chem., 243:556 (1968).
203. A. Stuart and H. G. Khorana, J. Biol. Chem., 239:3885 (1964).
204. A. Novogrodsky and J. Hurwitz, J. Biol. Chem., 241:2923 (1966).
205. R. Wu and A. D. Kaiser, Proc. Nat. Acad. Sci. USA, 57:170 (1967).
206. R. Weiss and C. C. Richardson, J. Mol. Biol., 23:405 (1967).
207. D. H. L. Bishop, N. R. Pace, and S. Spiegelman, Proc. Nat. Acad. Sci. USA, 58:1790 (1967).
208. R. Roblin, J. Mol. Biol., 31:51 (1968).
209. R. De Wachter, J. P. Verhassel, and W. Fiers, Biochem. Biophys. Acta, 157:195 (1968).
210. M. Watanabe and J. T. August, Proc. Nat. Acad. Sci. USA, 59:513 (1968).
211. D. A. Clayton and J. Vinograd, Nature, 216:652 (1967).
212. B. Hudson and J. Vinograd, Nature, 216:647 (1967).
213. L. Pikó, D. J. Blair, A. Tyler, and J. Vinograd, Proc. Nat. Acad. Sci. USA, 59:838 (1968).
214. Y. Hotta and A. Bassel, Proc. Nat. Acad. Sci. USA, 53:357 (1965).
215. P. A. Levene and L. W. Bass, Nucleic Acid, Chemical Catalog Co., New York (1931).

216. D. O. Jordan, Chemistry of the Nucleic Acids, Butterworth, London (1960).
217. A. Michelson, The Chemistry of Nucleosides and Nucleotides, Academic Press, New York – London (1963).
218. T. V. Venkstern, Uspekhi Biol. Khimii, 6:3 (1964).
219. R. H. Hall, Biochemistry, 4:661 (1965).
220. J. W. Littlefield and D. B. Dunn, Biochem. J., 70:642 (1958).
221. R. H. Hall, Biochem. Biophys. Res. Comm., 12:361 (1963).
222. M. N. Lipsett, J. Biol. Chem., 240:3975 (1965).
223. J. T. Madison and R. W. Holley, Biochem. Biophys. Res. Comm., 18:152 (1965).
224. A. W. Lis and W. E. Passarge, Arch. Biochem. Biophys., 114:593 (1966).
225. M. W. Gray and B. G. Lane, Biochemistry, 7:3441 (1968).
226. Y. Iwanami and G. M. Brown, Arch. Biochem. Biophys., 124:472 (1968).
227. L. Baczynsky, K. Biemann, and R. H. Hall, Science, 159:1481 (1968).
228. J. A. Carbon, L. Hung, and D. S. Jones, Proc. Nat. Acad. Sci. USA, 53:979 (1965).
229. J. A. Carbon, H. David, and M. H. Studier, Science, 161:1146 (1968).
230. W. E. Cohn, J. Biol. Chem., 235:1488 (1960).
231. D. B. Dunn, Biochim. Biophys. Acta, 38:176 (1960).
232. G. L. Cantoni, H. V. Gelboin, S. W. Luborsky, and H. H. Richards, Biochim. Biophys. Acta, 61:354 (1962).
233. U. L. RajBhandary, R. D. Faulkner, and A. Stuart, J. Biol. Chem., 243:575 (1968).
234. H. Feldmann, D. Dütting, and H. G. Zachau, Z. Physiol. Chem., 347:236 (1966).
235. H. G. Gassen and H. Witzel. Biochim. Biophys. Acta, 95:244 (1965).
236. D. B. Dunn, Biochim. Biophys. Acta, 46:198 (1961).
237. R. H. Hall, Biochem. Biophys. Res. Comm., 13:394 (1963).
238. K. Biemann, S. Tsunakawa, J. Sonnenbichler, H. Feldmann, H. Dütting and H. G. Zachau, Angew. Chem., 78:600 (1966).
239. M. J. Robins, R. H. Hall, and R. Thedford, Biochemistry, 6:1837 (1967).
240. R. H. Hall, L. Csonka, H. David, and B. McLennan, Science, 156:69 (1967).
241. F. Harada, H. J. Gross, F. Kimura, S. H. Chang, S. Nishimura, and U. L. RajBhandary, Biochem. Biophys. Res. Comm., 33:299 (1968); W. J. Burrows, D. J. Armstrong, F. Skoog, S. M. Hecht, J. T. A. Boyle, N. J. Leonard, and J. Occolowitz, Science, 161:691 (1968).
242. R. H. Hall, Biochemistry, 3:769 (1964).
243. R. H. Hall and G. B. Chedda, J. Biol. Chem., 240:PC 2754 (1965).
244. J. B. Smith and D. B. Dunn, Biochem. J., 72:294 (1959).
245. D. B. Dunn, Biochem. J., 86:14p (1963).
246. J. D. Smith and D. B. Dunn, Biochim. Biophys. Acta, 31:573 (1959).
247. R. H. Hall, Biochemistry, 3:876 (1964).
248. J. L. Nichols and B. G. Lane, Canad. J. Biochem., 44:1633 (1966); Biochim. Biophys. Acta, 166:605 (1968).
249. G. R. Wyatt and S. S. Cohen, Nature, 170:1072 (1952).
250. I. R. Lehman and E. A. Pratt, J. Biol. Chem., 235:3542 (1960).
251. S. Kuno and I. R. Lehman, J. Biol. Chem., 237:1266 (1962).
252. R. G. Kallen, M. Simon, and J. Marmur, J. Mol. Biol., 5:248 (1962).
253. I. Takahashi and J. Marmur, Nature, 197:794 (1963).
254. B. F. Vanyushin, Uspekhi Sovr. Biol., 65:163 (1968).
255. R. G. Wyatt, Biochem. J., 48:581 (1951).
256. D. B. Dunn and J. B. Smith, Biochem. J., 68:627 (1958).
257. G. Unger and H. Venner, Z. Physiol. Chem., 344:280 (1966).

258. A. Bendich, in: Methods in Enzymology, Vol. 3, S. P. Colowick and N. O. Kaplan (editors), Academic Press, New York – London (1957), p. 715.

259. W. E. Cohn, in: Methods in Enzymology, Vol. 3, S. P. Colowick and N. O. Kaplan (editors), Academic Press, New York – London (1957), p. 724.

260. J. D. Smith, in: Methods in Enzymology, Vol. 12, Part A, L. Grossman and K. Moldave (editors), Academic Press, New York – London (1967), p. 350.

261. R. Markham, in: Methods in Enzymology, Vol. 3, S. P. Colowick and N. O. Kaplan (editors), Academic Press, New York – London (1957), p. 743.

262. K. Randerath and E. Randerath, in: Methods in Enzymology, Vol. 12, Part A, L. Grossman and K. Moldave (editors), Academic Press, New York – London (1967), p. 323.

263. E. I. Budovskii and L. M. Klebanova, Vopr. Med. Khimii, 13:299 (1967).

264. E. Chargaff, Experientia. 6:201 (1950).

265. A. N. Belozerskii and A. S. Spirin, in: The Nucleic Acids, Vol. 3, E. Chargaff and J. N. Davidson (editors), Academic Press, New York (1960).

266. A. N. Belozerskii, Proceedings of the 5th International Biochemical Congress, Symposium III [in Russian], Izd. AN SSSR (1962), p. 123.

267. A. S. Antonov, Uspekhi Sovr. Biol., 60:161 (1965).

268. D. Elson and E. Chargaff, Biochim. Biophys. Acta, 17:367 (1955).

269. B. D. Hall and S. Spiegelman, Proc. Nat. Acad. Sci. USA, 47:137 (1961).

270. E. T. Bolton and J. McCarthy, Proc. Nat. Acad. Sci. USA, 48:1390 (1962); J. Mol. Biol., 8:201 (1964).

271. E. K. F. Bautz and B. D. Hall, Proc. Nat. Acad. Sci. USA, 48:400 (1962).

272. H. B. Strack and A. D. Kaiser, J. Mol. Biol., 12:36 (1965).

273. L. A. MacHattie, D. S. Ritchie, C. A. Thomas, and C. C. Richardson, J. Mol. Biol., 23:355 (1967).

274. J. Abelson and C. A. Thomas, J. Mol. Biol., 18:262 (1966).

275. J. Josse, A. D. Kaiser, and A. Kornberg, J. Biol. Chem., 236:864 (1961).

276. S. B. Weiss and T. Nakamoto, Proc. Nat. Acad. Sci. USA, 47:1400 (1961).

277. J. Hurwitz, J. J. Furth, M. Adrews, and A. Evans, J. Biol. Chem., 237:3752 (1962).

278. A. A. Baev, in: Nucleases [in Russian], V. S. Shapot (editor), Meditsina (1968), p. 164.

279. U. L. RajBhandary and A. Stuart, Ann. Rev. Biochem., 35:759 (1966).

280. K. Burton, in: Essays in Biochemistry, Vol. 1, P. N. Campbell and D. Greville (editors), Academic Press, New York – London (1965), p. 57.

281. M. Schütt, Z. Chem., 8:12 (1968).

282. P. Whitfeld, Biochem. J., 58:390 (1954).

283. D. M. Brown, M. Fried, and A. R. Todd, J. Chem. Soc., 2206 (1955).

284. C.-T. Yu and P. C. Zamecnik, Biochim. Biophys. Acta, 45:148 (1960).

285. H. C. Neu and L. A. Heppel, J. Biol. Chem., 239:2927 (1964).

286. P. Whitfeld, Biochim. Biophys. Acta, 108:202 (1965).

287. A. Steinschneider and H. Fraenkel-Conrat, Biochemistry, 5:2735 (1966).

288. H. L. Weith and P. T. Gilham, J. Am. Chem. Soc., 89:5473 (1967).

289. G. P. Moss, C. B. Reese, K. Schofield, R. S. Shapiro, and A. R. Todd, J. Chem. Soc., 1149 (1963).

290. T. Gabriel and A. L. Nussbaum, Abstr. Papers of XXIst Intern. Congress Pure Appl. Chem. Nucleic Acid Components, Prague, 51 (1967); T. Gabriel, W. Y. Chen, and A. L. Nussbaum, J. Am. Chem. Soc., 90:6833 (1968).

291. R. W. Holley, J. T. Madison, and A. Zamir, Biochem. Biophys. Res. Comm., 17:389 (1964).

292. S. K. Vasilenko, V. P. Demushkin, É. I. Budovskii, and D. G. Knorre, Dokl. Akad. Nauk SSSR, 162:694 (1965).

293. C. R. Bayley, K. W. Brammer, and A. S. Jones, J. Chem. Soc., 1903 (1961).

294. C. Tamm, M. E. Hodes, and E. Chargaff, J. Biol. Chem., 195:49 (1952).

295. A. S. Jones, J. R. Tittensor, and R. T. Walker, Nature, 209:296 (1966).

296. S. Takemura, Biochim. Biophys. Acta, 29:447 (1958); Bull. Chem. Soc. Japan, 32:920 (1959).

297. E. Chargaff and H. S. Shapiro, Biochemistry, 5:3012 (1966).

298. N. K. Kochetkov, E. I. Budowsky, M. F. Turchinsky, and N. A. Simukova, Biochem. Biophys. Res. Comm., 19:49 (1965).

299. N. K. Kochetkov, E. I. Budowsky, V. P. Demushkin, M. F. Turchinsky, N. A. Simukova, and E. D. Sverdlov, Biochim. Biophys. Acta, 142:35 (1967).

300. K. T. Beers, J. Biol. Chem., 235:2393 (1960).

301. T. A. Moskvitina and E. I. Budovskii, Dokl. Akad. Nauk SSSR, 171:999 (1966).

302. H. Witzel, Progress in Nucleic Acid Research, Vol. 2, J. N. Davidson and W. E. Cohn (editors), Academic Press, New York (1963), p. 221.

303. G. H. Gassen and H. Witzel, Europ. J. Biochem., 1:36 (1967).

304. F. Egami, K. Takahashi, and T. Uchida, in: Progress in Nucleic Acid Research, Vol. 3, J. N. Davidson and W. E. Cohn (editors), Academic Press, New York (1964).

305. R. I. Tatarskaya, N. M. Abrosimova-Amel'yanchik, V. D. Aksel'rod, A. I. Korenyako, T. V. Venkstern, A. D. Mirzabekov, and A. A. Baev, Dokl. Akad. Nauk SSSR, 157:725 (1964); R. I. Tatarskaya, N. M. Abrosimova-Amel'yanchik, V. D. Aksel'rod, A. I. Korenyako, N. Ya. Niedra, and A. A. Baev, Biokhimiya, 31:1017 (1966).

306. J. C. Lee, N. W. Y. Ho, and P. T. Gilham, Biochim. Biophys. Acta, 95:503 (1965).

307. O. I. Ivanova, D. G. Knorre, and É. G. Malygin, Molekul. Biol., 1:335 (1967).

308. N. K. Kochetkov, É. I. Budovskii, M. F. Turchinskii, and V. P. Demushkin, Dokl. Akad. Nauk SSSR, 152:1005 (1963).

309. N. K. Kochetkov, É. I. Budowsky, N. E. Broude, and M. Klebanova, Biochim. Biophys. Acta, 134:492 (1967).

310. J. Goldstein, J. Mol. Biol., 25:23 (1967).

311. M. Staehelin, in: Progress in Nucleic Acid Research, Vol. 2, J. N. Davidson and W. E. Cohn (editors), Academic Press, New York (1963).

312. M. Laskowski, in: Methods in Enzymology, Vol. 12, Part A, L. Grossman and K. Moldave (editors), Academic Press, New York – London (1967), p. 281.

313. N. B. Furlong, ibid., p. 318.

314. F. Sanger and G. G. Brownlee, ibid., p. 361.

315. K. Miura and Y. Hayashi, ibid., p. 390.

316. G. W. Rushizky, ibid., p. 295.

317. G. M. Tener, ibid., p. 398.

318. R. W. Holley, J. Apgar, G. A. Everett, J. T. Madison, M. Marquisee, S. H. Merril, J. H. Penswick, and A. Zamir, Science, 147:1462 (1965).

319. J. P. Penswick and R. W. Holley, Proc. Nat. Acad. Sci. USA, 53:543 (1965).

320. H. G. Zachau, D. Dütting, and H. Feldman, Z. Physiol. Chem., 347:212 (1966).

321. J. T. Madison, G. A. Everett, and H. Kung, Science, 153:531 (1966).

322. A. A. Baev, T. V. Venkstern, A. D. Mirzabekov, A. I. Krutilina, L. Li, and V. D. Aksel'rod, Molekul. Biol., 1:754 (1967).

323. U. L. RajBhandary, S. H. Chang. A. Stuart, R. D. Faulkner, R. M. Hoskinson, and H. G. Khorana, Proc. Nat. Acad. Sci. USA, 57:751 (1967).

324. H. M. Goodman, J. Abelson, A. Landy, S. Brenners, and J. D. Smith, Nature, 217:1019 (1968).

325. S. K. Dube, K. A. Marcker, B. F. C. Clark, and S. Cory, Nature, 218:232 (1968).

326. S. Takemura, T. Mizutani, and M. Miyazaki, J. Biochem., 63:227 (1968).
327. M. Staehelin, H. Rogg, B. C. Baguley, T. Ginsberg, and W. Wehrli, Nature, 219:1363 (1968).
328. B. G. Forget and S. H. Weissman, Science, 158:1695 (1967); J. Biol. Chem., 244:3148 (1969).
329. G. G. Brownlee, F. Sanger, and B. G. Barrell, Nature, 215:735 (1967); J. Mol. Biol., 34, 379 (1968).
330. M. Sugiura and M. Takanami, Proc. Nat. Acad. Sci. USA, 58:1595 (1967).
331. M. Takanami, J. Mol. Biol., 23:135 (1967); 29:323 (1967).
332. F. Sanger, G. G. Brownlee, and B. G. Barrell, J. Mol. Biol., 13:373 (1965).
333. N. Delihas, Biochemistry, 6:3356 (1967).
334. H. J. Gound, J. Mol. Biol., 29:307 (1967).
335. S. Mandelles, J. Biol. Chem., 242:3103 (1967).
336. R. de Wachter and W. Fiers, J. Mol. Biol., 30:507 (1967).
337. D. L. H. Bishop, D. R. Miles, and S. Spiegelman, Biochemistry, 7:3744 (1968).
338. S. Mandeles, J. Biol. Chem., 243:3671 (1968).
339. D. S. May and C. A. Knight, Virology, 25:502 (1965).
340. H. S. Shapiro, in: Methods in Enzymology, Vol. 12, Part A, L. Grossman and K. Moldave (editors), Academic Press, New York – London (1967), p. 205.
341. K. Burton, ibid., p. 222.
342. H. S. Shapiro, ibid., p. 212.
343. M. R. Lunt and K. Burton, Biochim. Biophys. Acta, 55:1005 (1962).
344. D. D. Brown and C. S. Weber, J. Mol. Biol., 34:661, 681 (1968).
345. H. G. Khorana, Some Recent Developments in the Chemistry of Phosphate Esters of Biological Interest, Wiley, New York (1961).
346. S. M. Zhenodarova, Uspekhi Khimii, 34:82 (1965).
347. S. M. Zhenodarova and M. I. Khabarova, Uspekhi Khimii, 35:1265 (1966).
348. F. Cramer, Angew. Chem., 78:186 (1966).
349. H. G. Khorana, J. P. Vizsolyi, and R. K. Ralph, J. Am. Chem. Soc., 84:414 (1962).
350. T. M. Jacob and H. G. Khorana, J. Am. Chem. Soc., 86:1630 (1964).
351. P. T. Gilham and H. G. Khorana, J. Am. Chem. Soc., 80:6212 (1958).
352. R. Lohrmann and H. G. Khorana, J. Am. Chem. Soc., 88:829 (1966).
353. G. Wiemann and H. G. Khorana, J. Am. Chem. Soc., 84:4329 (1962).
354. R. L. Letsinger and K. K. Ogilvie, J. Am. Chem. Soc., 89:4801 (1967); R. L. Letsinger and K. K. Ogilvie, J. Am. Chem. Soc., 91:3350 (1969); R. L. Letsinger, K. K. Ogilvie, and P. S. Miller, J. Am. Chem. Soc., 91:3360 (1969).
355. F. Eckstein and I. Rizk, Angew. Chem., 79:684, 939 (1967); Chem. Ber., 102: 2362 (1969).
356. R. L. Letsinger and V. Mahadevan, J. Am. Chem. Soc., 87:3526 (1965); 88: 5319 (1966).
357. F. Cramer, R. Hellig, H. Hettler, K. H. Scheit, and H. Seliger, Angew. Chem., 78:640 (1966).
358. H. Hayatsu and H. G. Khorana, J. Am. Chem. Soc., 88:3182 (1966); 89:3881 (1967).
359. L. R. Melby and D. R. Strobach, J. Am. Chem. Soc., 89:450 (1967).
360. G. M. Blackburn, M. J. Brown, and M. R. Harris, J. Chem. Soc., C, 2438 (1907).
361. G. M. Tener, H. G. Khorana, R. Markham, and E. H. Pol, J. Am. Chem. Soc., 80:6223 (1958).
362. H. G. Khorana and J. P. Vizsolyi, J. Am. Chem. Soc., 83:675 (1961).
363. A. Talahashi, J. Adler, and H. G. Khorana, J. Biol. Chem., 238:3080 (1963).

364. H. G. Khorana, A. F. Turner, and J. P. Vizsolyi, J. Am. Chem. Soc., 83:686 (1961).

365. R. K. Ralph and H. G. Khorana, J. Am. Chem. Soc., 83:2926 (1961).

366. R. K. Ralph, W. J. Connors, H. Schaller, and H. G. Khorana, J. Am. Chem. Soc., 85:1983 (1963).

367. A. F. Turner and H. G. Khorana, J. Am. Chem. Soc., 81:4651 (1959).

368. E. Ohtsuka, M. W. Moon, and H. G. Khorana, J. Am. Chem. Soc., 87:2956 (1965).

369. G. Wiemann, H. Schaller, and H. G. Khorana, J. Am. Chem. Soc., 85:3835 (1963).

370. S. A. Narang, T. M. Jacob, and H. G. Khorana, J. Am. Chem. Soc., 89:2167 (1967).

371. F. Cramer, W. Frolke, and H. Matzura, Angew. Chem., 79:580 (1967).

372. T. M. Jacob, S. A. Narang, and H. G. Khorana, J. Am. Chem., Soc., 89:2177 (1967).

373. A. M. Michelson and A. R. Todd, J. Chem. Soc., 2632 (1955).

374   T. M. Jacob and H. G. Khorana, J. Am. Chem. Soc., 87:368 (1965).

375. T. M. Jacob and H. G. Khorana, J. Am. Chem. Soc., 87:2971 (1965).

376. S. A. Narang and H. G. Khorana, J. Am. Chem. Soc., 87:2981 (1965).

377. S. A. Narang, T. M. Jacob, and H. G. Khorana, J. Am. Chem. Soc., 87:2988 (1965).

378. H. Kösell, H. Büchi, and H. G. Khorana, J. Am. Chem. Soc., 89:2185 (1967).

379. R. L. Letsinger, M. H. Caruthers, and D. M. Jerina, Biochemistry, 6:1379 (1967).

380. R. L. Letsinger, M. H. Caruthers, P. S. Miller, and K. K. Ogilvie, J. Am. Chem. Soc., 89:7146 (1967).

381. H. Kössel, M. W. Moon, and H. G. Khorana, J. Am. Chem. Soc., 89:2148 (1967).

382. E. Ohtsuka and H. G. Khorana, J. Am. Chem. Soc., 89:2195 (1967).

383. H. Schaller and H. G. Khorana, J. Am. Chem. Soc., 85:3828, 3841 (1963).

384. G. Weimann, H. Schaller, and H. G. Khorana, J. Am. Chem. Soc., 85:3835 (1963).

385. E. Ohtsuka, M. W. Moon, and H. G. Khorana, J. Am. Chem. Soc., 87:2956 (1965).

386. F. Eckstein, Chem. Ber., 100:2236 (1967).

387. A. Franke, F. Eckstein, K. H. Scheit, and F. Cramer, Chem. Ber., 101:944 (1968).

388. F. Kathawala and F. Cramer, Ann., 709:185 (1967).

389. F. Kathawala and F. Cramer, Ann., 712:195 (1968).

390. S. A. Narang, S. K. Dheer, and Michniewicz, J. Am. Chem. Soc., 90:2702 (1968).

391. R. Lohrmann, D. Söll, H. Hayatsu, E. Ohtsuka, and H. G. Khorana, J. Am. Chem. Soc., 88:819 (1966).

392. A. Holý and J. Smrt, Coll. Czech. Chem. Comm., 31:3800 (1966).

393. Y. Lapidot and H. G. Khorana, J. Am. Chem. Soc., 85:3852 (1963).

394. J. Smrt, Coll. Czech. Chem. Comm., 29:2049 (1964); J. Smrt and F. Śorm, Coll. Czech. Chem. Comm., 29:2971 (1964).

395. C. Coutsogeorgopoulos and H. G. Khorana, J. Am. Chem. Soc., 86:2926 (1964).

396. M. Smith, D. H. Rammler, I. H. Goldberg, and H. G. Khorana, J. Am. Chem. Soc., 84:430 (1962).

397. F. Cramer, R. J. Rhaese, S. Rittner, and K. H. Scheit, Ann., 683:199 (1965).

398. J. Smrt and F. Śorm, Coll. Czech. Chem. Comm., 27:73 (1962); 28:61 (1963).

399. S. Chladec and J. Smrt, Coll. Czech, Chem. Comm., 29:214 (1964).

400. D. H. Rammler and H. G. Khorana, J. Am. Chem. Soc., 84:3112 (1962).

401. J. Zemlička, S. Chladek, A. Holý, and J. Smrt, J. Coll. Czech. Chem. Comm., 31:3198 (1966).

402. J. Smrt, Coll. Czech, Chem. Comm., 33:1462 (1968).

403. A. Holý, Coll. Czech. Chem. Comm., 33:223 (1968).

404. D. Söll and H. G. Khorana, J. Am. Chem. Soc., 87:360 (1965).

405. F. Cramer, K. H. Scheit, and H. J. Rhaese, Ann., 693:244 (1966).

406. N. S. Tikhomirova-Sidorova, A. P. Kavunenko, and É. A. Pyaivinen, Zh. Obshch. Khimii, 37:1923 (1967).

407. D. Söll and H. G. Khorana, J. Am. Chem. Soc., 87:350 (1965).

408. J. Smrt and F. Šorm, Coll. Czech. Chem. Comm., 28:2415 (1963).

409. H. J. Rhaese, W. Siehr, and F. Cramer, Ann., 703:215 (1967).

410. B. E. Griffin, M. Jarman, and C. B. Reese, Tetrahedron, 24:639 (1968).

411. B. E. Griffin and C. B. Reese, Tetrahedron, 24:2537 (1968).

412. L. A. Heppel and P. A. Whitfeld, Biochem. J., 60:1 (1955).

413. L. A. Heppel, P. A. Whitfeld, and R. Markham, Biochem. J., 60:8 (1955).

414. M. R. Bernfield, J. Biol. Chem., 241:2014 (1966); 240:4753 (1965).

415. N. S. Tikhomirova-Sidorova, G. E. Ustyuzhanin, and E. M. Kogan, Zh. Obshch. Khimii, 36:2219 (1966); Biokhimiya, 32:867 (1967).

416. K. Sato-Asano and F. Egami, Biochim. Biophys. Acta, 29:655 (1958); J. Biochem. Japan, 48:284 (1960).

417. H. Hayashi and F. Egami, J. Biochem. Japan, 53:176 (1963).

418. K. H. Scheit and F. Cramer, Tetrahedron Letters, 2975 (1964).

419. D. Grünberger, A. Holý, and F. Šorm, Coll. Czech. Chem. Comm., 33:286 (1968).

420. N. M. Abrosimova-Amel'yanchik, R. I. Tatarskaya, and A. A. Baev, Molekul. Biol., 1:307 (1967).

421. S. M. Zhenodarova and E. M. Sedelnikova, Biochim. Biophys. Acta, 169:559 (1968).

422. F. J. Bolum, in: Progress in Nucleic Acid Research, Vol. 1, J. N. Davidson and W. E. Cohn (editors), Academic Press, New York (1963).

423. H. M. Keir, Progr. Nucl. Acid Res., 4:82 (1965).

424. D. Elson, Ann. Rev. Biochem., 34:449 (1965).

425. J. S. Krakow, C. Coutsogeorgopoulos, and E. S. Cannellanis, Biochim. Biophys. Acta, 55:639 (1962).

426. H. M. Keir and S. M. Smith, Biochim. Biophys. Acta, 68:589 (1963).

427. F. J. Bollum, E. Groeniger, and M. Yoneda, Proc. Nat. Acad. Sci. USA, 51: 853 (1964).

428. V. S. Tongur, in: Biosynthesis of Protein and Nucleic Acids [in Russian], A. S. Spirin (editor), Nauka (1965), p. 277.

429. M. Grunberg-Manago, in: Progress in Nucleic Acid Research, Vol. 1, J. N. Davidson and W. E. Cohn (editors), Academic Press, New York (1963).

430. R. M. S. Smellie, in: Progress in Nucleic Acid Research, Vol. 1, J. N. Davidson and W. E. Cohn (editors), Academic Press, New York (1963).

431. J. Hurwitz and J. T. August, in: Progress in Nucleic Acid Research, Vol. 1, J. N. Davidson and W. E. Cohn (editors), Academic Press, New York (1963).

432. L. A. Osterman, Uspekhi Biol. Khimii, 7:116 (1965).

433. I. Haruna, K. Nozu, Y. Ohtaka, and S. Spiegelman, Proc. Nat. Acad. Sci. USA, 50:905 (1963).

434. I. Haruna and S. Spiegelman, Proc. Nat. Acad. Sci. USA, 54:579 (1965).

435. C. Weissman, P. Borst, R. H. Burdon, M. A. Billeter, and S. Ochoa, Proc. Nat. Acad. Sci. USA, 51:682, 690 (1964).

436. M. Gellert, Proc. Nat. Acad. Sci. USA, 57:148 (1967).

437. B. Weiss and C. C. Richardson, Proc. Nat. Acad. Sci. USA, 57:1021 (1967).

438. B. M. Olivera and I. R. Lehman, Proc. Nat. Acad. Sci. USA, 57:1426 (1967).

439. N. K. Gupta, E. Ohtsuka, H. Weber, S. H. Chang, and H. G. Khorana, Proc. Nat. Acad. Sci. USA, 60:285 (1968).

440. F. N. Hayes, V. E. Mitchell, R. L. Ratliff, and D. L. Williams, Biochemistry, 6:2488 (1967).

441. F. J. Bollum, E. Groeniger, and H. Yoneda, Proc. Nat. Acad. Sci. USA, 51:853 (1964).

442. C. M. Radding, J. Josse, and A. Kornberg, J. Biol. Chem., 237:2869 (1962).

443. C. C. Richardson, C. L. Schildkraut, H. V. Aposhian, and A. Kornberg, J. Biol. Chem., 239 (1964).

444. R. B. Inman and B. L. Baldwin, J. Mol. Biol., 8:452 (1964).

445. H. K. Schachman, J. Adler, C. M. Radding, I. R. Lehman, and A. Kornberg, J. Biol. Chem., 235:3242 (1960).

446. S. Lee-Huang and L. Cavallieri, Proc. Nat. Acad. Sci. USA, 51:1022 (1964).

447. M. Riley, B. Maling, and M. J. Chamberlin, J. Mol. Biol., 20:359 (1966).

448. C. Byrd, E. Ohtsuka, M. W. Moon, and H. G. Khorana, Proc. Nat. Acad. Sci. USA, 53:79 (1965).

449. R. D. Wells and J. E. Blair, J. Mol. Biol., 27, 273 (1967).

450. R. D. Wells, E. Ohtsuka, and H. G. Khorana, J. Mol. Biol., 14:221 (1965).

451. R. D. Wells, T. M. Jacob, S. A. Narang, and H. G. Khorana, J. Mol. Biol., 27:237 (1967).

452. R. D. Wells, H. Büchi, H. Kösell, E. Ohtsuka, and H. G. Khorana, J. Mol. Biol., 27, 265 (1967).

453. M. J. Bessman, I. R. Lehman, J. Adler, S. B. Zimmerman, E. S. Simms, and A. Kornberg, Proc. Nat. Acad. Sci. USA, 44:633 (1958).

454. M. Chamberlin and R. L. Baldwin, J. Mol. Biol., 7:334 (1963).

455. A. G. Lezius and K. H. Scheit, Europ. J. Biochem., 3:85 (1967).

456. R. L. Ratliff, D. E. Hoard, D. G. Ott, and F. N. Hayes, Biochemistry, 6:851 (1967).

457. T. L. Ratliff, A. W. Schwartz, V. N. Kerr, D. L. Williams, D. G. Ott, and F. N. Hayes, Biochemistry, 7, 412 (1968).

458. K. Kato, J. M. Concalves, G. E. Houts, and F. J. Bollum, J. Biol. Chem., 242: 2780 (1967).

459. N. K. Gupta, E. Ohtsuka, V. Sgaramella, H. Büchi, A. Kumar, H. Weber, and H. G. Khorana, Proc. Nat. Acad. Sci. USA, 60:1388 (1968).

460. N. K. Gupta and H. G. Khorana, Proc. Nat. Acad. Sci. USA, 61:215 (1969).

461. R. Thach and P. Doty, Science, 147:1310 (1965); 148:632 (1965).

462. P. Leder, M. F. Singer, R. C. L. Brimacombe, Biochemistry, 4:1516 (1965).

463. M. R. Bernfield and M. W. Nirenberg, Science, 147:479 (1965).

464. M. Grunberg-Manago, P. J. Ortiz, and S. Ochoa, Biochim. Biophys. Acta, 20:269 (1956).

465. M. N. Thang, M. Graffe, and M. Grunberg-Manago, Biochim. Biophys. Acta, 108:125 (1965).

466. F. Pochon, A. M. Michelson, M. Grunberg-Manago, W. E. Cohn, and L. Dondon, Biochim. Biophys. Acta, 80:441 (1964).

467. B. E. Griffin and A. R. Todd, Proc. Nat. Acad. Sci. USA, 44:1123 (1958).

468. W. Szer and D. Shugar, Acta, Biochim. Polon, 8:235 (1961).

469. R. L. C. Brimacombe and C. B. Reese, J. Mol. Biol., 18:529 (1966).

470. B. E. Griffin, W. J. Haslam, and C. B. Reese, J. Mol. Biol., 10:353 (1964).

471. A. M. Michelson, J. Dondon, and M. Grunberg-Manago, Biochim. Biophys. Acta, 55:528 (1962).

472. A. M. Michelson and C. Monny, Biochim. Biophys. Acta, 149:88 (1967).
473. A. M. Michelson, C. Monny, R. A. Laursen, and N. J. Leonard, Biochim. Biophys. Acta, 119:258 (1966).
474. F. B. Howard, J. Frazier, and H. T. Miles, J. Biol. Chem., 241:4293 (1966).
475. D. A. Smith, R. L. Ratliff, D. L. Williams, and A. M. Martinos, J. Biol. Chem., 242:590 (1967).
476. J. S. Krakow and M. Karstadt, Proc. Nat. Acad. Sci. USA, 58:2094 (1967).
477. R. Haselkorn and C. F. Fox, J. Mol. Biol., 13:780 (1965).
478. B. D. Melhotra and H. G. Khorana, J. Biol. Chem., 240:1750 (1965).
479. S. Nishimura, T. M. Jacob, and H. G. Khorana, Proc. Nat. Acad. Sci. USA, 52:1494 (1964).
480. S. Nishimura, D. S. Jones, and H. G. Khorana, J. Mol. Biol., 13:302 (1965).
481. P. Roy-Burman and S. Roy-Burman, Biochim. Biophys. Acta, 142:355 (1967).
482. M. Chamberlin, R. L. Baldwin, and P. Berg. J. Mol. Biol., 7:334 (1963).
483. S. Nishimura, F. Harada, and M. Ikehara, Biochim. Biophys. Acta, 129:301 (1966).
484. M. Ikehara, K. Murao, F. Harada, and S. Nishimura, Biochim. Biophys. Acta, 155:82 (1968).
485. J. J. Pene, E. Knight, and J. E. Darnell, J. Mol. Biol., 33:609 (1968).
486. R. A. Weinberg and S. Penman, J. Mol. Biol., 38:289 (1968).
487. J. L. Hodnett and H. Busch, J. Biol. Chem., 243:6334 (1968).
488. A. C. Peacock and C. W. Dingman, Biochemistry, 6:1818 (1967).
489. C. W. Dingman and A. C. Peacock, Biochemistry, 7:659 (1968).
490. T. Nakamura, A. W. Prestayko, and H. Buch, J. Biol. Chem., 243:1368 (1968).
491. A. W. Prestayko and H. Busch, Biochim. Biophys. Acta, 169:327 (1968).
492. R. A. Jacobson and J. Bonner, Biochem. Biophys. Res. Comm., 33:716 (1968).
493. D. Bernhardt and J. E. Darnell, J. Mol. Biol., 42:43 (1969).
494. H. M. Munro and A. Fleck, Methods of Biochemical Analysis, 14:113 (1966).
495. T. I. Tikhonenko, The Biochemistry of Viruses [in Russian], Meditsina (1966), p. 53.
496. J. H. Strauss, R. B. Kelly, and R. L. Sinsheimer, Biopolymers, 6:793 (1968).
497. H. Boedtker, J. Mol. Biol., 35:61 (1968).
498. C. Corneo, E. Ginelli, C. Soave, and G. Bernardi, Biochemistry, 7:4373 (1968).
499. J. E. Darnell, Bacter. Rev., 32:262 (1968).
500. M. P. Schweizer, G. B. Chheda, L. Banczynskyj and R. H. Hall, Biochemistry 8:3283 (1969).
501. D. Dütting, Fortschr, Chem. Org. Naturst., 26:356 (1968).
502. J. T. Madison, Ann. Rev. Biochem., 37:131 (1968).
503. M. M. K. Nass, Science, 165:25 (1969).
504. E. Wimmer and M. E. Reichman, Science, 160:1452 (1968).
505. R. C. Merril, Biopolymers, 6:1727 (1968).
506. S. Cory, K. A. Marcker, S. K. Dube, and B. F. C. Clark, Nature, 220:1039 (1968).
507. B. G. Barrell and F. Sanger, FEBS Letters, 3:275 (1969).
508. M. Yaniv, B. G. Barrell, Nature, 222:278 (1969).
509. B. P. Doctor, J. E. Loebel, M. S. Sodd, and D. B. Winter, Science 163:693 (1969).
510. B. S. Dudock and G. Katz, J. Biol. Chem., 244:3069 (1969); B. S. Dudock, G. Katz, E. K. Taylor, and R. W. Holley, Proc. Nat. Acad. Sci. USA, 62:941 (1969).

511. S. Takemura, M. Murakami, and M. Miyazaki, J. Biochem. Japan, 65:489, 553 (1969).
512. S. Hashimoto, M. Miyazaki, and S. Takamura, J. Biochem. Japan, 65:659 (1969).
513. F. Amaldi and G. Attardi, J. Mol. Biol., 33:737 (1968).
514. P. Fellner and F. Sanger, Nature, 219:236 (1968).
515. D. G. Glitz, A. Bradley, and H. Fraenkel-Conrat, Biochim. Biophys. Acta, 161:1 (1968).
516. J. E. Dahlberg, Nature, 220:548 (1968).
517. R. De Wachter and W. Fiers, Nature, 221:233 (1969).
518. D. G. Glitz, Biochemistry, 7:927 (1968).
519. R. De Wacher, J. P. Verhassel, and W. Fiers, FEBS Letters, 1:93 (1968); Biochim. Biophys. Acta, 157:195 (1968).
520. J. M. Adams, P. G. N. Jeppesen, F. Sanger, and B. G. Barrel, Nature, 223: 1009 (1969).
521. R. W. Holley, Progr. Nucl. Acid. Res., 8:37 (1968).
522. H. G. Zachau, Angew. Chem., 81:645 (1969).
523. G. M. Blackburn, M. J. Brown, M. R. Harris, and D. Shire, J. Chem. Soc., (C), 676 (1969).
524. F. Cramer and H. Köster, Angew. Chem., 80:488 (1968).
525. L. R. Melby and D. R. Strobach, J. Org. Chem., 34:421, 427 (1969).

Chapter 2
Conformation of Nucleosides and Nucleotides

I. Introduction

Much of our understanding of the factors determining the chemical behaviour and biological specificity of nucleic acids depends on information concerning the distribution of electron density and the spatial arrangement of the various structural elements of these compounds.

Rotation around ordinary $C-C$, $C-N$, $C-O$, and so on, bonds in organic compounds is not always completely free, and of the infinite number of theoretically possible conformations only a certain number, characterized by a relative minimum of potential energy, actually occur. The stability of a given molecular conformation depends, on the one hand, on the Van der Waals' forces of repulsion between neighbouring atoms unlinked by covalent bonds, and on the other hand, on forces of attraction between functional groups on account of, for example, dipole – dipole interactions, hydrogen bonds, etc., which draw them together. A qualitative examination of the factors stabilizing a given conformation is usually possible on molecular models of the compound. In some cases this enables reasonably confident predictions to be made concerning the most stable conformation of that compound. When several factors acting in opposite directions are present, the quantitative estimation of the stability of various conformations becomes a more difficult problem; nevertheless, certain empirical principles have been deduced which, in some cases, allow the most stable conformation of a given compound to be selected.

---

* This problem is reviewed by Preobrazhenskaya, N. N. and Shabarova, Z. A., Uspekhi Khimii 38:222 (1969)

To determine the conformation of nucleosides and nucleotides, which are complex molecules, important information can be obtained by x-ray structural analysis and by the study of nuclear magnetic resonance and optical rotatory dispersion. Under favourable conditions investigation of the x-ray diffraction pattern from a crystal of a compound enables the position of the atoms composing the compound to be located to an accuracy, in the best work of this type, of a few thousandths of an Angström unit, and thus enables the conformation of the molecule in the crystal to be determined. Extensive computations are necessary in order to obtain such a pattern, however, and the problem can be appreciably simplified if one or several atoms with a high atomic number are present in the crystal. Only a few nucleosides and nucleotides have so far been investigated by x-ray structural analysis; this is due, not only to the laboriousness of the method, but also to the necessity of having a monocrystal of the substance measuring about 0.1 mm, and in the case of natural nucleosides and nucleotides, obtaining such a crystal can present considerable difficulties*.

When the nuclear magnetic resonance (NMR)† method is used, most information on the conformation of a molecule is usually obtained from the $J_{1,2}$ value: the constant of spin – spin interaction between two protons belonging to neighbouring carbon atoms. The $J_{1,2}$ value depends primarily on the magnitude of the dihedral angle $\Phi_{1,2}$ between the planes $C1 - C2 - H$ (at C1) and $C1 - C2 - H$ (at C2), in accordance with the equation:

$$J_{1,\,2} = J_0 \cos^2 \Phi_{1,\,2} - 0.28$$

---

* For further details of the method of x-ray structural analysis, see: Kitaigorodskii, A. N., The Theory of Structural Analysis [in Russian], Nauka (1957); Guinier, A., Radiocrystallo-graphie [Russian translation], Fizmatgiz (1961); Bokii, G. B. and Porai-Koshits, M. A., X-Ray Structural Analysis [in Russian], Moscow University Press (1964); Stout, G. H. and Jensen, L. H., X-Ray Structure Determination. Macmillan, Co, New York and London (1968).

The application of x-ray structural analysis to the investigation of nucleic acids and their components is surveyed by Davies, D. R:, Ann. Rev. Biochem. 36:321 (1967).

† For further details on NMR spectroscopy, see: Roberts, J.D., Nuclear Magnetic Resonance, McGraw-Hill, New York, (1959); Roberts, J.D., Introduction to Analysis of High-Resolution NMR Spectra, McGraw-Hill, New York (1963); Pople, J.A., Schneider, W.G., and Bernstein, H.J., High-Resolution Nuclear Magnetic Resonance, McGraw-Hill, New York (1959); Konrai, G., Uspekhi Organicheskoi Khimii 2:255 (1964); Stozers, G.B., in: Determination of the Structure of Organic Compounds by Physical and Chemical Methods [Russian translation], Vol. 1, Khimiya (1967), p. 204; Bhacca, N. S. and Williams, D. H., The Applications of NMR Spectroscopy in Organic Chemistry, Holden-Day, San Francisco (1966); Emsley, J.W., Feeney, J., and Sutcliffe, L.H., High-Resolution Nuclear Magnetic Resonance Spectroscopy, Oxford (1965-1966); Ionin, B. I, and Ershov, B. A. NMR Spectroscopy in Organic Chemistry [in Russian], Khimiya (1967); Bible, R. H., Interpretation of NMR Spectra, Plenum Press, New York (1965).

However, the spin – spin interaction constant also depends on other factors: the angles C1 – C2 – H (at C2) or C2 – C1 – H (at C1), the length of the C1 – C2 bond, and the electronegativity of atoms linked to the carbon atoms. Because of this, the dihedral angles can be calculated only approximately by the Karplus equation, and the conformation of a given compound can sometimes be determined unequivocally only by investigation of a series of model compounds in which the conformation is fixed.

Some information about the conformation of a compound can be obtained by comparing the chemical shifts of atoms in model compounds. The magnitude of the chemical shift is determined, inter alia, by the induced magnetic field created by neighbouring functional groups with magnetic anisotropy. Although the precise magnitude of this effect is difficult to estimate, determination of its sign is usually easy.

The important factors in the optical rotatory dispersion method, when used in stereochemical research, are the sign and amplitude of the Cotton effect: the characteristic extremum on the optical rotatory dispersion curve in the region of the UV-absorption band of the given compound. These values depend on asymmetry of the electrical and magnetic fields in which their chromophores lie, i. e., they are determined by the structure and conformation of the molecule. At present only the empirical approach can be usefully applied to the conformation analysis of complex organic compounds: by comparing curves of optical rotatory dispersion of a compound with the dispersion curves of compounds of known conformation. Essentially similar information can be obtained from the circular dichroism spectra of the investigated compound*

It must be emphasized that when these three basic methods are used to study the conformation of nucleosides, the substance tested must be in essentially different states. X-ray structural analysis obtains information about the conformation of a molecule in a crystal, where intermolecular interaction is strong. For the measurement of NMR signals, concentrated (0.2-0.5 M) solutions of the substance must be used, in which intermolecular interaction can still be considerable. Finally, optical rotatory dispersion curves in the region of the absorption band can only be obtained in dilute (10^{-3}-10^{-4} M) solutions, when intermolecular interaction can be virtually excluded and the conformation of the tested compound may differ appreciably from its conformation in a crystal or in concentrated solutions.

---

* For further details on the methods of optical rotatory dispersion and circular dichroism, see: Djerassi, C., Optical Rotatory Dispersion, McGraw-Hill, New York (1960); Klain, V., Uspekhi Org. Khimii 1:261 (1963); Velluz, L., Legrand, M., and Grosjean, M, Optical Circular Dichroism, Weinheim (1965); Crabbé, P., Optical Rotatory Dispersion and Circular Dichroism in Organic Chemistry, Holden-Day, San Francisco (1965); Djerassi, C., Optical Rotatory Dispersion and Circular Dichroism in Organic Chemistry, Heyden and Sons, London (1967).

Examination of the structural formulae of nucleosides shows that free rotation around ordinary bonds in these compounds is largely limited by the existence of two cyclic structures in the molecule: the heterocyclic ring and the pentofuranose residue. Accordingly, conformation analysis of nucleosides must include the examination of the conformation of the heterocyclic ring and of the carbohydrate residue as well as the mutual spatial arrangement of these two cyclic systems.

II. Conformation of the components of nucleic acids

1. Conformation of the heterocyclic rings

The heterocyclic bases which are usual components of nucleic acids are quasiaromatic compounds (see Chapter 3), and an essential condition for effective interaction between the π electrons in their ring is coplanarity of its component atoms. Accordingly, residues of the heterocyclic bases of nucleic acids can, as a first approximation, be regarded as two-dimensional, and this assumption is usually made in contemporary spatial models of nucleic acid and nucleotide structure.

However, precise x-ray structural investigations have demonstrated a definite curvature of the purine and pyrimidine rings, slight yet outside the limits of experimental error, as well as appreciable deviation of exocyclic substituents in some tested compounds from the plane of the heterocyclic ring. Values of the deviation of atoms of the heterocyclic ring from the mean plane, and interatomic distances and valence angles are given in Table 2.1. It is clear that the cytosinium cation (the heterocyclic ring in cytidine-3'-phosphate) exists in crystals in the form of an extremely shallow boat with C4 and N1 atoms slightly displaced to either side; these deviations in derivatives of uracil and thymine do not go beyond the limits of experimental error. Exocyclic substitutents C1' and O at C2 and C4 or N at C4, as well as the methyl group at C5, are usually an appreciable distance in front of or behind the plane of the heterocyclic ring; their linkages with the corresponding carbon atoms form an angle of 1.5–8° with the plane of the ring. Marked divergence of the shape of the pyrimidine ring from a regular hexagon and the similarity in length between the C – N bonds in the ring and the C – N bond with the exocyclic amino group also are noteworthy.

In the case of purine derivatives, deviation of atoms C5 and N1 from the plane of the ring can also be observed in the adeninium cation in adenosine-5'- and adenosine-3'-phosphates and of the N9 atom of adenine in the 2'-deoxyadenosine residue.

The significance of these slight curvatures of the heterocyclic rings in nulceosides and nucleotides in crystalline form is not yet clear, but these observations must obviously be taken into account when more accurate models of nucleic acids are constructed *

---

* Mean values of bond lengths and valence angles of the heterocyclic bases of nucleic acids, obtained from the latest x-ray structural data, are given in [68].

The heterocyclic bases of most minor components of nucleic acids are evidently similar in their spatial structure to the heterocyclic bases of the corresponding ordinary components; no x-ray structural data are available for bases of the minor components.

The only evident exception to this analogy is 5,6-dihydrouridine, in which the heterocyclic base must have a conformation which is markedly different from flat. Studies of NMR spectra of dihydrouracil (I) and its methyl derivatives (II, III) have led to the conclusion that these compounds have a half-chair conformation [8]:

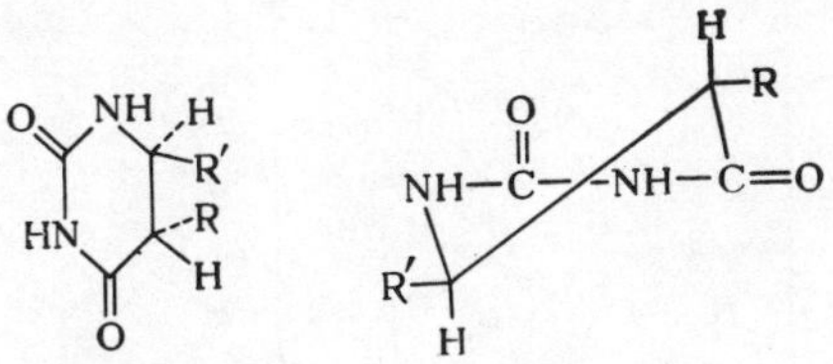

I (R = R'=H)

II (R= CH₃, R'=H)

III (R=H, R'=CH₃)

This conclusion has been confirmed by x-ray structural analysis of dihydrothymine and by studies of the kinetics of acid and alkaline hydrolysis of a number of dihydrouracil derivatives [10]. A similar conformation of the heterocyclic ring is also predominant in dihydrouracil, as x-ray structural analysis has shown [67].

2. Conformation of carbohydrate residues

In the case of cyclopentane derivatives, the most stable conformations are not planar, but those in which one carbon atom is slightly displaced from the plane (the C_S conformation), or in which two neighbouring carbon atoms are displaced from the plane on opposite sides (the C_2 conformation), [11].

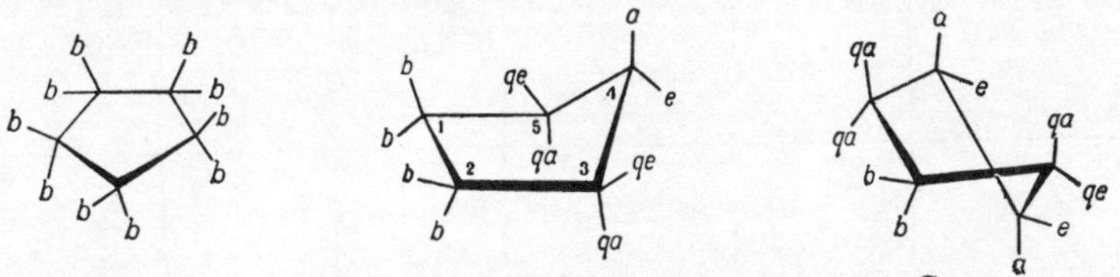

Planar　　　　　C_S　　　　　C_2

Substituents at carbon atoms lying in the plane of the ring occupy symmetrical (bissectrial) positions b. In the case of atoms displaced away from the plane, these positions are not equivalent. The bond between carbon and the substituent occupying the equatorial position e is approximately parallel to the plane of the ring, while the bond with the axial substituent a is approximately perpendicular to it. Finally, the arrangement of the substituents at carbon atoms next to an atom displaced from the plane is intermediate in

TABLE 2.1. Structure of Heterocyclic Bases of the Components of Nucleic Acids (Results of X-Ray Structural Analysis)

Compound	Deviation of atoms from mean plane of ring, Å	Interatomic distances, Å	Valence angles, deg
Cytosinium cation in cytidine-3'-phosphate [1]	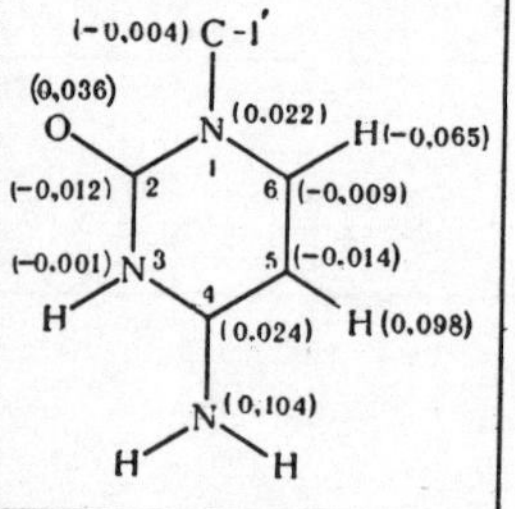 (−0.004) C-1′ (0.036) O (−0.012) 2 (−0.001) N3 H N(0.022) H(−0.065) 6(−0.009) 5(−0.014) (0.024) H(0.098) N(0.104) H H	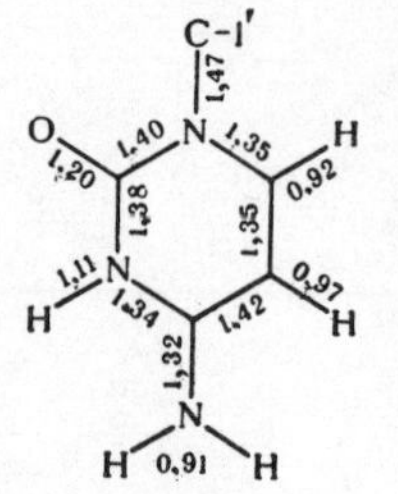 C-1′ 1.47 O 1.20 1.40 N 1.35 H 0.92 1.38 1.35 1.11 N 0.97 1.34 1.42 H H 1.32 N H 0.91 H	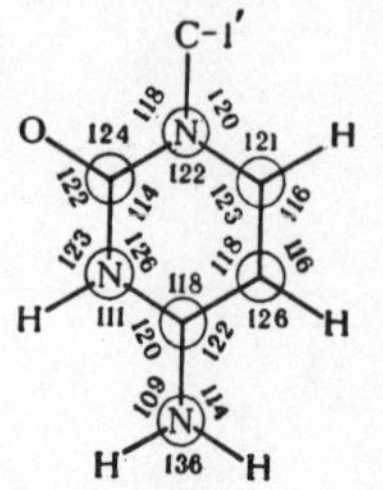 C-1′ 124 118 120 121 H 122 114 122 123 116 123 118 118 118 116 H 111 122 126 H 109 114 N 136 H H
Cytosinium cation in cytidine-3'-phosphate [2]	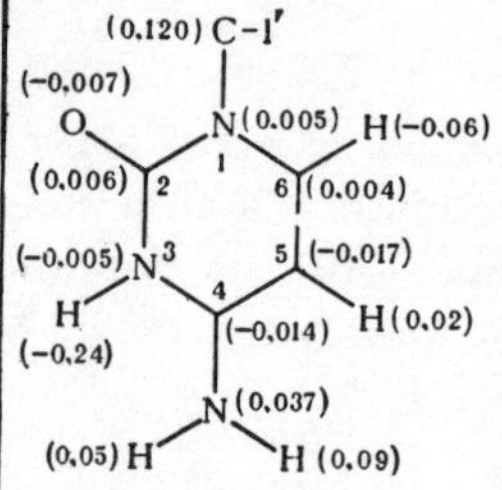 (0.120) C-1′ (−0.007) O (0.006) 2 (−0.005) N3 H (−0.24) N(0.005) H(−0.06) 6(0.004) 5(−0.017) (−0.014) H(0.02) N(0.037) (0.05) H H(0.09)	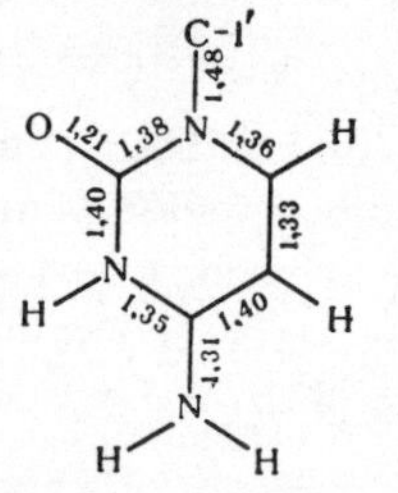 C-1′ 1.48 O 1.21 1.38 N 1.36 H 1.40 1.33 N 1.35 1.40 H 1.31 N H H	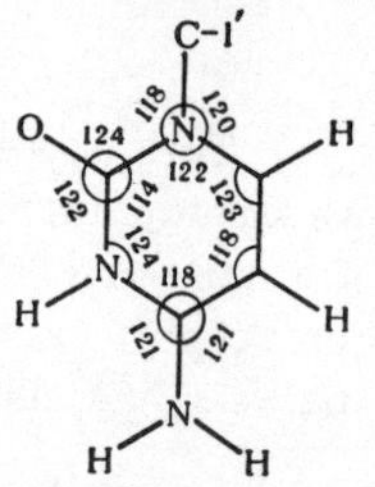 C-1′ 124 120 O N H 122 114 122 123 124 124 118 118 121 N 121 H H N H
Thymine in thymidine-5'-phosphate [3]	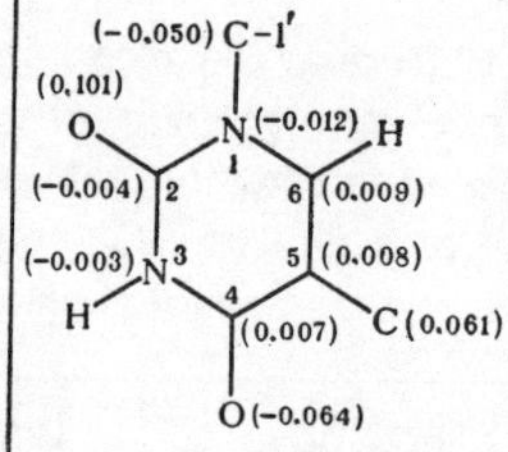 (−0.050) C-1′ (0.101) O (−0.004) 2 (−0.003) N3 H N(−0.012) H 6(0.009) 5(0.008) (0.007) C(0.061) O(−0.064)	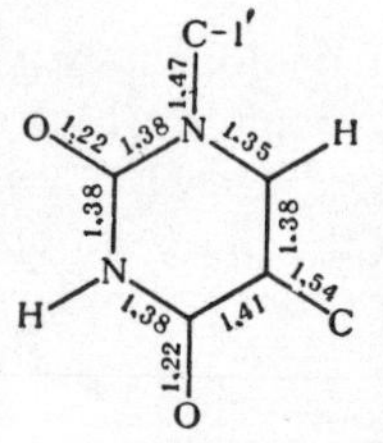 C-1′ 1.47 O 1.22 1.38 N 1.35 H 1.38 1.38 N 1.38 1.54 H 1.41 C 1.22 O	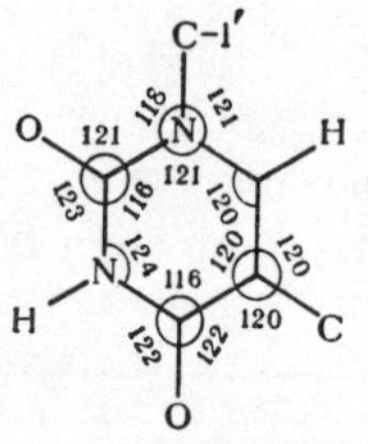 C-1′ 121 116 121 O N H 123 116 121 120 124 116 120 C 122 122 120 O
Uracil in uridine-5'-phosphate [4]	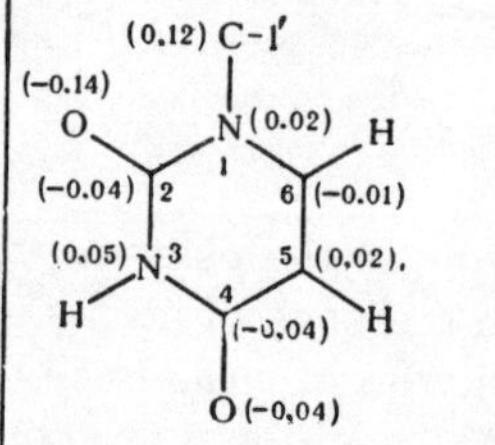 (0.12) C-1′ (−0.14) O (−0.04) 2 (0.05) N3 H N(0.02) H 6(−0.01) 5(0.02), (−0.04) H O(−0.04)	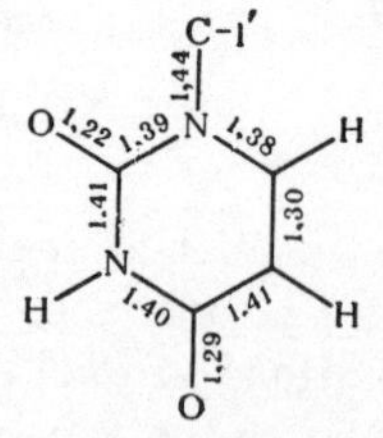 C-1′ 1.44 O 1.22 1.39 N 1.38 H 1.41 1.30 N 1.40 1.41 H 1.29 O	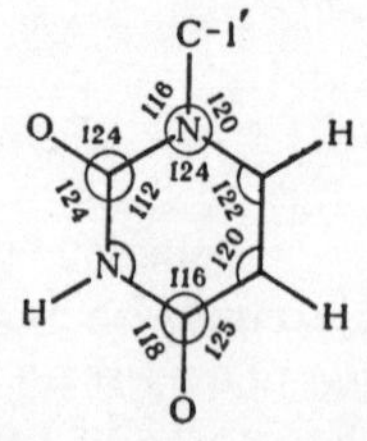 C-1′ 124 116 120 O N 124 H 124 112 124 122 N 116 H 118 125 H O

TABLE 2.1 (Continued)

Compound	Deviation of atoms from mean plane of ring, Å	Interatomic distances, Å	Valence angles, deg
Adenine * in 2'-deoxy-adenosine [5]	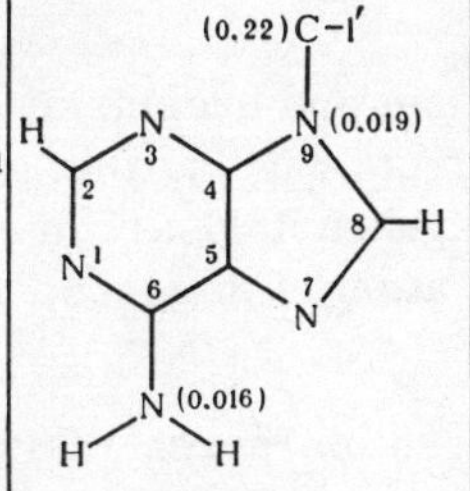	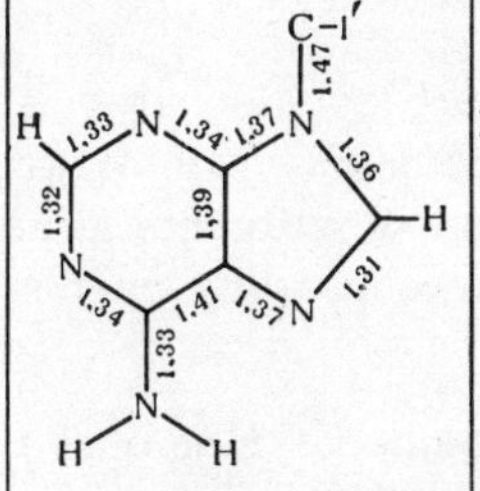	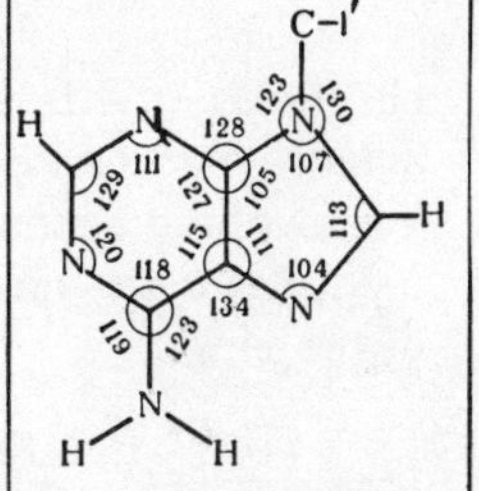
Adeninium* cation in adenosine-5'-phos-phate [6]	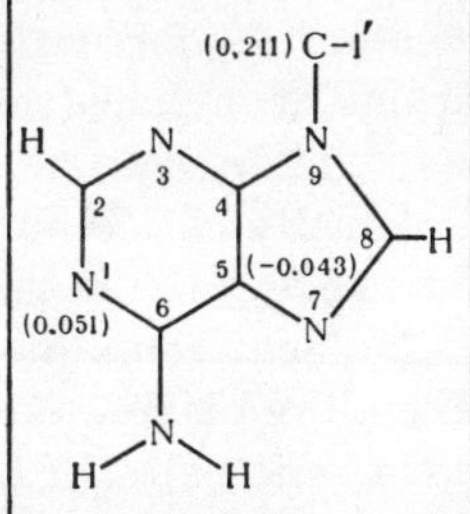	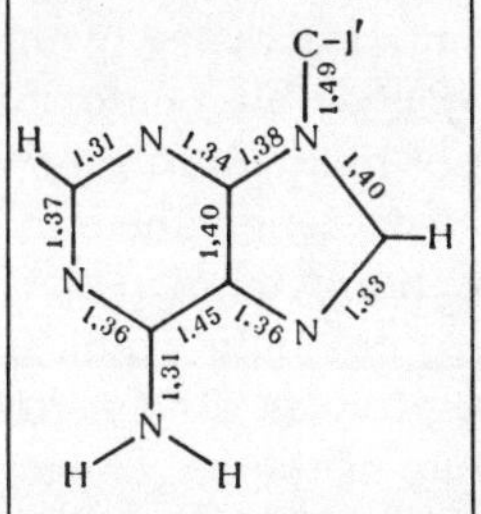	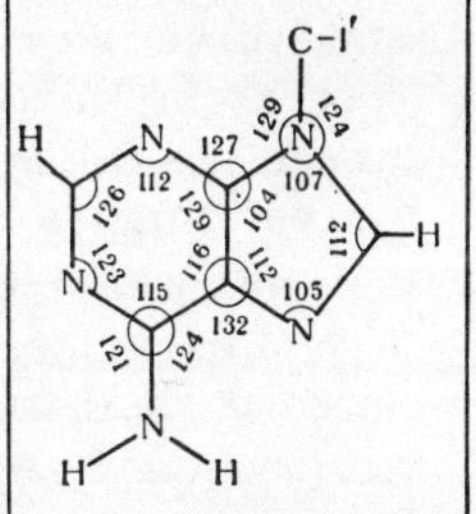
Adeninium* cation in adenosine-3'-phos-phate [7]	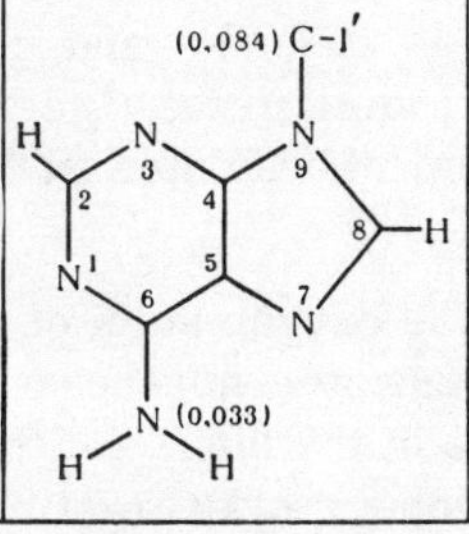	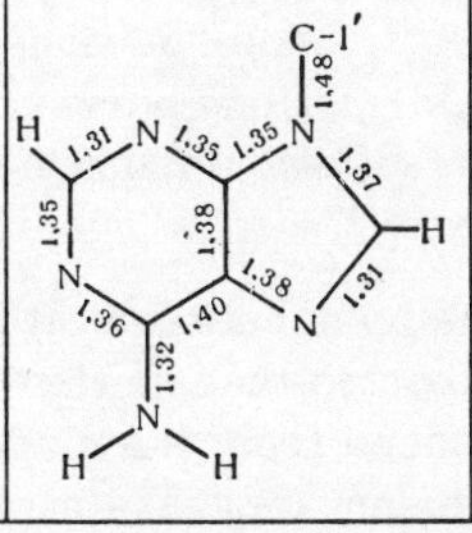	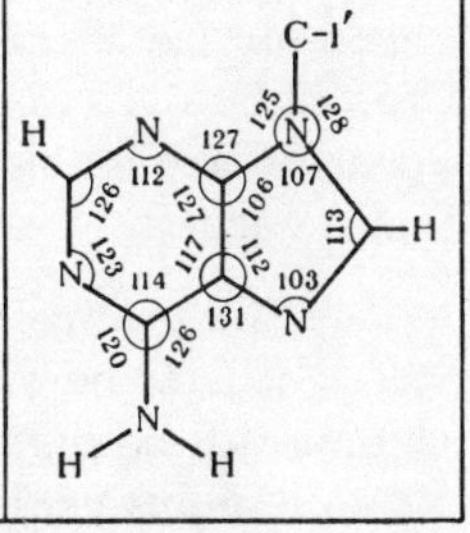

* Only the highest values of deviations of atoms from the plane are indicated.

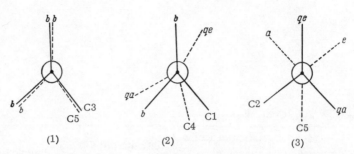

Fig. 2.1. Newman projections in the C_s conformation of cyclopentane along bonds: C1—C2 (1); C2—C3 (2); C3—C4 (3); numbering of the atoms in the ring as above.

character: the terms quasiequatorial (qe) and quasiaxial (qa) positions of the substituents can be used. It is clear from Fig. 2.1 that interaction between noncovalent-bonded substituents at neighbouring carbon atoms is considerably modified if the corresponding carbon atom is displaced away from the plane of the ring.

The five-membered ring of the pentofuranose residue can assume conformations similar to those of cyclopentane. Because of asymmetry of the tetrahydrofuran ring, for every cyclopentane conformation there are five different theoretically possible conformations of tetrahydrofuran and ten different conformations of the pentofuranoses. Conformations analogous to the C_s conformation of cyclopentane are generally called envelope conformations, and represented by the symbol [12] V^n, where n is the number of the carbon atom of the pentose molecule which is displaced from the plane; the index is placed above if the atom is displaced from the plane on the same side as the hydroxymethyl group of the sugar residue, and below if the carbon atom is displaced to the opposite side. Another abbreviation which is sometimes used for the V^n and V_n conformations is C_n-endo- and C_n-exo- conformations respectively. Conformations corresponding to the C_2 conformation of cyclopnetane are called twist-conformations, abbreviated to T^n_{n+1} and T^{n+1}_n or C_n-endo- and C_{n+1}-exo-.

Since there is no substituent at the oxygen atom of the pentofuranose ring, it is to be expected that neither this atom nor the neighbouring C1' or C4' atoms will be displaced from the plane of the ring, but the C2' or C3' atoms. This is more convenient from the standpoint of energetics.

$$\text{HOCH}_2 \quad O \quad B$$
$$\text{HO} \quad R$$

B denotes base residue; $R = H$ or OH

The corresponding conformations of the ribose or deoxyribose residues in nucleosides are given in Table 2.2; the positions occupied by substituents in the ring (CH_2OH at C4', base residue at C1', hydroxyl group at C3' and, in the case of ribose, the hydroxyl group at C2') in these conformations are given in the same table.   In every case at least one of the bulky substituents occupies the disadvantageous bissectrial or quasiaxial position, and the various conformations are evidently very similar in their stability.   Table 2.2 also shows the only possible conformation of the ribose or deoxyribose residue in nucleosides in which the bulkiest substituents occupy quasiequatorial positions, namely the V^O (O-endo-)X conformation, when the oxygen of the ring is outside the plane.   However, in this conformation the hydroxyl groups at C2' and C3' of the ribose residue occupy extremely disadvantageous (see Fig. 2.1) bissectrial positions.

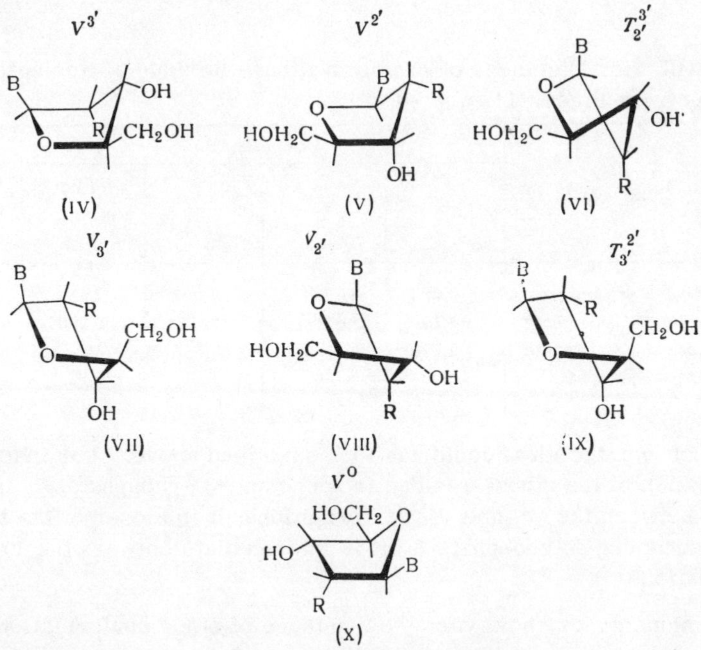

B denotes base residue; R = H or OH

The results of x-ray structural analysis experiments show that the ribose residue in nucleosides and nucleotides usually has the $V^{3'}$ conformation. This type of conformation of the carbohydrate residue has been demonstrated, for example, by x-ray structural analysis of crystals of cytidine [13, 14], uridine [14], uridine-5'-phosphate, adenosine [14], and the adenosine·5-bromouridine complex [15].   More precise x-ray structural investigations (Table 2.3) have shown that besides considerable deviation (0.5–0.6 Å) of the 3' carbon atom of the ribose residue from the plane of the ring, the neigbouring (C2') carbon atom is also 0.10–0.15 Å out of this plane, on the same or opposite side. The ribose residue in this case thus has a conformation which is intermediate

TABLE 2.2. Conformations of Nucleosides and Position of Substituents of Ribofuranose Residue

Conforma-tion	Formula	Position of substituents			
		CH_2OH	heterocyclic base B	OH (at C3')	R (at C2')
$V^{3'}$	IV	qe	b	e	qa
$V^{2'}$	V	b	qe	qa	e
$V_{3'}$	VII	qa	b	a	qe
$V_{2'}$	VIII	b	qa	qe	a
$T_{2'}^{3'}$	VI	qe	qa	e	a
$T_{3'}^{2'}$	IX	qa	qe	a	e
V^{o}	X	qe	qe	b	b

TABLE 2.3. Distances of Atoms in Ribose Residue of Nucleotides away from Plane C1' − C4' − O

Compound	Deviation, Å				Literature cited
	C2'	C3'	C5'	N (of heterocyclic ring)	
Cytidine-3'-phosphate	0.434	−0.146	1.236	0.889	[1]
Adenosine-5'-phosphate	−0.120	0.573	0.722	1.273	[6]
Adenosine-3'-phosphate	0.075	0.620	−	−	[7]
Uridine-5'-phosphate	0.52	−0.02	1.16	0.92	[4]

in type between the ideal conformations examined above. For instance, the conformation of the ribose residue in adenosine-3'-phosphate [7] is intermediate between the $V^{3'}$ and $V^{2'}$ conformations (it is closer to the former), and in adenosine-5'-phosphate [6] it is intermediate between the $V^{3'}$ and $T_{3'}^{2}$ conformations.

In some cases, however, the existence of other conformations of the ribose residue has been proved. In the case of cytidine-3'-phosphate [1], for instance, the sugar residue has an intermediate conformation between the $V^{2'}$ and $T_{3'}^{2}$ types (closer to the former), while in the case of 5-bromouridine [16] the conformation is of the $V_{2'}$ type.

An unusual conformation for the ribose residue, in which the C4' carbon atom projects from the plane of the ring has been discovered in crystals of adenosine-3',5'-cyclic phosphate (XI) [17]. This type of change of conformation is due to the formation of a condensed system containing a 6-membered and a 5-membered ring. In another nucleoside of similar structure which has been investigated (uridine-3',5'-cyclic phosphate; XII) [18], the sugar residue has the $V^{3'}$ conformation.

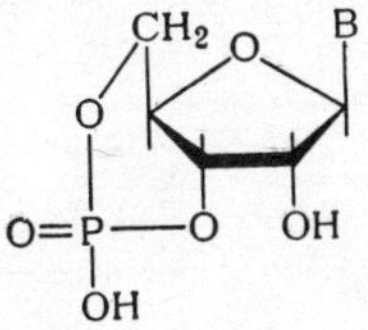

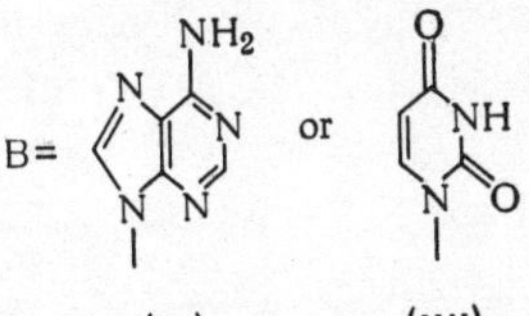

Attempts have been made to use the NMR method to determine the conformation of the carbohydrate residue of nucleosides and nucleotides in solution, but the results so far obtained do not allow an unequivocal interpretation, and the problem remains unresolved.

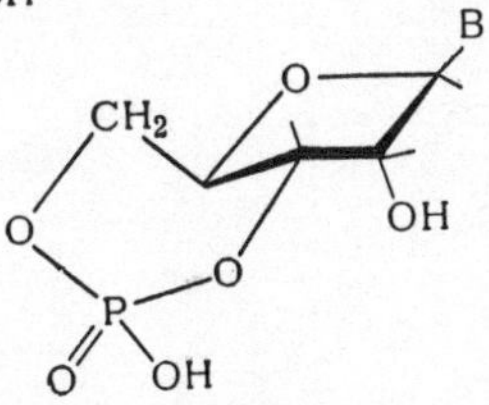

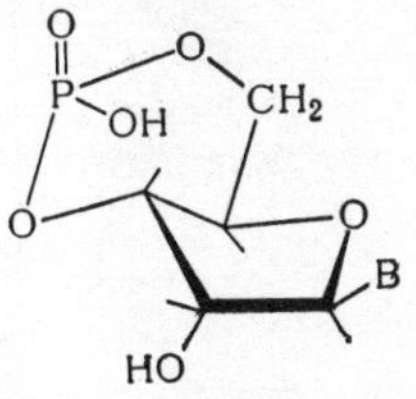

Investigation of the conformation of the ribose residue in nucleosides and nucleotides by means of NMR spectroscopy is usually based on analysis of changes in the coupling constant of the proton at C1' in different nucleosides ($J_{1', 2'}$), because the signal of this proton differs appreciably in the magnitude of its chemical shift from signals of other protons of the nucleosides, and its splitting can easily be observed. Examination of molecular models shows that the $V^{3'}$ conformation corresponds to a dihedral angle $\Phi_{1', 2'}$ = 115°, and the $V^{2'}$ conformation to an angle $\Phi_{1', 2'}$ = 150°, which correspond to values of the decoupling constants $J_{1', 2'}$ of 1.7 and 6.9 Hz respectively*. According to the experimental data (Table 2.4), values of $J_{1', 2'}$ in the nucleosides and nucleotides lie within the range 3-6.5 Hz and increase in the order:

cytidine < uridine < adenosine < guanosine

In an earlier investigation [23] of the NMR spectra of nucleosides and nucleotides it was concluded that the $V^{2'}$ conformation is predominant in purine nucleosides and the $V^{3'}$ conformation in pyrimidine nucleosides. However, this conclusion is not sufficiently strict, because a change in $J_{1', 2'}$ may be due to other factors also, notably to changes in the electronegativity of the substituent at C1' (see Chapter 3). It has also been shown that bond lengths and valence angles in ribofuranose derivatives differ appreciably from the standard values [24], and the usual molecular models evidently cannot be used for precise determination of the angle $\Phi_{1', 2'}$. The values observed for nucleosides can evidently be regarded as agreeing sufficiently closely both

---

* These values were obtained by means of the initial parameters of Karplus's equation (see page 100). More recently several other empirical equations connecting the value of the spin-spin interaction constant with the dihedral angle have been suggested.

TABLE 2.4. Experimentally Determined Values of the Constant of Spin — Spin Interaction $J_{1',2'}$ for Nucleosides and Nucleotides

Compound	$J_{1',2'}$ in D_2O, Hz		
	data of [20]	data of [21]	data of [22]
Guanosine...............	6.4	6.4	—
Adenosine...............	5.0	5.5	5.7
Adenosine-5'-phosphate......	4.5	—	—
Adenosine-3'-phosphate......	4.2	—	5.2
	(data of [19])		
Uridine	—	3.3	4.0
Uridine-3'-phosphate	—	—	4.2
Cytidine...............	3.0	3.0	—

with the $V^{3'}$ conformation (which predominates in crystals, as x-ray structural investigations have shown), and with the $V^{2'}$ conformation.

X-ray structural analysis of nucleosides and nucleotides (of the ribo- and deoxyribo- series) demonstrates a close analogy between the conformation of the 2'-deoxyribose and ribose residues in them. The most accurate data, in the writer's opinion, for deviation of atoms in monosaccharide residues of deoxyribonucleosides and deoxyribonucleotides from the $C1' - C4' - O$ plane are summarized in Table 2.5.

These results show that the conformation of 2'-deoxyadenosine can be regarded as intermediate between $V_{3'}$ and $T_3^{2'}$ conformations; the conformation of thymidine-5'-phosphate as close to the conformation of adenosine-3'-phosphate (intermediate between $V^{3'}$ and $V^{2'}$ conformations), while 5-fluoro-2'-deoxyuridine has a conformation close to that of cytidine-3'-phosphate (intermediate between $V^{2'}$ and $T_3^{2'}$ conformations). In other investigations, the 2'-deoxyribose residue has been ascribed the $V^{2'}$ conformation; it is found in 5-bromo-2'-deoxyuridine [26], 5-iodo-2'-deoxyuridine [27], and in nucleosides present in the complex of 2'-deoxyguanosine with 5-bromo-2'-deoxycytidine [28].

Investigation of the conformation of 2'-deoxynuleosides in solution with the aid of NMR spectroscopy has led to the conclusion that the 2'-deoxyribose residue has a V^O conformation [29]. This conclusion is based on the assumption that Karplus's equation is quantitatively satisfied and on measurement of the dihedral angles $\Phi_{1',2'}$ and $\Phi_{2',3'}$ in atomic models of different conformations of 2'-deoxyribose; this approach, however, can no longer be accepted as valid and is in need of revision.

On the whole, therefore, these results show that the sugar residue has different conformations in different natural nucleosides and nucleotides. This indicates that the energy values of the different conformations must be very close.

TABLE 2.5. Deviations of Atoms in Monosaccharide Residues of Deoxyribonucleosides and Deoxyribonucleotides from C1' − C4' − O Plane

Compound	Deviation, Å				Literature cited
	C2'	C3'	C5'	N (of heterocyclic ring)	
2'-Deoxyadenosine	0.60	−0.504	1.358	1.187	[5]
Thymidine-5'-phosphate	0.098	0.602	0.698	1.103	[13]
5-Fluoro-2'-deoxyuridine	0.494	−0.170	1.281	1.183	[25]

3. Mutual arrangement of the carbohydrate residues and heterocyclic rings

For the mutual arrangement of the sugar residue and heterocyclic base in nucleosides to be unequivocally determined, not only the length of the C − N glycoside bond and the angle formed by it with the plane of the heterocyclic base (see Table 2.1), but the values of two further angles must be determined.

One of these is the dihedral angle formed by the planes of the base and of the pentofuranose residue; this angle can be determined only by x-ray structural analysis. The values of the dihedral angle found in this way lie between 65 and 80°; when it is determined it is usual to take as the pentofuranose plane, not the C1' − C4' − O plane, but the "mean" plane (the plane such that the sum of the squares of deviations of all atoms of the pentofuranose ring from it is a minimum), because the values obtained may to some extent reflect differences in the conformation of the carbohydrate residue. Most values determined experimentally, however, fall within the range 70-75°.

Mutual rotation of the sugar residue and heterocyclic base around the C1'− N glycoside bond is characterized by yet another angular value. This rotation is usually designated as the angle of rotation $\Phi_{C, N}$ formed by projections of the C1' − O and N1 − C6 bonds in pyrimidine (or C1' − O and N9− C8 bonds in purine) residues on the plane perpendicular to the C1' − O bond, as shown on the Newman projections illustrated below:

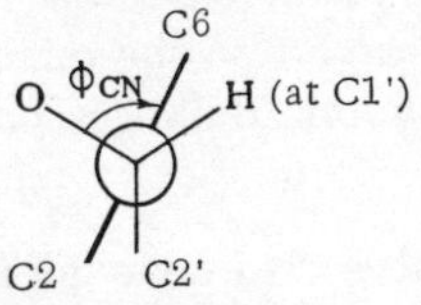

For pyrimidine nucleosides 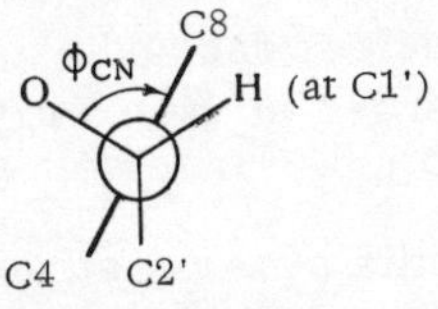   For purine nucleosides

The angle $\Phi_{C, N}$ is conventionally regarded as positive if measured in a clockwise direction.

Examination of molecular models shows that free rotation around the N-glycoside bond in nucleosides is impeded. Particularly severe limitations are observed in pyrimidine nucleosides because of interaction between the oxygen atom at C2 or the hydrogen atom at C6 of the heterocyclic ring and substituents at C2' and C3' of the sugar residue, and also the oxygen atom of the furanose ring. Less strong interactions evidently exist also in purine nucleosides, although in this case the N3 atom of the purine ring may come into close contact with the cyclic oxygen atom of the sugar residue and the hydrogen atom at C2' or C3' as the bond. Weakest interactions are observed if $\Phi_{C,N}$ is $-30°$ or $+150°$; the conformations corresponding to these values of the angle of rotation are customarily called anti- and syn-conformations of the nucleosides [30]. In the case of uridine and adenosine, for example:

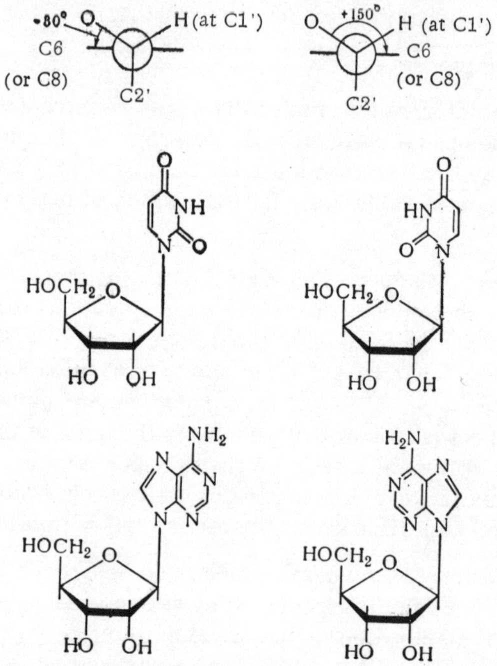

This qualitative conclusion obtained by examination of molecular models has been confirmed by more precise analysis of the change in distances between atoms during rotation of the ribose residue around the N-glycoside bond [31] in nucleosides and nucleotides, and also the quantum-chemical calculation of changes in the energy of the system during a change in the angle $\Phi_{C,N}$ [32].

The results of x-ray structural analysis of several ribonucleosides and ribonucleotides show that, as a rule, in all cases studied the conformation of these compounds in the crystal is very close to the anti-conformation (Table 2.6), although for pyrimidine derivatives the absolute value of angle $\Phi_{C,N}$ is usually slightly greater than for purine derivatives.

TABLE 2.6. Values of Angle of Rotation $\Phi_{C,N}$ for Several Ribo-nucleosides and Ribonucleotides

Compound	$\Phi_{C,N}$, deg	Literature cited
Pyrimidine derivatives		
Uridine-5'-phosphate...............	−43	[4]
Cytidine-3'-phosphate...............	−42	[1]
Cytidine......................	−18	[13]
5-Bromouridine (in complex with adenosine)....	−20	[15]
Purine derivatives		
Adenosine-5'-phosphate...............	−18	[6]
Adenosine-3'-phosphate...............	−4	[7]
Adenosine (in complex with 5-bromouridine) ...	−10	[15]

TABLE 2.7. Amplitude of Cotton Effect (a) for Nucleosides, Nucleotides, and Some Model Anhydro Compounds of the Pyrimidine Series [33]

Compound	Formula	$a \cdot 10^{-2}$, deg	$\Phi_{C,N}$ (approximate value), deg
Uridine.......................	−	+117	−
Cytidine......................	−	+152	−
Cytidine-3'-phosphate..............	−	+204	−
Cytidine-5'-phosphate..............	−	+137	−
5',6-exo-O-Cyclo-6-hydroxyuridine.....	XIII	+490	−30
2',2-exo-O-Cyclouridine............	XIV	+266	−60
3',2-exo-O-Cyclouridine	XV	+43	+105
2',3'-O-Isopropylidene-5',2-exo-O-cyclo-uridine......................	XVI	−539*	+150
2-O-Ethyl-2',3'-O-isopropylideneuridine ..	XVII	+227	−

* Value understated in absolute terms.

The exceptions are crystals of adenosine-3',5'-cyclic phosphate [17]: the elementary lattice of the crystal contains two molecules of the compound, one of which is in the anti-conformation ($\Phi_{C,N}$ −50°), while the other is in the syn-conformation ($\Phi_{C,N}$ +102°). It has recently been shown [69] that 4-thiouridine has the syn-conformation in crystals ($\Phi_{C,N}$ +83°).

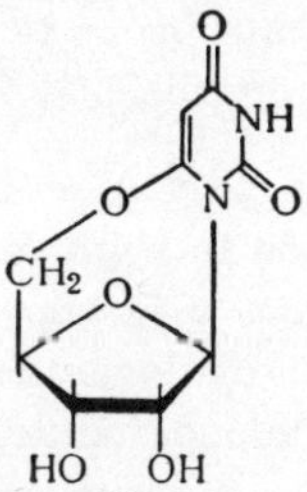

XIII 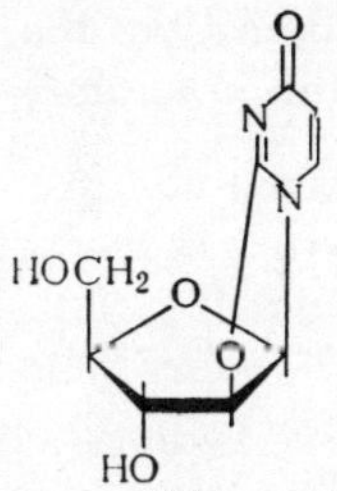   XIV 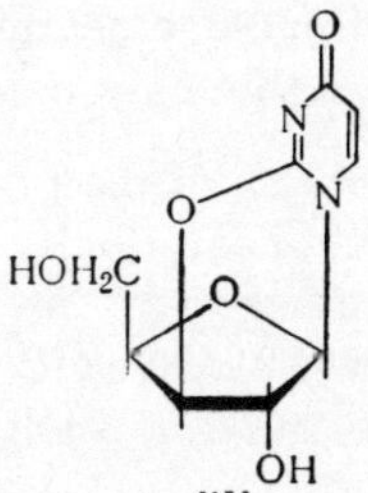   XV

E

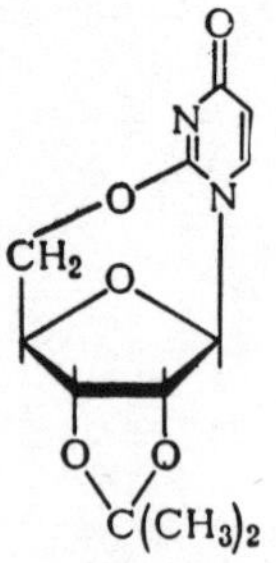

XVI 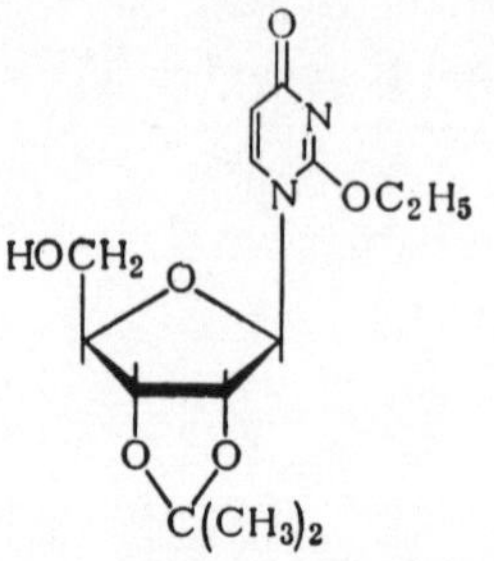   XVII

The anti-conformation is the preferable conformation for pyrimidine nucleosides in solutions also, as a study of their optical rotatory dispersion has shown. These compounds have a positive Cotton effect with high amplitude [33-36] (Table 2.7). A study of the optical rotatory dispersion curves of model anhydro compounds (XIII-XVI) [33] with a fixed position of their ribose residue and heterocyclic base has shown that the amplitude of the Cotton effect is highly dependent on their mutual orientation.

An increase in the angle of rotation $\Phi_{C, N}$ leads to a decrease in amplitude of the positive Cotton effect; for compound XVI with a fixed syn-conformation a high negative Cotton effect is observed; the change in sign of the Cotton effect, moreover, is unrelated to a change in the chromophore during the conversion from uridine to its 2-O-substituted derivative, as is clear from the value of the amplitude for 2-O-ethyl-2',3'-O-isopropylideneuridine (XVII). These results clearly show that the preferential conformation for pyrimidine nucleosides in solution is the anti-conformation, although rotation about the C1' − N-glycosidic linkage is nevertheless possible; this leads to a decrease in amplitude of the Cotton effect compared with compounds with a fixed conformation. The anti-conformation of pyrimidine nucleosides in solution is confirmed by NMR findings. This has been demonstrated for solutions of acetates of nucleosides in dimethyl sulphoxide, where the change in chemical shift of the protons of the acetyl group at C2' during the conversion from uridine derivatives to derivatives of 5,6-dihydrouridine indicates an appreciable effect of magnetic anisotropy of the double bond and, consequently, the close spatial proximity of the methyl protons of the acetyl group at C2' and the double bond [37]. A study of the relationship between the chemical shift of the proton at C6 of the pyrimidine ring in nucleoside-5'-phosphates and pH indicates the close spatial proximity of the phosphate group to this proton, and this is possible only in the case of an anti-conformation [38, 71].

The problem of the preferential conformation of purine nucleosides in solution has not been finally solved. These compounds in solution are characterized by a negative Cotton effect [34, 35]. A negative Cotton effect is also observed in the case of 2',3'-O-isopropylidene-3,5-cyclonucleosides of the purine series [36, 39], such as, for example, in the adenosine derivative XVIII.

XVIII   XIX

On the other hand, model compounds in which the ribose residue is
fixed in the anti-conformation, such as 8,2'-anhydro-8-hydroxy-9-β-D-ara-
binofuranosyladenine (XIX), have a positive Cotton effect [40, 41], from which
it can be concluded that purine nucleosides have the syn-conformation in solution. This conclusion is confirmed [42] by the change in sign of the Cotton
effect of adenosine derivatives if bulky substituents are introduced at the C5'
position, making the existence of the compound in the syn-conformation impossible *.

Measurements of the NMR spectra of nucleoside-5'-phosphates show,
however, that ionization of the phosphate group has an appreciable effect on
the chemical shift of the proton at C8 but not at C2, which is possible only in
the case of an anti-conformation [38]. Finally, as recent work has shown,
with a change in pH there is a change in sign of the Cotton effect in guanosine
derivatives [43]; this can be interpreted as indicating the possibility of reversible interconversions between syn- and anti-conformations of a nucleoside†.

The predominant conformation of deoxyribonucleosides and deoxyribonucleotides is evidently analogous to the conformation of the corresponding
ribose derivatives. The results of x-ray structural analyses so far undertaken (Table 2.8) show that in most cases the conformation of deoxyribonucleosides and deoxyribonucleotides in the crystal form is close to the anti-conformation, although in one case evidence of a syn-conformation was obtained.

Optical rotatory dispersion curves of pyrimidine deoxyribosides show
a positive Cotton effect [34, 44]; for a model compound with fixed syn-conformation (5',2-exo-O-cyclothymidine) a negative Cotton effect has been observed,

---

* The hypothesis that adenosine derivatives have an anti-conformation is confirmed by the results
of approximate calculations of the rotatory power of these compounds [70].

† Investigation of the NMR spectra of 2',3'-O-isopropylideneadenosine, 2',3'-O-isopropylidene-
guanosine, and 3,5'-cyclo-2',3'-O-isopropylideneadenosine by the double resonance method has
shown that reversible interconversions are possible between syn- and anti-conformations, and that
the anti-conformation is less advantageous for adenosine derivatives than for guanosine derivatives
[72].

TABLE 2.8. Angles of Rotation $\Phi_{C,N}$ for Several Deoxyribonucleosides and Deoxyribonucleotides

Compound	$\Phi_{C,N}$, deg	Literature cited
Pyrimidine derivatives		
Thymidine-5'-phosphate..........................	−48	[3]
2'-Deoxy-5-bromocytidine (in complex with 2'-deoxy-guanosine)...............................	−61	[28]
2'-Deoxy-5-fluorouridine	−62	[26]
2'-Deoxy-5-bromouridine........................	−43	[26]
2'-Deoxy-5-iodouridine	−65	[27]
Purine derivatives		
2'-Deoxyadenosine	−3	[5]
2'-Deoxyguanosine (in complex with 2'-deoxy-5-bromo-cytidine)................................	+138	[28]

from which it can be postulated that anti-conformers are predominant in solutions of compounds in which rotation around the glycoside bond can take place. Purine deoxyribosides are analogous to purine ribosides, as can be judged from the character of their optical rotatory dispersion curves (negative Cotton effect) and the change in their NMR spectrum with a change in pH; it can accordingly be concluded that the conformations of these compounds are closely similar.

4. Intramolecular interactions

In the earlier discussion of the theoretically possible conformations of nucleosides (see page 106) examination of the chief factors determining conformation was limited to Van der Waals' forces of repulsion. Yet molecules of nucleosides and nucleotides contain considerable numbers of functional groups which may interact with each other. Intramolecular interactions of this type may have a substantial influence on the conformation of nucleosides and nucleotides in dilute solutions. Although a detailed examination of this problem is impossible at the present time, a brief discussion of existing data concerning intramolecular interactions in nucleosides and nucleotides is of the greatest value.

There is an extensive literature on the question of the formation of intramolecular hydrogen bonds between the hydroxyl group at C2' of the ribose residue and the carbonyl group at C2 of the pyrimidine base:

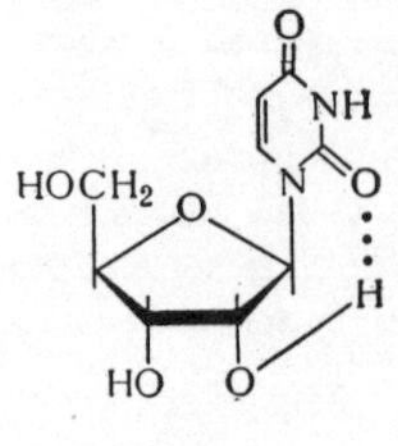

The possibility of the formation of such hydrogen bonds can be demonstrated easily on molecular models of nucleosides with an anti-conformation. Interaction along these lines was first postulated in order to explain differences in the UV-spectra of uridine and uridine-3'-phosphate, on the one hand, and uridine-2'-phosphate, on the other hand, in a strongly alkaline medium [45]. No direct proof of the formation of a hydrogen bond of this type has yet been obtained. The distance between the 2-exo-O and 2'-exo-O oxygen atoms in crystals of pyrimidine ribonucleosides and ribonucleotides, obtained by x-ray structural analysis, is greater than the distance necessary for the formation of a hydrogen bond. However, this is not surprising, because both oxygen atoms in the crystal lattice are in the immediate vicinity of functional groups of neighbouring nucleotide molecules; in that case, intermolecular interactions of this type can be stronger than intramolecular interactions existing in dilute solutions.

The available indirect evidence of the existence of a hydrogen bond between the carbonyl group at C2 of the pyrimidine ring and the hydrogen atom of the hydroxyl group at C2' of the ribose residue can be divided into two groups. The first group consists of facts relating to the difference in physical properties or chemical reactivity of pyrimidine nucleosides and their derivatives in which a hydroxyl group is present at C2' (for example, uridine, uridine-3'-phosphate, uridine-5'-phosphate) and derivatives in which this hydroxyl group is absent (for example, 2'-deoxyuridine, uridine-2'-phosphate, alkyluracils). Besides the data on ultraviolet spectra of uridine derivatives already mentioned above, other relevant data here include those concerning the difference in ionization constants of cytosine derivatives of these two groups [46], differences between the NMR spectra of pyrimidine nucleotides [38], and differences between the velocities of certain reactions of ribo- and deoxyribonucleosides (for example, the photochemical hydration of cytidine derivatives [47], and the catalytic hydrogenation [48] and hydroxylaminolysis [49] of uridine derivatives).

The other group of indirect evidence consists of data relating to the change in reactivity of functional groups of the sugar or phosphate residue depending on the nature of the heterocyclic base. Differences of this type have often been observed, but detailed kinetic investigations in this field are virtually absent. One example of such work is that of Witzel [50], who studied the kinetics of hydrolysis of dinucleoside monophosphates containing residues of pyrimidine nucleoside-3'-phosphates and their comparison with the corresponding compounds containing residues of purine nucleoside-3'-phosphates. The kinetic differences observed can be explained by increased nucleophilicity of the hydroxyl group at C2' of the pyrimidine nucleosides because of hydrogen bonding with the heterocyclic nucleus. Interaction between the carbonyl group at C2 of the pyrimidine ring with the hydroxyl group at C2' of the ribose residue has been postulated to explain the mechanism of action of pancreatic ribonuclease [51, 52]. Other data, which can be regarded as indirect evidence that intramolecular hydrogen bonding can also take place in purine nucleosides, with the participation of N3 of the heterocyclic ring and the hydroxyl group at C2' of the ribose residue, have also been published. They include descriptions of differences in UV-spectra [51] and NMR spectra [38, 53] of purine ribo- and 2'-deoxyribonucleosides.

Considerable factual evidence of differences in the physicochemical characteristics of oligonucleotides and polynucleotides containing ribo- and deoxyribonucleotides has now been obtained (for further details see Chapter 4). The logical cause of these differences is evidently some role of the hydroxyl group at C2' of the ribose residue in stabilizing the conformation of the polynucleotide chain in the case of the ribopolymers. Besides the formation of a hydrogen bond with the carbonyl group of the pyrimidine base (or N3 atom of the purine base), the formation of hydrogen bonds with oxygen atoms of the phosphate group has also been postulated in order to explain the observed properties [54]:

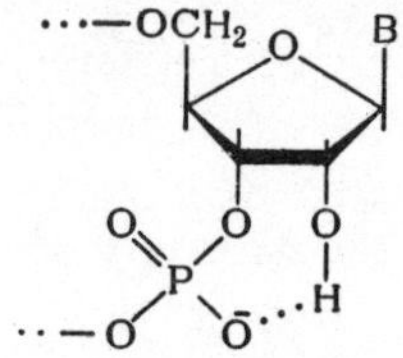

B— base residue

This type of interaction can evidently take place also in monomeric nucleoside-3'-phosphates.

Intramolecular interactions due to hydrogen bonds can also exist in more complex low molecular weight nucleotide derivatives, such as nucleoside diphosphate sugar [55], and nucleoside-5'-phospho(P – N)-amino acid and its peptide [56].

Intramolecular interactions of another type can arise in those nucleotide derivatives which contain residues of heterocyclic bases (with aromatic properties) as substituent: interaction between two aromatic systems, analogous to interactions observed in polynucleotides (see Chapter 4). Intramolecular interactions of this type have been reliably proved for nicotinamide-adenine dinucleotide [57-62], flavin-adenine dinucleotide [63-66], and their derivatives.

Strong interactions between two heterocyclic bases are observed in P^1, P^2-dinucleoside-5'-pyrophosphates and in dinucleoside monophosphates. Interactions of this type play an important role in the stabilization of the conformation of oligonucleotides and polynucleotides. They will be examined in more detail in Chapter 4.

Bibliography

1. M. Sundaralingam and L. H. Jensen, J. Mol. Biol., 13:914 (1965).
2. C. E. Bugg and R. E. Marsh, J. Mol. Biol., 25:67 (1967).
3. K. N. Trueblood, P. Horn, and V. Luzzati, Acta Cryst., 14:965 (1961).
4. E. Shefter and K. N. Trueblood, Acta Cryst., 18:1067 (1965).
5. D. G. Watson, D. J. Sutor, and P. Tollin, Acta Cryst., 19:111 (1965).
6. J. Kraut and L. H. Jensen, Acta Cryst., 16:79 (1963).
7. M. Sundaralingam, Acta Cryst., 21:495 (1966).
8. P. Rouillier, J. Delmau, and C. Nofre, Bull. Soc. Chim. France, 3515 (1966).
9. S. Furberg and L. H. Jensen, J. Am. Chem. Soc., 90:470 (1968).
10. B. Kurtev, I. Pojarliev, I. Blagoeva, and I. Burgudjiev, Abstr. Papers XXIst Intern. Congr. Pure. Appl. Chem., Nucl. Acid Components, No. 2, Prague (1967).
11. J. E. Kilpatrick, K. S. Pitzer, and R. Spitzer, J. Am. Chem. Soc., 69:2483 (1947).
12. L. D. Hall, Chem. and Ind., 950 (1963).
13. S. Furberg, C. S. Peterson, and C. Romming, Acta Cryst., 18:313 (1965).
14. S. Furberg, Acta Chem. Scand., 4:751 (1950).
15. A. E. Haschemeyer and H. M. Sobell, Acta Cryst., 18:525 (1965).
16. J. Iball, C. H. Morgan, and H. R. Wilson, Proc. Roy. Soc., A295:320 (1966).
17. K. Watenpaugh, J. Dow, L. H. Jensen, and S. Furberg, Science, 159:206 (1968).
18. C. L. Coulter, Science, 159:888 (1968).
19. C. D. Jardetzky, J. Am. Chem. Soc., 84:62 (1962).
20. C. D. Jardetzky and O. Jardetzky, J. Am. Chem. Soc., 82:222 (1960).
21. L. Catlin and J. C. Davis, J. Am. Chem. Soc., 84:4464 (1962).
22. H. P. M. Fromageot, R. E. Griffin, C. B. Reese, J. E. Sulston, and D. R. Tretham, Tetrahedron, 22:705 (1965).
23. C. D. Jardetzky, J. Am. Chem. Soc., 82:229 (1960).
24. M. Sundaralingam, J. Am. Chem. Soc., 87:599 (1965).
25. D. R. Harris and W. M. Macintyre, Biophys. J., 203 (1964).
26. J. Iball, C. H. Morgan, and H. R. Wilson, Nature, 209:1230 (1966).
27. N. Camerman and J. Trotter, Acta Cryst., 18:203 (1965).
28. A. E. Haschemeyer and H. M. Sobell, Acta Cryst., 19:125 (1965).
29. C. D. Jardetzky, J. Am. Chem. Soc., 83:2919 (1961).
30. J. Donohue and K. Trueblood, J. Mol. Biol., 2:363 (1960).
31. A. E. Haschemeyer and A. Rich, J. Mol. Biol., 27:369 (1967).
32. F. Jordan and B. Pullman, Theoret. Chim. Acta, 9:242 (1968).
33. T. R. Emerson, R. J. Swan, and T. L. V. Ulbricht, Biochemistry, 6:842 (1967).
34. J. T. Yang, T. Samejima, and P. K. Sarkar, Biopolymers, 4:623 (1966).
35. T. Nishimura and B. Shimizu, Biochim. Biophys. Acta, 157:221 (1968).
36. D. W. Miles, R. K. Robins, and H. Eyring, Proc. Nat. Acad. Sci. USA, 57:1138 (1967).
37. R. J. Cushley, K. A. Watanabe, and J. J. Fox, J. Am. Chem. Soc., 89:394 (1967).
38. M. P. Schweizer, A. D. Broom, P. O. P. Ts'o and D. P. Hollis, J. Am. Chem. Soc., 90:1042 (1968).
39. A. Hampton, J. Org. Chem., 32:1688 (1967).
40. M. Ikehara, M. Kaneko, K. Muneyama, and K. Tanaka, Tetrahedron Letters, 3977 (1967).
41. M. Ikehara, M. Kaneko, and M. Sagai, Chem. Pharm. Bull., 16:1151 (1968).
42. W. A. Klee and S. H. Mudd, Biochemistry, 6:988 (1967).

43.   V. Gushelbauer and Y. Courteis, FEBS Letters, 1:183 (1968).

44.   T. L. V. Ulbricht, J. P. Jennings, P. M. Scopes, and M. Klyne, Tetrahedron Letters, 695 (1964).

45.   J. J. Fox, L. F. Cavalieri, and N. Chang, J. Am. Chem. Soc., 75:4315 (1953).

46.   S. Lewin and D. A. Humphreys, J. Chem. Soc. (B), 210 (1966).

47.   K. L. Wierzchowsky and D. Shugar, Biochim. Biophys. Acta, 25:355 (1957); Acta Biochim. Polon., 8:219 (1961).

48.   N. K. Kochetkov, É. I. Budovskii, V. N. Shibaev, and G. I. Eliseeva, Dokl. Akad. Nauk SSSR, 159:609 (1964).

49.   N. K. Kochetkov, É. I. Budovskii, V. N. Shibaev, and G. I. Eliseeva, Dokl. Akad. Nauk SSSR, 172:603 (1967).

50.   H. Witzel, Ann., 635:182 (1960).

51.   H. Witzel, in: Progress in Nucleic Acid Research and Molecular Biology, Vol. 2, D. N. Davidson and W. E. Cohn (editors), Academic Press, New York (1963), p. 221.

52.   H. G. Gassen and H. Witzel, Europ. J. Biochem., 1:36 (1967).

53.   A. D. Broom, M. P. Schweizer, and P. O. P. Ts'o, J. Am. Chem. Soc., 89: 3612 (1967).

54.   J. Brahms, J. C. Maurizot, and A. M. Michelson, J. Mol. Biol., 25:481 (1967).

55.   N. K. Kochetkov, É. I. Budovskii, and V. N. Shibaev, Biokhimiya 28:741 (1963).

56.   N. I. Sokolova, E. A. Stumbravichute, P. P. Purygin, Z. A. Shabarova, and M. A. Prokof'ev, Dokl. Akad. Nauk SSSR, 174:722 (1967).

57.   G. Weber, J. Chim. Phys., 55:878 (1958).

58.   S. F. Velick, J. Biol. Chem., 233:1455 (1958).

59.   W. L. Meyer, H. R. Mahler, and B. R. Baker, Biochim. Biophys. Acta, 64:353 (1962).

60.   O. Jardetzky and N. G. Wade-Jardetzky, J. Biol. Chem., 241:85 (1966).

61.   G. Pfleiderer, K. Woenckhaus, K. Scholz, and H. Feller, Ann., 675:205 (1964).

62.   G. Pfleiderer and C. W. Woenckhaus, Ann., 690:170 (1965).

63.   O. A. Bessey, O. H. Lowry, and R. H. Love, J. Biol. Chem., 180:755 (1949).

64.   G. Weber, Biochem. J., 47:114 (1950).

65.   B. M. Chassy and D. B. McCormick, Biochemistry, 4:2612 (1965).

66.   I. C. M. Tsibris, D. B. McCormick, and L. B. Wright, Biochemistry, 504 (1965).

67.   D. Rohrer and M. Sundaralingam, Chem. Comm., 746 (1968).

68.   J. Donohue, Arch. Biochem. Biophys., 128:591 (1968).

69.   W. Saenger and K. H. Scheit, Angew. Chem., 81:121 (1969).

70.   D. W. Miles, S. J. Hahn, R. K. Robins, M. J. Robins, and H. Eyring, J. Phys. Chem., 72:1483 (1968).

71.   I. Feldman and R. P. Agarwal, J. Am. Chem. Soc., 90:7329 (1968).

72.   P. A. Hart and J. P. Davis, J. Am. Chem. Soc., 91:512 (1969).

Chapter 3
Electronic Structure and Reactivity of the Monomer Components of Nucleic Acids

I. Introduction

The functional specificity of nucleic acids and nucleotide coenzymes is determined by the reactivity of the nucleoside components and, in particular, of the heterocyclic bases incorporated in them. The term reactivity must always be understood in its widest sense, to include not only interactions leading to the formation or rupture of covalent bonds, but also interactions of other types: with neighbouring bases in the same polynucleotide chain or with the complementary bases of another polynucleotide, with proteins (histones, virus membrane proteins, enzymes of nuclear metabolism, and so on), ions of metals, etc.

The whole arsenal of classical and modern organic chemistry is used to study the chemical properties of the components of nucleic acids: theoretical quantum-chemical calculations and empirical methods based on the study of reactivity of analogues. In this chapter, we shall examine from this point of view the principal characteristics of the components of nucleic acids and nucleotide coenzymes responsible for their chemical specificity in the basic electronic state. The properties of components of nucleic acids in an excited state will be examined in Chapter 12.

II. Distribution of electron density in the heterocyclic bases of nucleic acids

The properties of molecules are determined by the character of their electronic structure, and because of this, before discussing the reactivity of the heterocyclic bases composing nucleic acids, we must examine the distribution of electron density in the molecules of these bases. We shall examine

E*

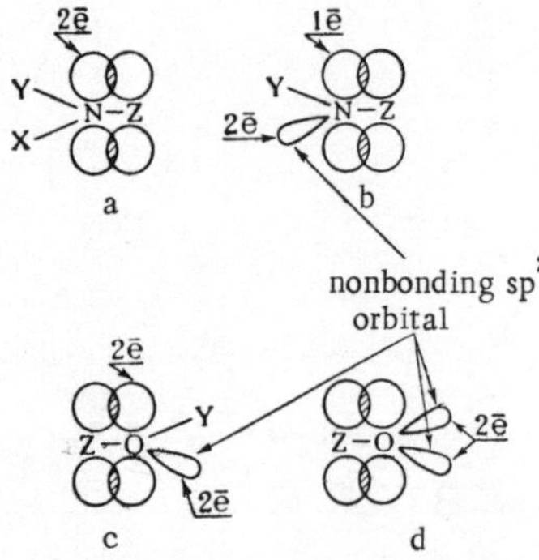

Fig. 3.1. Ways in which nitrogen and oxygen atoms of heterocyclic bases take part in conjugation: a) the nitrogen atom forms three σ bonds with atoms X, Y, and Z. In the p orbital of the nitrogen atom, which overlaps with the p orbital of atom Z, there are two electrons. The corresponding nitrogen atoms are marked δ+ in Fig. 3.3. b) The nitrogen atom forms two σ bonds with atoms Y and Z. In the p orbital of the nitrogen atom, overlapping with the p orbital of the Z atom, there is one electron. In the n orbital, making an angle of 120° with the N – Y and N – Z bonds, there are two nonbonding electrons. The corresponding nitrogen atoms are marked δ – in Fig. 3.3. c) The oxygen atom forms two σ bonds with atoms Z and Y. In the p orbital of the oxygen atom, overlapping with the p orbital of the Z atom, there are two electrons. In the nonbonding n orbital, making an angle of 120° with the Z – O and O – Y bonds, there are two nonbonding electrons. The corresponding oxygen atoms are marked δ+ on Fig. 3.3 (the sulphur atom forming two σ bonds plays a similar role in conjugation). d) The oxygen atom forms one σ bond. In the p orbital of the oxygen atom, overlapping with the p orbital of the Z atom, there is one electron. In the two nonbonding n orbitals, making an angle of 120° with each other and with the O – Z bond, there are two n electrons in each case. The corresponding atoms in the bases in Fig. 3.3 are marked δ – (the sulphur atom forming one σ bond also contributes one electron to the conjugated system).

the theoretical aspect of the problem and, of necessity more briefly, the few experimental attempts which have been made to confirm the theoretical assumptions. Since the tautomeric equilibrium of the bases is usually displaced very strongly toward one of the forms (see pages 137 et seq.), the one which is evidently responsible for the properties of the compound, we shall examine in this chapter the question of the distribution of electrons principally in tautomeric forms of the bases which are the most stable under ordinary conditions (25°C, 1 atm). In addition, although protonation or deprotonation, with the formation of the corresponding cation or anion, can appreciably modify the electron distribution in molecules of the bases, in this section we shall examine only neutral bases, because nearly all the theoretical calculations are in fact carried out on this form. Examination of neutral molecules gives considerable information pertinent to the study of the chemical properties of the bases, including those in an ionized state.

1. Theoretical considerations

Heterocyclic bases and conjugated systems. Purine and pyrimidine bases of nucleic acids are cyclic systems composed of interlinked trigonal (sp^2 hybrid) atoms, the p electron atomic orbitals of which intersect to form π electron molecular orbitals. In accordance with the generally accepted views [1–3] regarding methods of conjugation of sp^2 hybrid heteroatoms, the nitrogen atom which in the compounds under review is linked to three other atoms contributes two electrons to the general π electron system and has no

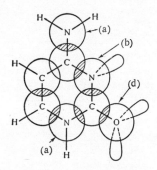

Fig. 3.2. The cytosine molecule as an example of a conjugated system with heteroatoms
of types a, b, and d (see Fig.
3.1).

nonbonding or n orbitals (Fig. 3.1). Examples of such atoms are the nitrogen atoms of exocyclic amino groups in cytosine, adenine, or guanine.

The nitrogen atom, which is linked to only two other atoms, i. e., which has two σ bonds, contributes only one π electron to the general system, and there are two electrons in its nonbonding orbital. Examples of such atoms are the N3 atom of cytosine (Fig. 3.2) and the N7 atom of guanine and adenine.

In the same way, the oxygen atom, which forms two σ bonds, contributes two p electrons to the system and has two electrons in one nonbonding orbital; the carbonyl oxygen atom, forming one σ bond, contributes only one electron to the general system and has two electrons in each of two nonbonding orbitals. The carbon atom participating in conjugation contributes one electron to the common π electron system. Naturally the properties of the compounds are determined by the distribution of the π, n, and σ electrons. In many cases, however, notably in processes taking place without rupture of σ bonds, the more labile π electrons play the principal role. It must be remembered that they interact in a specific manner with σ electrons, although as a first approximation this interaction can be disregarded and the π and σ electron systems can be considered separately.

Even the purely qualitative examination of the role of individual atoms in conjugation leads to some important conclusions regarding the chemical properties of the bases (Fig. 3.3). In particular, atoms which contribute two electrons to the common π electron system (designated types a and c in Fig. 3.1) must have a certain deficiency of electrons or, in other words, a partial positive charge $\delta+$ must be concentrated on them. Atoms contributing one p electron to the system and having two electrons in the nonbonding orbital (designated types b and d in Fig. 3.1) must possess a partial negative charge $\delta-$ (Fig. 3.3). Accordingly, the first class of atoms must interact far less readily with electrophilic agents than the second.

Distribution of π electron density. A qualitative picture of the distribution of electron density in the molecule can be obtained by making mesomeric structures [4]. This shows, in particular, that electron density in pyrimidine bases is higher at C5 than at C6, and that the C8 atom of purine bases must possess some deficiency of electrons. Although even qualitative conclusions are often extremely useful for the interpretation of the chemical behaviour of molecules, preference naturally lies with a quantitative, or even semiquantitative analysis, which can be undertaken by calculation. For

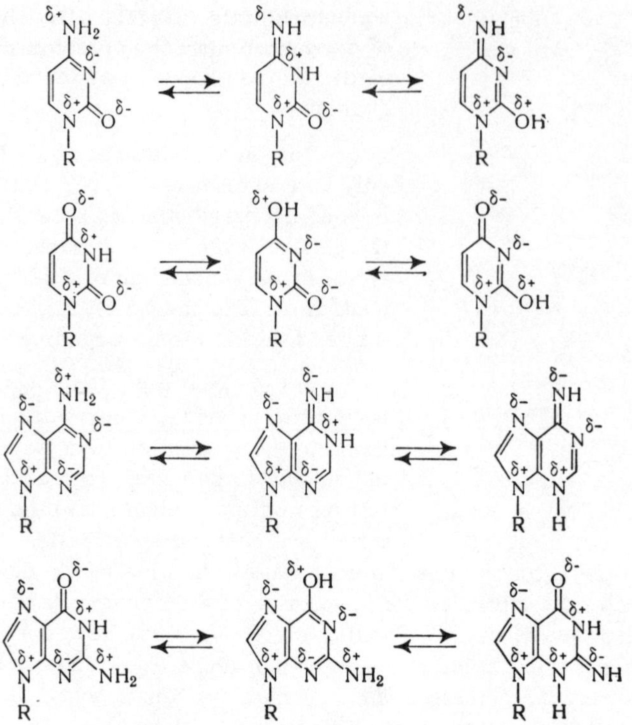

Fig. 3.3. Conjugation of heteroatoms in tautomeric forms of derivatives of heterocyclic bases of nucleic acids.

application to the bases of nucleic acids, usually various forms of the molecular orbital method are used for the calculations, in the approximation of the linear combination of atomic orbitals. Even quite recently the only available data were the results obtained by the use of the simplest method, that of Hückel [2, 5, 6]. This is a semi-empirical method, because parameters chosen beforehand in accordance with the experiment or on the basis of general considerations are used in the calculations, and the results of the calculations are highly dependent on their magnitude which, in turn, is frequently chosen purely arbitrarily. Although, as a rule, the most fundamental conclusions reached by different workers on the basis of their calculations are in agreement, not only quantitative but also qualitative disagreements are frequently found in the results concerning more special problems. These calculations yield information concerning the density of π electrons on individual atoms participating in conjugation, the orders of the bonds between two linked atoms (these orders characterize the density of the electron cloud between the atoms), and the distribution of the energy levels of the molecule. The results of the calculation are presented as a molecular diagram. Along the bonds on the skeleton of the molecule are indicated their bond orders,

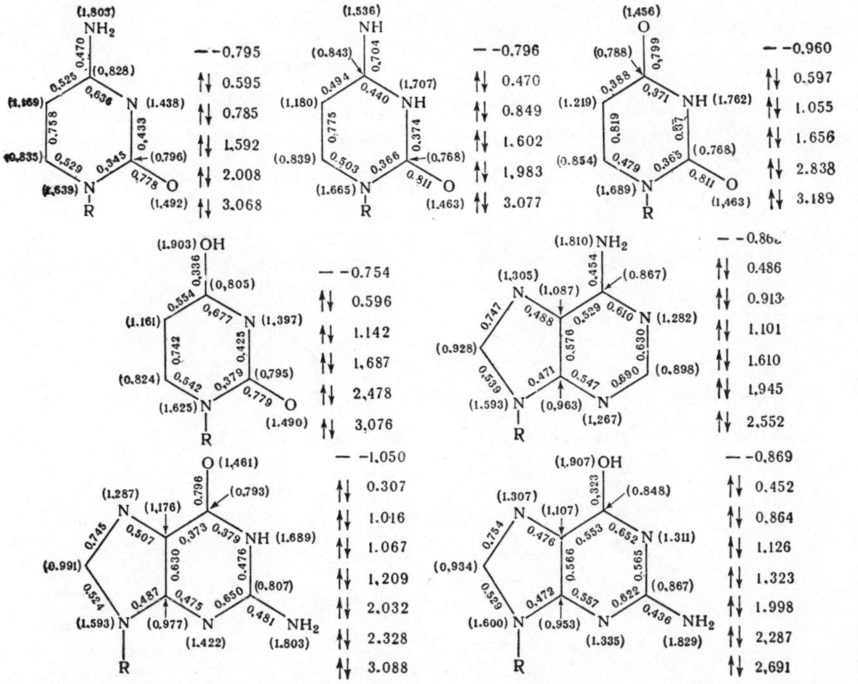

Fig. 3.4. Molecular diagrams of some tautomeric forms of bases of nucleic acids (free bases and bases incorporated in various derivatives). Schemes showing the distribution of π electrons by molecular energetic levels in conventional units of energy β (the difference in energies between two resolved energetic levels in ethylene is 2β) are given alongside the diagrams. Empty electron orbitals are indicated by "-" and electron orbitals containing two electrons with opposite spins are denoted by "⥮".

while by the side of the skeletal and exo-atoms the π electron density around them is shown in parentheses.

Molecular diagrams of some of the most important tautomeric forms of heterocyclic bases of nucleic acids (free bases and bases incorporated in various derivatives), obtained by Pullman [2] using Hückel's method*, are shown in Fig. 3.4. By using the electron densities on the atoms given on the molecular diagrams, it is easy to calculate the partial charges ($\delta+$ or $\delta-$) on the corresponding atoms by subtracting from the number of electrons supplied by a particular atom to the general π system the magnitude of the π electron density at that atom. In this way a clearer picture is obtained of the role of each atom as an electropositive or electronegative centre in the molecule. Corresponding diagrams obtained by calculation by Huckel's method are given in Fig. 3.5; diagrams obtained by the self-consistent field method are shown in Fig. 3.6. It can easily be seen that the qualitative predictions

---

* Hückel's method does not take into account interaction between electrons in the molecule. This drawback is partly removed in improved methods, such as the self-consistent field method [7, 8].

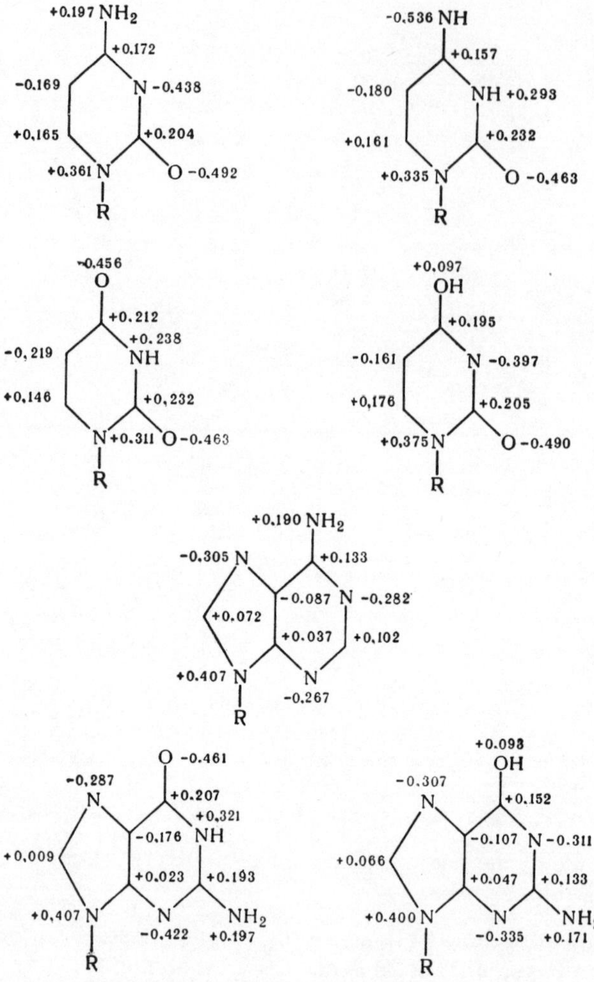

Fig. 3.5. Distribution of partial ("net") charges on skeletal and exo-atoms of some tautomeric forms of nucleic acid bases (free bases and their derivatives), calculated by the molecular orbital method in Hückel's approximation [2].

made on the basis of mesomeric structure building are concerned in theoretical calculations by Hückel's approximation of the molecular orbital method.

Among the carbon atoms of the central rings of the pyrimidine bases, C2, C4, and C6 possess a partial positive charge, whereas a considerable partial negative charge is concentrated on C5. Partial positive charges of considerable magnitude are located on C2 and C6, and a substantially smaller charge on C8 of purine bases. Groups contributing two π electrons to the system (for example, NH_2, OH, etc.) carry partial positive charges, while

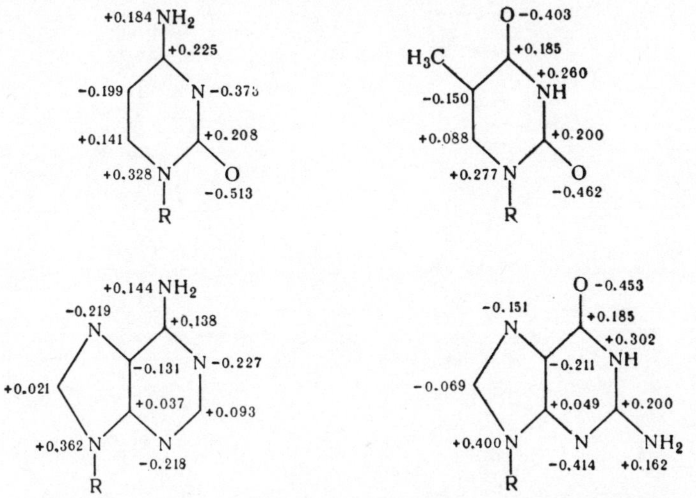

+0.184 NH$_2$
+0.225
-0.199
N -0.375
+0.141
+0.208
+0.328 N
O
-0.513
R

O -0.403
H$_3$C
+0.185
+0.260
-0.150
NH
+0.088
+0.200
+0.277 N
O
-0.462
R

+0.144 NH$_2$
-0.219 N
+0.138
-0.131
N -0.227
+0.021
+0.037
+0.093
+0.362 N
N
-0.218
R

O -0.453
-0.151 N
+0.185
+0.302
-0.211 NH
-0.069
+0.049
+0.200
+0.400 N
N
NH$_2$
-0.414 +0.162
R

Fig. 3.6. Distribution of partial ("net") charges of π electrons in skeletal and exo-atoms of ketoamino forms of nucleic acid bases (free bases and their derivatives), calculated by the self-consistent field method [7].

heteroatoms supplying one π electron carry considerable partial negative charges. These conclusions regarding the distribution of electron density are confirmed by calculations made by other workers using the same method [5, 6] and also more refined methods [7, 8]*. The conclusions drawn from theoretical calculations are of great value for predicting the reactivity of molecules of bases.

Nevertheless, considerable discrepancies still occur in the calculations made by different workers. This is true both of the simplified molecular orbital method and also of more sophisticated methods. Although in the latter case more satisfactory energetic characteristics of the molecule are evidently obtained, the electron densities in calculations employed by different workers remain different in both absolute numerical and relative values. As an example, the order of increase in electron density on nitrogen atoms of the pyridine type in the amino form of adenine obtained by calculation can be compared:

Simplified molecular orbital method [13]...........	N7>	N1>	N3>
Self-consistent field method, after Kuprievich [10].....	N7>	N1>	N3>
The same, after Pullman [11]	N7>	N3>	N1
The same, after Nagata et al. [12]...............	N1>	N3>	N7
The same, after Giessner-Prettre [7]	N1>	N7>	N3>

* Mention must be made of some investigations (for example [9]) in which, because the wrong empirical parameters were chosen for calculations by methods still in the developmental stage, the results agree neither with those calculated by other methods nor with the experimental data.

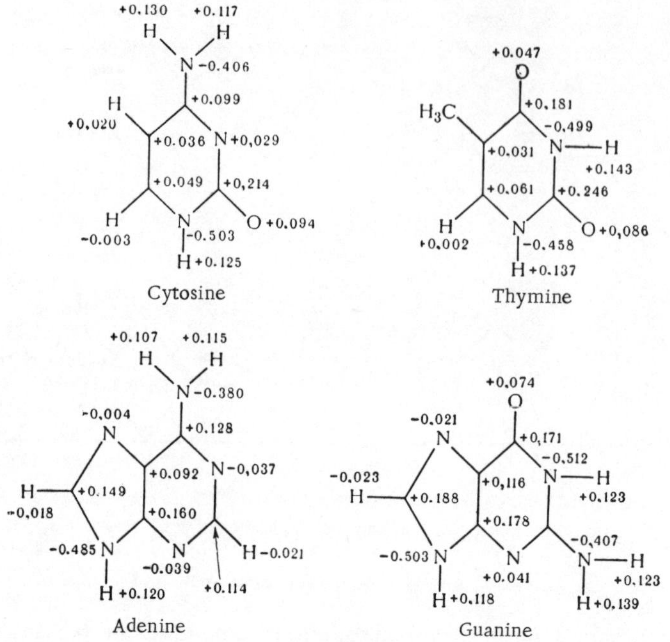

Fig. 3.7. Distribution of σ electron charges for ketoamino forms of nucleic acid bases [7].

These and other observed discrepancies are evidently due to the imperfect character of the method of calculation and to the arbitrary method used in the selection of the original parameters. It is to be hoped that the development of theoretical concepts and computing techniques will ultimately result in more reliable predictions of the distribution of electron density, but for the time being we must evidently be content with the information available through calculations made by different methods. This principle will be followed during the subsequent account of the chemical properties of the bases.

Allowance for σ electrons and net electron density distribution. Many of the physical properties of bases, such as dipole moments, depend on both the π electron and σ electron characteristics of the molecule. This evidently applies also to the chemical properties of the bases associated with rupture of σ bonds. Knowledge of the distribution of σ electrons in the molecule is therefore extremely important. Electron density distribution in nucleic acid bases for σ electrons is shown in Fig. 3.7 and the net charge distribution is shown in 3.8 [7]. The qualitative charge distribution remains the same after allowance is made for the σ electrons in the calculations as if only π electrons are considered (Fig. 3.6). For example, the charges on the C5 and C6 atoms of pyrimidine bases calculated in terms of π electrons only and in terms of π and σ electrons can be compared. In both cases the electron density is higher at the C5 atom. The two calculations further predict that net positive charges are concentrated at C6 of adenine and cytosine.   It may be that calculations allowing for σ electrons are more suitable for the prediction of the chemical properties of the bases, but at the present time they are not readily available.

2. Experimental data and their comparison with calculated data

In order to choose the best method of calculation, a wide range of experimental data must be available so that the results can be verified.   The electron density distribution evidently determines such characteristics of molecules as bond lengths, dipole moments, chemical shifts (during NMR measurement), and the structure of EPR signals and nuclear quadrupole resonance signals.

X-Ray structural analysis.   Bond lengths in the molecules of chemical compounds can be determined from the results of x-ray structural analysis of their crystals.   Bond lengths determined in this manner can be compared with bond lengths calculated theoretically on the basis of bond orders [8].

As an example, the bond lengths in the molecule of unprotonated cytosine obtained experimentally [14] and by calculation [15] are compared in Table 3.1.

The qualitative agreement between the two is good.   This suggests that calculations provide a pattern of electron density distribution which reflects,

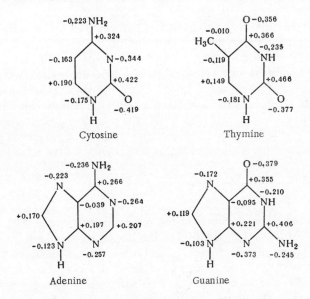

Fig. 3.8. Net π and σ charges for ketoamino forms of nucleic acid bases [7].

TABLE 3.1. Comparison of Theoretical [15] and Experimental [14] Bond Lengths in Molecule of Unprotonated Cytosine

Bond	Bond order	Theoretical bond length, Å	Bond length found, Å
N1 – C2	0.38	1.38	1.376
N1 – C6	0.53	1.36	1.361
C2 – 2-exo-O	0.78	1.24	1.260
C2 – N3	0.48	1.38	1.354
C4 – N3	0.64	1.33	1.351
C4 – C5	0.53	1.41	1.432
C4 – 4-exo-N	0.47	1.36	1.332
C5 – C6	0.76	1.88	1.348

at least qualitatively, the true distribution*. It is also clear from Table 3.1 that with an increase in bond order, bond length decreases.

NMR spectra. Chemical shifts of protons in the NMR spectra of different compounds depend on the electron density at the atoms with which they are associated. Fairly close correlation between π electron density calculated for a particular atom and the chemical shift of the proton associated with that atom is observed for benzoid systems [17-20]. Attempts to make a similar comparison between experimentally determined values of chemical shifts of the protons at positions C2, C6, and C8 of purine with the electron densities at these carbon atoms have proved much less successful [21-25]. This result is understandable, however, if it is remembered that in this particular case the magnitude of the chemical shift must be powerfully influenced by field effects of neighbouring heteroatoms possessing a high net negative charge, and also the fact that the influence of magnetic anisotropy of the ring is different for protons occupying different positions in the ring. If values of the chemical shifts are calculated from the calculated values of charges on the atoms, allowing for all these effects, the order of increase in magnitude of the chemical shift in the purine molecule can be correctly predicted: $\delta(C8) < \delta(C2) > \delta(C6)$ [25]. Nevertheless, NMR spectroscopy is evidently of little value as a method of investigation of electron density distribution in molecules of such complex compounds as the nucleic acid bases.

Investigation of dipole moments. Whereas conclusions regarding the state of individual atoms in the molecule can be drawn from the results of x-ray structural analysis and NMR spectroscopy, the dipole moment is determined by the electronic structure of the molecule as a whole [26]. Unfortunately, there is very little information in the literature on dipole moments of the heterocyclic bases of nucleic acids, and broad comparisons cannot therefore be made. However, dipole moments of several bases, calculated with allowance for the effect of σ electrons, agree sufficiently closely with the experimentally determined values (Table 3.2) [27].

---

* For x-ray structural data on other bases, see Chapter 2; for calculated values, see [2, 5-12, 16].

TABLE 3.2. Dipole Moments of Some Heterocyclic Compounds [27, 29, 30–32, 37]

| Compound | Dipole moments, D | | | |
| | calculated | | | found |
	μ_π	μ_σ	μ_{net}	
Pyrimidine.	0.87	1.50	2.37	2.42
9-Methylpurine.	3.53	0.73	4.26	4.3
9-Methyladenine.	2.70	0.37	3.06	3.0
1,3-Dimethyluracil	3.28	0.86	3.94	3.9
5-Bromouracil	3.20	1.45	3.98	4.5 ± 0.3

Note: Allowance for effect of σ bonds made as in [28].

The validity of electron density calculations for bases which are components of nucleic acids can thus be verified at present only indirectly. Tests show that the quantitative results do not agree very closely either with each other (when obtained by different methods of calculation) or with the experimental findings. Nevertheless, certain fundamental qualitative conclusions which can be drawn from the theoretical data agree fully with the observed results (for example, comparison of bond orders and lengths, prediction of reactivity, magnitude of dipole moments, and so on).

It is thus reasonably probable that the qualitative picture of electron distribution obtained by calculation is on the whole not far from the truth.

III. Energetic characteristics of the bases of nucleic acids

Unlike the pattern of distribution of electron density, the energetic characteristics of molecules can be determined fairly accurately by quantum-chemical calculations. The energetic characteristics of compounds which are of greatest interest to chemists are the resonance energy and the energy of the highest filled and lowest available levels, the meaning and importance of which will be briefly examined below.

It was pointed out above that calculations of electronic properties of the molecule by the molecular orbital method can yield a series of permitted energy levels on which π electrons can be located. Naturally in the basic state of the molecule, the electrons occupy levels with the lowest possible energy, and two electrons with opposite spins will be found in each molecular orbital corresponding to a given energy level (Fig. 3.4). The total energy of the π electrons will accordingly be twice the sum of the energies of the filled levels.

1. Resonance energy [33–35]

To calculate the resonance energy each π electron is first regarded as delocalized, i.e., as belonging to the molecule as a whole, and the total energy of these delocalized electrons is calculated. If, on the other hand, the molecule is regarded as a combination of ordinary and localized double bonds, not interacting with each other, as it is conventionally represented by the classical description of its formula, and the total energy of the π electrons of its individual double bonds is calculated, this energy will be appreciably higher than in the corresponding molecule with delocalized electrons. The difference between these energies, calculated for a given molecule, is called the resonance energy (the energy of delocalization), and it characterizes the stability of the molecule as a whole. Since the resonance energy depends on the number of electrons participating in conjugation, a more suitable characteristic of the increase in stability due to conjugation is the "reduced" resonance energy: the resonance energy calculated per delocalized electron [34]. Other conditions being equal, a molecule with a higher reduced resonance energy is more stable than one with a lower reduced resonance energy. The resonance energies of several nucleic acid bases, calculated by the two methods, and expressed per single π electron are given below [11]:

	By the molecular orbital method, β units	By the self-consistent field method, eV
Uracil	0.19	0.260
Cytosine	0.23	0.373
Adenine	0.32	0.904
Guanine	0.27	0.446

i.e., the resonance energy values decrease in the order:

adenine > guanine > cytosine > uracil

It thus follows that in any type of conversion taking place under the influence of physical factors or chemical agents, the adenine nucleus must be most stable; this is in fact observed in practice.

Values of the resonance energy can be used to estimate the relative stability of different tautomeric forms (see page 135), and also to study equilibrium conversions. Unfortunately, however, the calculated values of the resonance energy of nucleic acid bases cannot yet be subjected to direct experimental verification [36, 37].

2. Energy of the highest filled and lowest empty orbitals [38, 39]

The energy of the highest filled orbital is directly related to the ease of removal of electrons from the molecule: the higher the energy, the easier it is to remove an electron with the formation of a cation. Conversely, the energy of the lowest empty orbital determines the ease of entry of an electron: the lower the corresponding energetic level, the easier it is for the molecule to accept an electron and become an anion. Calculations made by different methods give the following orders of values for the energy of the highest filled level:

after Pullman [39]:

> guanine > adenine > thymine > cytosine > uracil

after Fernandez-Alonso [6]:

> guanine > thymine > adenine > uracil > cytosine

after Nagata [12]:

> guanine > adenine > cytosine > uracil > thymine

after Kuprievich [10] and Giessner-Prettre [7]:

> guanine > adenine > cytosine > thymine > uracil

These calculated values for the energies of the highest filled level thus show that of all the bases guanine must be the best electron donor and uracil and thymine the worst; in general, the purines are better donors than the pyrimidines.

Values of the energy of the highest filled level can be compared with experimentally determined values of the ionization potential (Table 3.3).

Another way of experimentally verifying the calculated energies of the highest filled and lowest empty electron orbitals is by polarographic investigation of the bases. During polarography, electrons are transferred from the compound to the electrode (oxidation) or, conversely, from the electrode to the compound (reduction). The ease with which these processes take place can be compared directly with the height of the highest filled and lowest empty electronic levels respectively. Experience shows that polarographic oxidation also gives results which agree basically with those predicted on theoretical grounds. For instance, all purine bases except purine itself give an oxidative wave on a graphite electrode, and guanine is more readily oxidized than adenine [39]. The electron-donor properties of bases of the nucleic acids can thus evidently be predicted relatively well theoretically on the basis of modern concepts, especially if sophisticated methods of calculation are used.

Less concordant results are obtained by estimation of the energy of the lowest empty electron level of bases, which determines their electron-acceptor properties. Calculations by various methods give the following orders for the energy of the lowest empty level:

after Pullman [11, 39]:

> cytosine < adenine < thymine < uracil < guanine

after Pullman [11] (improved method):

> adenine < cytosine < uracil < thymine < guanine

TABLE 3.3. Energy of the Highest Filled Level and Ionization Potentials of Some Purine and Pyrimidine Bases [40]

Compound	Ionization potential, eV	Energy of highest filled level, eV	Compound	Ionization potential, eV	Energy of highest filled level, eV
6-Azauracil.	10.18	9.65	Xanthine.	9.30	8.82
Uracil	9.82	9.15	Hypoxanthine.	9.17	8.00
Purine	9.68	8.87	Adenine	8.91	7.92
Thymine.	9.43	8.80	Cytosine.	8.90	8.16

after Kuprievich [10]:

cytosine < uracil < thymine < guanine < adenine

after Nagata [12]:

thymine < uracil < cytosine < guanine < adenine.

These orders differ so much that it is impossible as yet to draw any conclusions or to make any predictions regarding the electron-acceptor properties of the bases.

The results of quantum-chemical calculations obtained by different methods thus agree in the most important conclusions, but in many small details they frequently yield conflicting results. The value of modern theoretical methods as a basis for prediction must not therefore be overestimated. At the same time, they can certainly serve as a reference point for experimental research and as a useful guide to the interpretation of results.

IV. Tautomerism of the bases of nucleic acids

One of the most important aspects of the chemistry of nuclear bases is their tautomerism. For instance, one of the most widely accepted theories of spontaneous mutation is based on the possible existence of bases in different tautomeric forms. It can be expected, for example, that cytosine in the amino form must, because of its electronic structure, form a complementary pair with guanine, whereas in its imino form it is complementary to adenine; thymine (uracil), in the diketo form, must pair with adenine, while in the tautomeric 4-hydroxy form it must pair with guanine. The same picture must also be observed for derivatives of the bases.

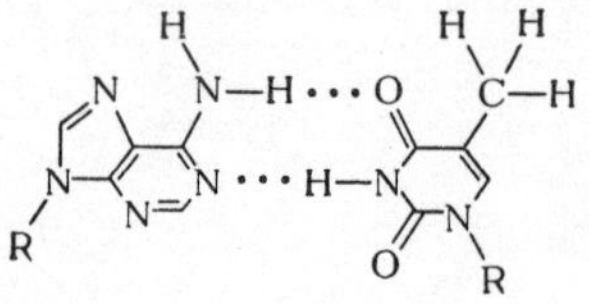

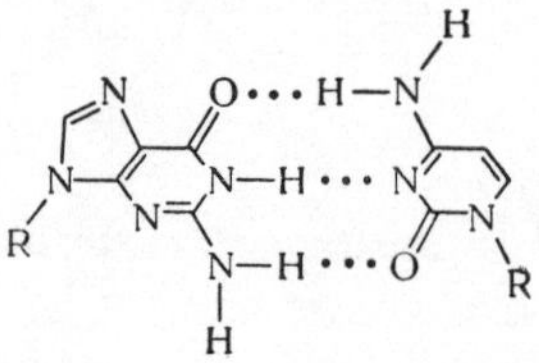

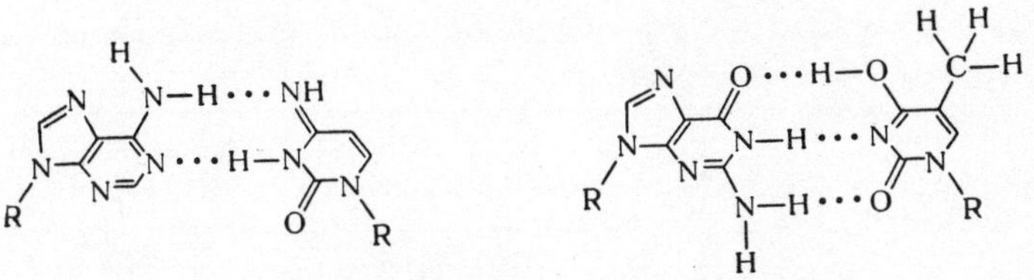

R denotes hydrogen atom or various radicals

This suggestion is still nothing more than an hypothesis, but at the same time it is very attractive and it indicates one way in which the problem of template synthesis can be studied: the search for correlations between the tautomeric form of a base and its role in biosynthesis.

The different tautomeric forms of the bases must of course differ in their reactivity. If the reactivity of true (pure) tautomeric forms of compounds can be judged to some degree from the reactivity of their models, the sharp difference in behaviour of the various tautomeric forms becomes obvious. As an example, the behaviour of uridine can be compared with that of 1-(β-D-ribosyl)-4-ethoxydihydropyrimidone-2 (a model of the rare tautomeric 4-hydroxy form of uridine) toward nucleophilic agents. Whereas uridine in a neutral medium is fairly inert, its analogue reacts readily with a wide variety of agents [41, 42] and, because of this, it is an extremely useful intermediate product in the synthesis of nucleoside analogues.

In this section we shall briefly examine the tautomerism of nucleic acid bases, by first considering the theoretical aspect of the problem and then examining the experimental evidence for the existence of tautomeric forms of the bases and their equilibrium. We shall take advantage of the fact that the general problems of tautomerism of heterocyclic compounds have been fully examined by Katritzky in his admirable survey [43], and we shall therefore concentrate on the more special problems concerned with tautomerism of the nucleic acid bases as components of nucleosides, reverting to fundamentals only when it is necessary to confirm the conclusions drawn for a nucleoside or when no data are available for nucleosides.

1. Theoretical considerations

Every base which is a component of a nucleic acid can, generally speaking, exist in several tautomeric forms, the number of which depends on the number of exocyclic groups in the purine or pyrimidine ring.

Heterocyclic hydroxy and amino compounds are known to exist (under ordinary conditions) chiefly in the keto and amino tautomeric forms [43, 44]. As will be shown below, the bases of nucleic acids are no exception to this rule. The a priori examination of the problem of which tautomeric form is more stable under particular conditions requires the consideration of several factors determining the free energy of transformation of the compound from one tautomeric form to the other and, consequently, the equilibrium constant of this transition. First, during a change in tautomeric form the system of

TABLE 3.4. Combined Energy of Isolated Bonds and Resonance Energy for Enol and Keto Forms of Bases

Base	Combined energy of isolated bonds [45], kcal/mole			Resonance energy*, kcal/mole		
	for enol form	for keto form	Δ	for enol form	for keto form	Δ
Hypoxanthine.................	1123	1140	17	75	68	7
Guanine	1257	1275	18	83	77	6
Cytosine..................	1035	1053	18	53	46	7
Uracil	998 (4-Hydroxy form)	1015.8	18	43 (4-Hydroxy form)	38	5

* Calculated from data in [34].

σ bonds of the molecule is reorganized. Second, the structure of the π electron system in conjugated molecules is modified and, consequently, so also is their resonance energy. Third, the solubilizing power of the molecule is altered, if the tautomeric equilibrium is examined in solution, the most interesting situation for the investigation of chemical and biochemical problems. Fourth and finally, during transformation of the compound from one tautomeric form to the other the system of intramolecular hydrogen bonds is modified. The combined action of these factors determines the relative stability of the different tautomeric forms under the particular conditions concerned. As a preliminary guide, the relative stability of the various tautomeric forms can be estimated theoretically, disregarding differences in their degree of solvation. For each tautomeric form the energy of the molecule as a whole can be calculated from experimental data for the energy of the isolated bonds represented in the canonical formula, and the resonance energy calculated by a quantum-chemical method can then be added to it, or a complete quantum-chemical calculation can be carried out, taking both π and σ bonds into account. The conclusion reached from estimation of the resonance energy values calculated for tautomeric enol and keto forms under standard conditions (25°C, 1 atm) is that the hydroxy forms are more stable (Table 3.4). On the other hand, however, calculation of the energy of the molecule contributed by the isolated bonds of the corresponding tautomeric forms shows that, under standard conditions, keto forms are much more stable than the enol forms and the difference between the combined energies of the isolated bonds for the keto and enol forms is greater than the difference between the corresponding values of the resonance energy. From the point of view of bond energy in tautomeric equilibrium of the nucleic acid bases under standard conditions, keto forms must therefore predominate over enol.

In the case of amine — imine equilibrium, the energy of the σ bonds in the two tautomeric forms can be taken as equal; in that case the stability of

TABLE 3.5. Calculated Energy of Transformation of Tautomeric Forms of Some Nucleic Acid Bases [46]

Base	Type of tautomeric equilibrium	ΔH, eV	Base	Type of tautomeric equilibrium	ΔH, eV
Adenine	Amine—imine	1.477	Thymine	Keto—enol	0.345
Cytosine	Amine—imine	0.550	Uracil	4-Keto—4-hydroxy	0.329
Guanine	Keto—enol	0.185			

these forms can be determined by the ratio between their resonance energies. Calculations made by Hückel's approximation of the molecular orbital method [34] show that the resonance energy of the amino form is higher than the resonance energy of the imino form, and the former is thus more stable*.

Calculations taking into account the influence of σ electrons [46] also enable the choice to be made in favour of keto and amino forms (under standard conditions).

Values of the energy of transition from the keto to the enol form and from the amino to the imino form for the bases of nucleic acids are given in Table 3.5. These results show that the tautomeric equilibrium is shifted most strongly toward the amino form in the case of adenine; the shift toward the more stable form for amino — imino equilibrium is more strongly expressed than the corresponding shift for keto — enol tautomerism. These conclusions are in agreement with the experimental data, which are presented below for several bases.

2. Experimental data

Uridine and Thymidine

Theoretically both these compounds can exist as free tautomeric forms:

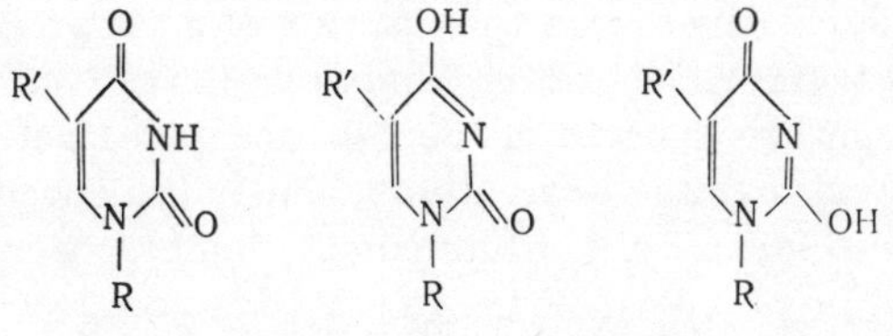

R denotes ribose or 2'-deoxyribose residue, $R' = H$ or CH_3

X-ray analysis has shown that uracil in the crystalline state exists in the diketo form. However, the results obtained for crystalline substances cannot be directly extrapolated to their state in solution; for this reason, the experimental data concerning solutions must now be examined.

---

* In some cases calculations by the simple molecular orbital method have given conflicting results [5]. The results of calculations by this method are considerably influenced by the choice of parameters. This disadvantage can largely be avoided by the use of more refined methods of calculation (see, for example, [10-12]).

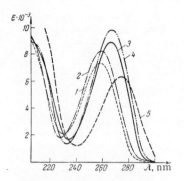

Fig. 3.9. UV-Spectra of uracil
derivatives (aqueous solutions,
pH 7.2) [48]: 1) 3-methyluracil;
2) uracil; 3) 1,3-dimethyluracil;
4) 1-methyluracil; 5) 1-methyl-
4-ethoxydihydropyrimidone-2.

UV-Spectra. Comparison of the UV-spectra of uracil, and 1-methyl-, 3-methyl-, and 1,3-dimethyluracils [48] in water at pH 7.2 (Fig. 3.9) shows that during the transition from uracil to 1-methyluracil and from 3-methyluracil to 1,3-dimethyluracil, an absolutely identical bathochromic shift and identical characteristic changes in the molar extinction coefficient are observed. The two methyl groups in 1,3-dimethyluracil have an additive action on the character of the spectrum. The differences between the absorption curves of these four compounds can be reduced simply to additive effects of the methyl groups. It thus follows that all four compounds exist in the same tautomeric form, i.e., the diketo form. This conclusion is confirmed by the sharp difference between the spectrum of these compounds and the spectrum of 1-methyl-4-ethoxydihydropyrimidone-2. In a similar way the UV-spectra of uridine and 3-methyluridine (in aqueous solutions) are extremely similar (λ_{max} 262 nm), whereas the spectrum of 4-ethoxy-1-glucopyranosyldihydropyrimidone-2 differs strongly from them (λ_{max} 273 nm) [49]. It thus follows from the character of their UV spectra that both uracil and uridine exist in the diketo form in aqueous solutions.

IR-Spectra. More accurate information still can be obtained by comparing the IR-spectra of the compound under investigation and of known model tautomeric forms. Miles [49, 50] has compared the IR-spectra of uridine, 3-methyl uridine, 1-(β-D-glucopyranosyl)-uracil, 1-(β-D-glucopyranosyl)-4-ethoxydihydropyrimidone-2, uridine-5'-phosphate, and the analogous thymine derivatives in heavy water (Table 3.6). He found that the spectra of uracil and thymine derivatives with substitution of a sugar residue in position 1 and the spectra of the corresponding 3-methyl derivatives are very similar, but differ sharply from the spectra of the 4-hydroxy derivatives. This means that all such uracil and thymine derivatives, under the chosen conditions in D_2O exist in the diketo form. A similar conclusion has been reached in later investigations [51-52] (see also [52] for crystalline compounds). Angell [51] states that the IR-spectra of nucleosides in the solid state and in aqueous solutions are virtually identical.

NMR Spectra. Investigation of NMR spectra in nonaqueous solvents can be used to determine the localization of mobile protons. In the NMR spectra of uridine and thymidine, obtained in dimethyl sulphoxide, a peak corresponding in area to one proton, and which is the resonance signal of the N−H proton, is observed in the region of chemical shifts of about 11 ppm. This means that under these conditions thymidine and uridine exist in the diketo form [53, 61].

TABLE 3.6. IR-Spectra of Uracil and Thymine Derivatives in D_2O [49, 50]

Compound	Characteristics of band, μ^*
Uridine.	5.92 (s); 6.05 (vs); 6.19 (m); 6.84 (s); 7.18 (m); 7.74 (m)
3-Methyluridine	5.90 (s); 6.04 (vs); 6.20 (s); 6.80 (s); 7.05 (w); 7.25 (w); 7.73 (w)
1-(β-D-Glucopyranosyl)-uracil	5.91 (vs); 6.04 (vs); 6.17 (m); 6.85 (s); 7.20 (m); 7.55 (w); 7.71 (m)
1-(β-D-Glucopyranosyl)-4-ethoxy-dihydropyrimidone-2	6.05 (vs); 6.13 (s); 6.31 (w); 6.49 (s); 6.74 (s); 6.85 (s); 7.21 (m); 7.27 (m); 7.61 (vs); 7.81 (w)
Uridine-5'-phosphate.	5.96 (s); 6.0 (vs); 6.21 (m); 6.85 (s); 7.07 (vw); 7.20 (m); 7.73 (m)
Thymidine	5.97 (s, sh); 6.03 (vs); 6.15 (s); 6.78 (m); 7.33 (m); 7.70 (m)
1-(β-D-Glucopyranosyl)-3-methyl-uracil	5.90 (s); 6.05 (vs); 6.19 (s)
Deoxyuridine.	5.91 (s, sh); 6.01 (vs)
3-Methylthymidine.	5.93 (s); 6.00 (s); 6.18 (vs)
1-(β-D-Glucopyranosyl)-thymine. . .	5.90 (s); 6.00 (vs); 6.10 (vs)
1-(β-D-Glucopyranosyl)-3-methyl-thymine	5.93 (s); 6.00 (s); 6.16 (vs)
1-(β-D-Glucopyranosyl)-4-ethoxy-5-methyldihydropyrimidone-2	— 6.00 (vs); 6.15 (s)

* Besides the wavelength of the band the characteristics of its intensity are given in the table in parentheses: vs) very strong band; s) strong; m) moderately strong; w) weak; sh) band appears as shoulder on another band; vw) very weak; s, sh) strong band appearing as a shoulder on another band.

The experimental data as a whole thus indicate that the diketo pyrimidine structure is predominant in uridine derivatives both in the crystalline state and in aqueous and nonaqueous solutions. However, there are no grounds for asserting that this tautomeric form is the only possible form for these compounds. The quantity of another tautomeric form (or forms) may be so small that physical methods, in their present state, are unable to detect it. Meanwhile, knowledge of the constants of tautomeric equilibrium between the possible tautomeric forms is required for the solution of many problems. Since the physical methods listed above do not provide information about these constants, the only possible method for their estimation is to compare the ionization constants of known models of tautomeric forms [43]. This approach can be used in cases when the tautomeric forms to be studied give a cation or anion of identical structure. Equilibrium between the two tautomeric forms and the common ionized form can be described as follows:

$$\xrightarrow{k_{XH}} HXH^+ \xrightarrow{k_{HX}}$$

$$H^+ + XH \xrightleftharpoons{K_\tau} HX + H^+ \qquad K_{XH} = \frac{[H^+][XH]}{[HXH^+]} \qquad K_{HX} = \frac{[H^+][HX]}{[HXH^+]} \qquad (1)$$

Here XH and HX represent two tautomeric forms differing in the position of the proton; K_T is the constant of tautomeric equilibrium. In this case the experimentally observed ionization constant is given by:

$$K_{obs} = \frac{[H^+]\,([XH] + [HX])}{[HXH^+]} \qquad (2)$$

Comparison of the expressions for the three ionization constants gives:

$$K_{obs} = K_{XH} + K_{HX}$$

It is evident from the definition $K_T = [XH]/[HX]$ that $K_T = K_{XH}/K_{HX}$. Consequently, if the ionization constant of the compound and the ionization constant of even one of the tautomeric forms are known, the tautomeric equilibrium constant can be determined. The ionization constant of one of the forms is usually determined with the aid of a model in which a certain tautomeric form is fixed by the introduction of a substituent at the corresponding atom. This can be done, for example, by introducing a methyl group (assuming that the methyl group does not modify the ionization constant very greatly by comparison with the true tautomeric form). In the case of uracil derivatives, however, it is more difficult to use this approach, because in an alkaline medium the O–alkyl derivatives of uridine used as models of its tautomeric forms do not dissociate, and its dissociation constant in the acid region is very low. The ionization constants of model methylated compounds were determined by Katritzky [54]. Values of pK_a' for a number of model pyrimidines, derivatives of uracil, measured in concentrated aqueous solutions of sulphuric acid are given below [54]:

	pK_a'
Uracil	−3.38
1-Methyluracil	−3.40
5-Bromouracil	−7.25
1-Methyl-5-bromouracil	−6.60
1,3-Dimethyluracil	−3.25
1,3-Dimethyl-5-bromouracil	−6.44
4-Ethoxydihydropyrimidone-2	+1.00
1-Methyl-4-methoxydihydropyrimidone-2	+0.65
1-Methyl-4-methoxy-5-bromodihydropyrimidone-2	−3.32

The value of the tautomeric equilibrium constant calculated from these data is $10^{4.0}$, which is in favour of the diketo form*. It is interesting to note

---

* This result can be interpreted only as approximate [54], since pK was determined in a medium of very concentrated sulphuric acid, and with the aid of Hammett's acidity function H_0 [55], which is determined by the equation $H_0 = \log (B/BH^+) - pK_a$. However, the experimentally determined gradient of the curve of $\log (B/BH^+)$ as a function of H_0 differed considerably from unity. Determination of the constant of tautomeric equilibrium by another method gives the value $10^{3.3}$ for K_T.

TABLE 3.7. UV-Spectra of Neutral Forms of Cytosine Derivatives [57, 58]

Compound	Formula	Medium	λ_{max}, nm	$\log \varepsilon_{max}$
1-Methylcytosine........	I (R = Me, R' = H)	pH 7.0	274	3.92
			230 (shoulder)	3.89
		Ethanol, 10^{-2} M NaOH	275	3.94
1,3-Dimethylcytosine.....	II (R = R" = Me, R' = H)	pH 11.5	273	3.94
			225	4.0
4-Imino-2-methoxy-1-methyl-1,4-dihydropyrimidine.	III (R = R" = Me, R' = H)	Ethanol, 10^{-2} M NaOH	275 (shoulder)	3.78
			246	4.16
1-Methyl-4-exo-N,N-dimethylaminodihydropyrimidone-2	I (R = R' = Me)	pH 7.0	282	4.08
			219 (shoulder)	4.04

that the introduction of an electronegative substituent such as bromine in the 5 position of the pyrimidine ring shifts the equilibrium appreciably toward the enol form. In this case, K_T is $10^{3.3}$. Despite their relative inaccuracy, these results show that under the conditions of investigation, equilibrium is shifted very strongly in favour of the diketo structure, and they make clear the impossibility (at present) of detection of minor tautomeric forms experimentally.

Cytidine

Cytidine can evidently exist as three tautomeric forms I–III (R = R' = R" = H):

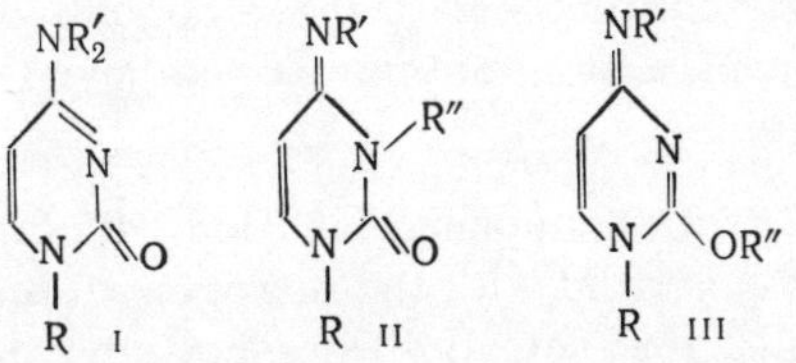

The results of x-ray structural analysis indicate that cytidine in the crystalline state exists in the amino form [56]. Investigation of the structure of cytidine in aqueous and nonaqueous solvents has been attempted by various methods.

UV-spectra. It is concluded from the comparison of the UV-spectra of neutral forms of 1-methylcytosine and 4-imino-2-methoxy-1-methyl-1,4-dihydropyrimidine, the model of tautomeric form III, that it is not the predominant form in tautomeric equilibrium (Table 3.7). However, it is impossible on the basis of the UV-spectra to decide between forms I and II, because the spectra of 1-methylcytosine, 1-methyl-4-exo-N,N-dimethylaminodihydropyrimidone-2 (the model compound of form I) and of 1,3-dimethylcytosine (the model compound of form II) are very similar.

TABLE 3.8. IR-Spectra of Glycoside Derivatives of Cytosine in D_2O [50]

Compound	Characteristic bands, μ		
Cytidine	6.08	6.21	—
Deoxycytidine	6.05	6.15	—
1-(β-D-Glucopyranosyl)-4-exo-N,N-di-methylaminodihydropyrimidone-2....	6.09	6.15	6.50

IR-Spectra. Solutions of cytidine, deoxycytidine, and 1-(β-D-gluco-pyranosyl)-4-exo-N,N-dimethylaminodihydropyrimidone-2 in D_2O have closely similar IR-spectra in the region of the carbonyl bands (Table 3.8) [50].

However, since the last compound has a known keto structure, cytidine must evidently also possess a keto structure. If the bands in the spectrum of cytidine and its derivatives are carefully related, bands associated with the amino group [51] and the carbonyl group can be distinguished, and cytidine and its analogues can be ascribed a keto – amino structure.

NMR Spectra. Cytidine keeps its keto – amino structure in nonaqueous solvents also, as is shown by the study of its NMR spectra. In dimethyl sulphoxide in the region $\delta \approx 7.5$ ppm, for instance, the characteristic peak of an aromatic amino group is observed, corresponding to two protons [53], whereas the peak of the N – H proton in the region 11 ppm is not observed. This is clearly seen if the NMR spectrum of cytidine is compared with the spectra of thymidine or uridine (see above). It thus follows that in dimethyl sulphoxide cytidine has an amino structure. The spectra of the base in deoxycytidine and cytidine are absolutely analogous [59, 60], so that the tautomeric structure of these two compounds is the same *.

Final proof of the correctness of these conclusions is given by comparison of the NMR spectra of 1-methylcytosine and 1-methyl-4-exo-^{15}N-cytosine [62] (Fig. 3.10).

In the spectrum of the compound labelled with heavy nitrogen, the peak corresponding to protons bound with the nitrogen atom (or atoms) of the ring is split into two symmetrical peaks. This shows beyond doubt that both protons belong to one exocyclic amino group and not to two imino groups: one cyclic and one exocyclic. It is thus firmly established that these various cytosine derivatives are present (virtually entirely) in the keto – amino form in the crystalline state and in solution.

---

* As the result of one investigation [61], different tautomeric structures (amino and imino respectively) have been ascribed to cytidine and deoxycytidine. However, as more recent work has shown [59, 60], the differences in the NMR spectra on which this conclusion was based were due to the fact that cytidine in a neutral form was compared with protonated deoxycytidine.

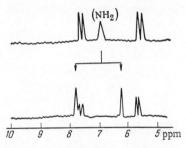

Fig. 3.10. NMR spectra of solutions*
of 1-methylcytosine (upper curve) and
1-methyl-4-exo-^{15}N-cytosine (lower
curve) in dimethyl sulphoxide at 23°C
[62].

Investigations have also shown that
the keto − amino structure is predominant
in cytidine analogues, notably 6-azacyti-
dine [63, 64] and 5-azacytidine [65].

Values of pK. The extent to which
the tautomeric keto − amino form of
cytidine derivatives in aqueous solutions
predominates can be estimated from data
for the basicity of models of the tautomeric forms and, in particular, from
the known [66] dissociation constants of
1,4-exo-N,N-trimethylcytosine (pK$'_a$
4.20) and 1,3-dimethylcytosine (pK$'_a$
4.29). Since the UV-spectra of the
cationic forms of these compounds indicate that their structure is identical,
their tautomeric equilibrium constant could be determined (see page 140); its
value is approximately 10^5. Tautomeric equilibrium constants calculated from
different models can be compared. Values of pK$'_a$ for methylated derivatives
of cytidine are compared in Table 3.9 with those for the corresponding 1-
methylcytosine derivatives. The tautomeric equilibrium constants of amino
and imino forms of cytosine can be determined from these data by comparing
the results for 1-methylcytosine and 1,3-dimethylcytosine (K$_T$ $10^{4.9}$) or for
1,3-dimethylcytosine and 1,4-exo-N,N-trimethylcytosine (K$_T$ 10^5), and the
tautomeric equilibrium constant of cytidine can be determined by comparing
the results for cytidine and 3-methylcytidine (K$_T$ $10^{4.6}$) and for 3-methyl
cytidine and 4-exo-N,N-dimethylcytidine (K$_T$ $10^{5.1}$). All the values of the
equilibrium constants obtained in this way are very similar and they indicate
a strong displacement of equilibrium toward the amino form.

Comparison of the tautomeric equilibrium constants of cytidine and
uridine shows that in the case of uridine the content of the minor form is
about 5-10 times higher than in the case of cytidine (in aqueous solutions
under ordinary conditions). This is in agreement with the results of quantum-
chemical calculations, predicting an easier tautomeric conversion in the case
of uridine (see page 137) than with cytidine.

The content of tautomeric form II in a solution of cytidine is thus very
low. However, the evidence so far adduced in favour of predominance of the
keto − amino form does not rule out the possibility of appreciable amounts of
the tautomeric form III, and there are in fact good reasons why this form
cannot be disregarded [58, 69, 70]. In particular, there is the identical
photochemical behaviour of 1-methyl-2-methoxy-4-iminodihydropyrimidine

---

* Here and subsequently, unless the solvent used is specifically mentioned, the compounds are in
aqueous solution.

TABLE 3.9. Values of pK_a' for Various Methyl Derivatives of Cytosine and Cytidine

Compound	pK_a'	Literature cited
1-Methylcytosine.	4.55	67
1,3-Dimethylcytosine.	9.4	68
	9.29	66
1,4-exo-N,N-Trimethylcytosine.	4.20	66
Cytidine	4.1	67
3-Methylcytidine.	8.7	68
4-exo-N,N-Dimethylcytidine.	3.62	67

(the model of form III) and of 1-methylcytosine in ethanol during UV-irradiation: both compounds undergo photochemical conversion, whereas 1,3-dimethylcytosine remains unchanged under these conditions. The most effective wavelength at which this photochemical conversion takes place is near to the wavelength of the absorption maximum in the spectrum of 1-methyl-2-methoxy-4-iminodihydropyrimidine (~240 nm) [58]. However, further evidence is necessary before a final conclusion can be drawn regarding the contribution of tautomeric form III.

Some very interesting work has been done on the analogues of natural nucleotides, because they can help to reveal factors influencing tautomeric equilibrium. The study of tautomeric equilibrium in the case of 4-exo-N-hydroxy- and 4-exo-N-aminocytosines must now be discussed [71].

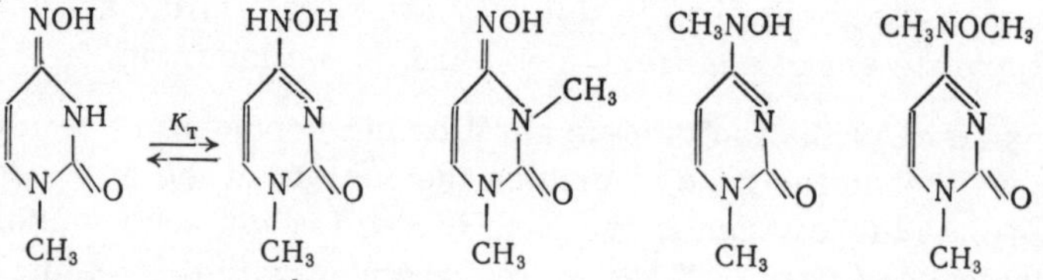

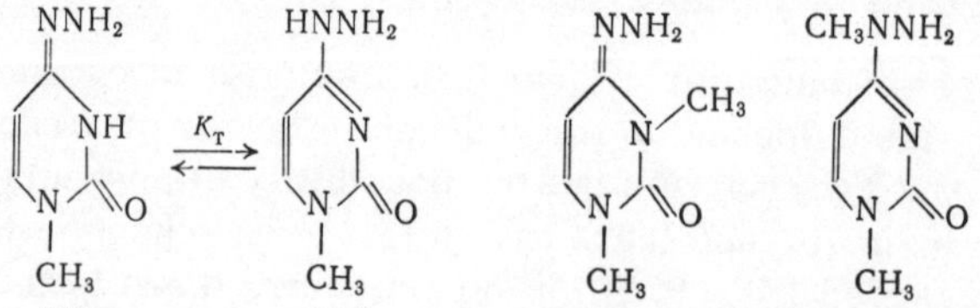

Compounds IV-VII and VIII-X give cations of identical structure, and the tautomeric equilibrium constants for these compounds can be estimated from the ionization constants for the corresponding tautomeric forms. Calculations have shown that K_T between IVa and IVb is about 10, and for the conversion between VIIIa and VIIIb the value of K_T is 1/30. The impression is gained that with an increase in electronegativity of the substituent at the nitrogen atom (H- < NH_2- < OH-) equilibrium is shifted toward the imino form. Modification of the pyrimidine ring by introduction of electronegative substituents in the 5 position of uridine (see page 141) or into the exocyclic amino group of cytidine thus appreciably shifts the tautomeric equilibrium toward minor forms.

Effect of saturation of the C5 – C6 bond in the molecule of pyrimidine bases on tautomeric equilibrium. An important role in the chemistry and biology of nucleic acids is played by saturation of the C5–C6 bond in the molecule of the pyrimidine bases, which takes place, for example, during irradiation of nucleic acids with UV-light (see Chapter 12), a procedure frequently used in functional and, in particular, in genetic investigations. Comparison of the characteristics of the bonds [50] in the IR-spectra of 5,6-dihydrouridine and 5,6-dihydrothymidine with the IR-spectrum of 3-methyl-5,6-dihydrouridine in D_2O gives the following picture:

$$\lambda, \mu$$

5,6-Dihydrouridine.	5.85;	5.98
5,6-Dihydrothymidine.	5.88;	6.00
3-Methyl-5,6-dihydrouridine	5.85;	6.02

It is evident from these figures that in this case also the diketo structure remains predominant. This conclusion is confirmed by x-ray structural investigations of dihydrothymine [72].

The study of tautomeric equilibrium of dihydrocytosine is a more difficult problem because this compound is unstable. Nevertheless, it is concluded from the results of NMR and UV-spectroscopy of compounds XI-XII [73] that tautomeric equilibrium for these compounds in aqueous solutions is shifted toward the amino form. The value of K_T between XIa and XIb can be estimated by comparing pK_a for compounds XI and XII; it is ~25. Comparison of this value with the corresponding value of K_T for cytidine shows that in the first case there is a strong shift toward the imino form. This conclusion is also in agreement with the results of quantum-chemical calculations [74] of the resonance energies of tautomeric forms of dihydrocytosine.

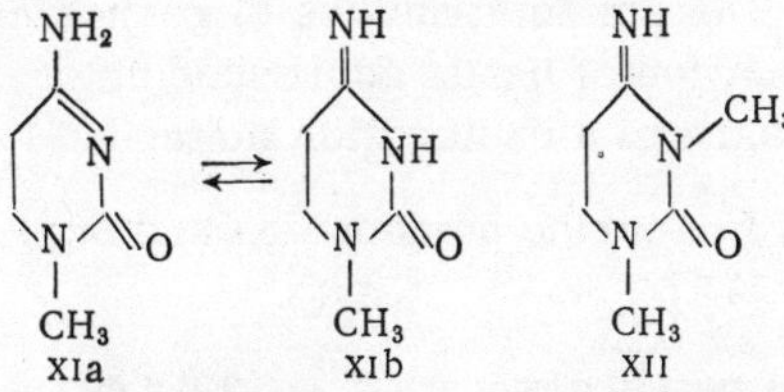

$$pK_a' 6.62 \qquad\qquad pK_a' 8.05$$

It can thus be concluded from such evidence as is at present available that the changes in the chemical structure of the natural bases shift the

TABLE 3.10. IR-Spectra of Guanosine and Inosine Derivatives in D_2O [76, 78]

Compound	Characteristic bands, μ		
Guanosine.....................	6.00	6.34	6.37
1,9-Dimethylguanine	5.98	6.28	6.46
9-(β-D-Ribofuranosyl)-2-amino-6-methoxy-purine......................	6.18	6.27	6.54
Inosine	5.97	–	–
1-Benzylinosine..................	5.97	–	–
9-(β-D-Ribofuranosyl)-6-methoxypurine ...	6.21	6.28	–

tautomeric equilibrium appreciably (under ordinary conditions) toward the minor forms. It is difficult at present to decide whether this is due to the biased character of the results so far obtained or whether it is a general rule. If it is a general rule, it provides weighty evidence in support of the hypothesis of the important role of tautomerism in the biogenesis of the nucleic acids. It may be that in the course of evolution bases which were too easily convertible from one tautomeric form into another were discarded because they led to an excessive degree of variation. From this point of view the investigation of tautomeric conversions is of particular interest.

Guanosine and Inosine

The use of UV-spectra for the study of tautomeric equilibrium of the purine bases of nucleic acids gives only minimal information because the spectra of the stabilized tautomeric forms differ only very slightly from each other [75].

IR-Spectra. Comparison of the spectra of guanosine, 1,9-dimethylguanine, and 9-(β-D-ribofuranosyl)-2-amino-6-methoxypurine in D_2O shows a close resemblance in the case of the first two compounds and an appreciable difference for the third compound, in whose spectrum the carbonyl band at $6-6.25\ \mu$ has disappeared (Table 3.10)*. A similar picture can be observed in the case of inosine: the presence of a band at $5.97\ \mu$ in the spectrum of inosine and its 1-benzyl derivative, and disappearance of this band with the conversion to the enol model indicates predominance of the keto structure in the case of ionosine. The amino structure of guanosine has been established on the basis of identification of bands corresponding to the NH_2 group in the crystalline state by means of a deuteration method [51].

Results obtained for purine nucleosides and nucleotides agree with those obtained for the bases [75].

---

* The band at $6.2\ \mu$ in the spectrum of guanosine and at $5.9\ \mu$ in that of inosine has been identified as carbonyl on the basis of the IR-spectra of 6-^{18}O-nucleosides [77].

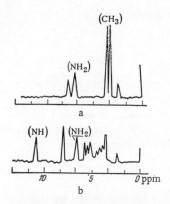

Fig. 3.11. NMR Spectra of 1,9-dimethylguanine (a) and guanosine (b) in deuterodimethyl sulphoxide.

NMR Spectra. Comparison of the NMR spectra of guanosine and 1,9-dimethylguanine (Fig. 3.11) in dimethyl sulphoxide leads to the conclusion that the single-proton signal in the region of chemical shifts of 11-12 ppm in the guanosine spectrum corresponds to the proton bound with the N1 nitrogen atom. The singlet double-proton signal in the region ~7 ppm corresponds to easily exchanged protons [53, 61]. The unsplit character of this signal means that the two protons are bound to the same atom, which can only be the exocyclic nitrogen atom. All this is evidence in support of a keto – amino structure for guanosine in dimethyl sulphoxide [76].

Unfortunately no data are available for the tautomeric equilibrium constants of guanosine, but on the basis of quantum-chemical calculations (page 137) it can be assumed that equilibrium in this case must be shifted more strongly toward the enolic form than in the case of uridine.

Adenosine

Adenosine can exist in two tautomeric forms: amino or imino. The similarity between the IR-spectra of 9-(β-D-ribofuranosyl)-6-exo-N,N-dimethylaminopurine and adenosine-5'-phosphate (the monosodium salt) [50] in D_2O suggests that in aqueous solution adenosine exists as the amino compound. In the crystalline state, adenosine preserves its amino structure, as is shown by x-ray structural data and the IR-spectra of the crystalline compound [51]. The NMR spectrum of adenosine and of deoxyadenosine in dimethyl sulphoxide contains a characteristic double-proton signal of an amino group in the region 7.5 ppm, but no signals are observed from other protons capable of exchange with the solvent [53, 61]. The means that adenosine exists in the amino form also in dimethyl sulphoxide. If it is assumed that the amino and imino forms of adenosine give the same cation on protonation, the tautomeric equilibrium constant in water can be calculated from the values of pK_a' for 6-exo-N,N-dimethylaminopurine riboside (3.62 with an ionic strength $\mu = 0.5$) and for 1-methyladenosine (8.25 with $\mu = 0.5$). This calculation gives a value of $K_T \approx 4 \cdot 10^4$.

The theoretical arguments leading to the conclusion that the bases of the nucleic acids have a keto – amino structure are thus fully confirmed experimentally. In all cases investigated, equilibrium is strongly shifted toward the predominant tautomeric form, so that minor forms of the compounds cannot be detected by optical methods.

V. Ionization constants of the bases of nucleic acids

1. General considerations

Even establishing that a particular tautomeric form is predominant is not sufficient to explain all the chemical properties of the bases. As a rule all reactions of bases take place in solution, and protonated (cationic) or deprotonated (anionic) forms of the bases take part in many of them. In an acid medium the bases may be protonated by the equilibrium conversion:

$$BH^+ \rightleftarrows B + H^+ \tag{3}$$

In that case the "basic" ionization constant will be described by the expression:

$$K_{1a}^0 = \frac{(B)(H^+)}{(BH^+)} \tag{4}$$

where (B), (H$^+$), and (BH$^+$) denote activities of the corresponding components.

In an alkaline medium the bases lose a proton in accordance with the equilibrium:

$$BH \xrightleftharpoons{K_a} B^- + H^+ \tag{5}$$

The "acid" dissociation constant is described in the form:

$$K_{2a}^0 = \frac{(B^-)(H^+)}{(BH)} \tag{6}$$

Ionization constants determined from activity are called true or thermodynamic constants, and they characterize the acid – base properties of bases in infinitely dilute solutions. In practice, the apparent ionization constants are determined: these are expressed through the concentration of the compound studied and they depend on the presence of other compounds in the solution. In particular, with the potentiometric and spectrophotometric methods of determination of ionization constants which are most frequently used in nucleic acid chemistry, the empirical values are determined by the following equations:

$$K_{1a}' = \frac{(H^+)[B]}{[BH^+]} = \frac{(B)(H^+)}{(BH^+)} \cdot \frac{f_{BH^+}}{f_B} \tag{7}$$

$$K_{2a}' = \frac{(H^+)[B^-]}{[BH]} = \frac{(B^-)(H^+)}{(BH)} \cdot \frac{f_{BH}}{f_{B^-}} \tag{8}$$

The values shown in square brackets in these equations correspond to concentrations of the components in equilibrium, while those in parentheses denote their activities; the f values are coefficients of activity.

To determine the true equilibrium constants, extrapolation must be carried out to infinite dilution. This problem is examined in rather more detail on page 159). The equilibrium constants are generally expressed in the form:

$$pK_a = -\log K_a$$

Some of the bases of nucleic acids possess strong basic properties and are protonated in a weakly acid medium, but are deprotonated only in a strongly alkaline medium; other bases, on the contrary, are weak acids and, while they form anions in a weakly basic medium, they are protonated only in a strongly acid medium. The first group includes cytosine and adenine; the values of pK_a for these compounds are associated with Equations (3) and (4). The second group includes thymine and uracil, and their usual pK_a values are associated with Equations (5) and (6). Hypoxanthine, xanthine, and guanine occupy an intermediate position; they are protonated at comparatively high pH values for these compounds in accordance with Equation (3) and they are deprotonated in accordance with Equation (5) at fairly low pH values for these compounds in the alkaline region. Accordingly, their acid – base properties are described by two values of pK_a, one of which is associated with the first process, the other with the second.

2. Localization of the attachment and detachment of protons in nucleosides and nucleotides

It is natural to expect that a proton would be attached to those atoms in the molecule on which the highest net negative charge due to π electrons is concentrated, i.e., to the nitrogen atoms forming the two σ bonds. These atoms, in addition, possess a lone electron pair which, in turn, must be readily protonated. In the case of the exocyclic nitrogen atoms forming three σ bonds, in the first place, as is shown in Fig. 3,5, they carry a net positive charge. Second, attachment of a proton to these atoms would require the withdrawal from conjugation of the pair of electrons located in the p orbital (see page 122 and Fig. 3.1), i.e., would require additional expenditure of energy.

In accordance with these considerations, the possible places of protonation (shown inside the circles) in the predominant tautomeric forms of the bases are N3 in cytosine; N1, N3, and N7 in adenine; N3 and N7 in guanine and hypoxanthine; and N7 in xanthine. This applies also to derivatives of these bases.

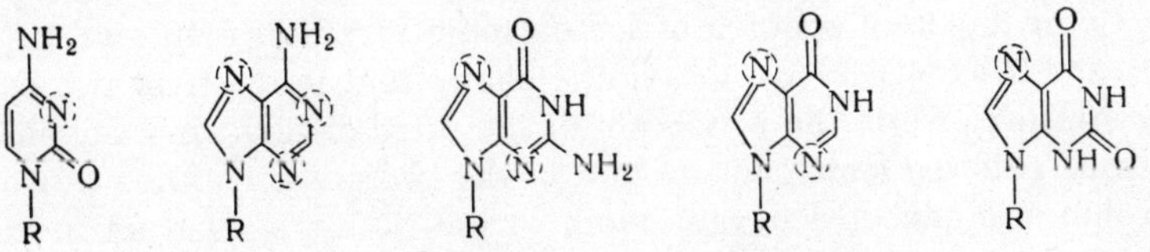

R denotes a hydrogen atom or various radicals

However, the choice of the predominant site (from the several possible sites) of protonation cannot be determined from the distribution of the π electron density only, because nonbonding lone pair electrons also participate in the process. Accordingly Pullman and Nakayama developed a method of calculation which takes into account the lone pair electrons [80]. This calculation correctly predicts that attachment of the proton must take place predominantly in the N1 position of adenine and the N7 of guanine and hypoxanthine.

This approach does not allow for the influence of the medium in which protonation takes place on ionization of the base, nor for the effect of many other factors associated with solvation and intramolecular interactions.

The results of x-ray structural analysis confirm the correctness of these conclusions regarding the site of proton attachment to the base moieties of crystalline protonated bases, nucleosides, and nucleotides.

They also show that in the adeninium cation as such [82], and also in this cation in adenosine-3'-phosphate [83] or adenosine-5'-phosphate [84], the proton is attached to the N1 nitrogen atom. In the case of guanine it has been concluded that the proton is attached to N7 [85], and finally, on the basis of results obtained for cytidine-3'-phosphate [86, 87] and 5-carboxymethylcytosine [88], the proton in the cytosinium cation is located at N3.

The correctness of the conclusions drawn from theoretical considerations and from the results of x-ray structural analysis is confirmed by spectrophotometric data for solutions of the compounds. The most convincing evidence is derived, as a rule, from synthesis of models and comparison of their spectral characteristics with those for the protonated molecule. On the basis of a study of the UV-spectra of the bases and their derivatives, Dekker (in his survey [89]), puts forward a number of arguments which enable the localization of protons in the ionized compounds to be established more precisely. The points which he makes will be briefly summarized below:

1. The UV-spectra of the cationic forms of 7-methylguanine and 7-methyldeoxyguanosine-5'-phosphate are similar to the spectra of protonated guanine and deoxyguanosine-5'-phosphate. It is therefore probable that protonation is located at N7 of guanine.

2. The spectra of 3-methyl- and 7-methyladenosines in the cationic form differ from the spectrum of protonated adenosine and, consequently, in the case of this last compound, attachment of the proton at N1 is most likely.

3. Comparison of spectra of the cationic forms of cytidine, 1,3-dimethylcytosine, and 2-exo-O-2'-cyclocytidine shows that in the first two cases they are very similar, while the spectrum of the third differs very considerably from them. This evidently means that of the two possible sites of attachment of the proton (the carbonyl oxygen atom or the N3 nitrogen atom) in the case of cytidine attachment is principally at N3.

Even more convincing than the results described by Dekker [89] are those obtained by comparing the IR-spectra of the cationic forms of compounds with a stabilized tautomeric structure with the spectra of the corresponding protonated compounds. In the monoprotonated form, guanosine in D_2O exhibits an IR-spectrum analogous to the IR-spectrum of 1,7,9-trimethylguanine, indicating a keto structure for the protonated form of guanosine. The spectra of 7,9-di-(β-hydroxyethyl)-guanine and of 1,7,9-trimethylguaninium iodide in the cationic form are similar to the spectrum of protonated guanosine, evidence of the protonation of guanosine in the N7 position [76].

The characteristic band at 1215 cm^{-1}, ascribed to $C-NH_2$ vibrations, is observed in both the neutral and the protonated forms of crystalline adenosine [90]. This is also true of the two bands at 3100 and 3300 cm^{-1}, associated with vibrations of the amino group [51]. This evidently means that the proton is attached in the N1, N3, or N7 positions, but not to the nitrogen of the exocyclic amino group.

Protonation does not affect the band at 1285 cm^{-1} observed in the spectrum of neutral cytidine. This band is ascribed to $C-NH_2$ vibrations, and the absence of a shift on protonation means that the NH_2 group is unaffected. Consequently, the most probable position of attachment of the proton [90] is N3. This conclusion is also confirmed by the character of the changes in frequencies corresponding to $C=O$ vibrations, on protonation of cytidine in D_2O and in the crystalline state [90].

Finally, arguments in favour of protonation of adenosine at N1 and of guanosine at N7, as well as evidence of the protonation of cytidine at N3, at least in nonaqueous solvents, are given by the results of NMR spectroscopy.

The magnitude of the chemical shift of the proton at C8 of the protonated form of adenosine in D_2O compared with the nonprotonated form is 0.2-0.3 ppm toward low field. At the same time, the chemical shift of the corresponding proton in protonated guanosine compared with the nonprotonated form of the nucleoside is very large, namely 0.95 ppm toward low fields [91-93]. This may be an argument in support of the protonation of adenosine in the pyrimidine ring and of guanosine in the imidazole ring.

The structure of the NMR spectra of the cations cytosinium and 1-methylcytosinium in dimethyl sulphoxide [94] and in trifluoroacetic acid [95] unquestionably proves that the proton in the protonated molecule is attached to N3.

The magnitude of the chemical shift of the proton in the C8 position of the inosine cation compared with the neutral form of this nucleoside [93] may indicate that in this particular case protonation takes place just as in guanosine at the N7 position of the imidazole ring.

It is thus now firmly established that on protonation the above-mentioned forms of adenine, guanine, and cytosine (or of their nucleosides and nucleotides) are the main products. However, the possibility is not complete-ly ruled out that, besides these forms, cations with a proton attached at other positions of the molecule which cannot be found because of the insufficient sensitivity of methods now used may be present although, admittedly, in much smaller amounts.

Arguments in support of the existence of other protonated forms of adenine, as a base, which differ from the basic form mentioned above (pro-tonation at N1) were recently put forward [96] on the basis of a study of the fluorescence spectra of methyl derivatives of adenine at pH 2. It was con-cluded that in this case, although protonation does take place at N1, this does not explain the phenomenon of fluorescence, since 1-methyladenine, 3-methyladenine, and 3,7-dimethyladenine do not exhibit fluorescence. It was postulated that in this case responsibility for the fluorescence rests with a twice protonated tautomer, protonated at the exocyclic amino group and also at N7. Nevertheless, in the case of adenosine, fluorescence can be explained by protonation at N1, because no evidence or arguments in support of the existence of other protonated tautomeric forms have been produced.

3. Localization of charges in the ions of bases

It can thus be accepted that, of the possible protonated forms, those whose existence has been predicted theoretically play the most important role. However, the site of attachment of the proton does not necessarily correspond to the location of the positive charge in the molecule. Since the bases of nucleic acids are conjugated molecules, the positive charge in them is largely delocalized, and a greater proportion of the charge may be concentrated on some other atom than that to which the proton is attached. Hence, despite the fact that in the protonated form of adenine the proton is attached at N1, the positive charge is localized to a much greater degree on the exocyclic nitrogen atom of the amino group and on the N9 atom [82].

These remarks also apply to the localization of the negative charge formed by detachment of a proton from nucleosides of the uridine or guanosine type. On the basis of the close analogy between the IR-spectra of guanosine and its O-methyl derivative in an alkaline medium [76] and disappearance of the characteristic band waves due to C = O group at high pH values [77], it is concluded that the negative charge is localized predominantly on the oxygen atom. The same conclusion can be drawn from the analogy between the UV-spectra of the anion of 9-ethylguanine and 2,6-diaminopurine [97]. These conclusions are also valid, evidently, for inosine. The analogy between the UV-spectra of the anion of uridine and of its 4-exo-O-alkyl derivative in-dicates that the negative charge is localized predominantly on the C4 oxygen atom [48, 98]. This conclusion is supported by the close similarity between the IR-spectra of uridine in an alkaline medium and of dihydropyrimidone-2 [99]. However, different structures of the anion have been suggested for thymidine: with the negative charge localized at O2 [99] and O4 [48, 98].

Some investigators (see [90], for example) have accepted structures for anions with charges on the nitrogen atom from which the proton has migrated, but it would seem that the notion that the greater proportion of the negative charge is localized on oxygen atoms rests on a firmer experimental basis.

To sum up, whereas the question of predominance of one or other tautomeric form and of localization of the proton in protonated forms can be regarded as solved, the problem of the distribution of electron density in the resulting cations and anions is still only partly clear. Its final solution will be of great importance to our understanding of many of the properties of these bases, including the influence of substituents on their acid—base properties.

4. Effect of various factors on the acid—base balance of the bases

The acid – base properties of the bases, like their tautomeric equilibria, are influenced above all by the structural features of the base molecules themselves (the presence of substituents, the various intramolecular interactions: hydrogen bonding, formation of zwitterions, and so on). In addition, external factors dependent on the experimental conditions (temperature, ionic strength, etc.), also play an important role.

Structural Factors

Role of the carbohydrate residue. The best known example of the influence of a substituent on acid – base properties is the difference between the ionization constants of the bases and of their ribo and deoxyribo derivatives. Values of the corresponding ionization constants are given in Table 3.11. The decrease in pK_a observed during the transition from the base to its deoxyribo derivative, and thereafter to the ribo derivative can be explained by the inductive effect $(-I)$ of the sugar residue, which is evidently higher in the case of ribose than in the case of deoxyribose. This is confirmed by the marked decrease in the value of pK'_a in the case of 2'-deoxy-2'-fluorocytidine compared with cytidine, since fluorine possesses much greater electronegativity than the hydroxyl group. However, this still does not explain the change in pK'_a in a series of stereoisomeric pentofuranosyl-thymines (Table 3.12), the carbohydrate residues of which contain equal numbers of hydroxyl groups and, consequently, must also possess closely similar inductive effects. It may be that in this case hydrogen bonding between the hydrogen of the 2'-hydroxyl group of the pentose and the oxygen atom of the carbonyl group at C2 of the pyrimidine ring plays a role of some importance. Arguments for and against the existence of such a hydrogen bond in nucleosides and nucleotides are examined in Chapter 2. It is impossible to distinguish between the inductive effect and the effect of hydrogen bonding on the basis of the available data, and the solution to this problem must await a systematic study of the influence of substituents in the carbohydrate residue on the value of pK_a.

F*

TABLE 3.11. pK$_a'$ Values for Bases, Deoxyribonucleosides, and Ribonucleosides

Derivatives	pK$_a'$ of base	Literature cited	pK$_a'$ of deoxyribonucleoside	Literature cited	pK$_a'$ of ribonucleoside	Literature cited
Uracil..................	9.5	[100]	9.3	100	9.25	[100]
Cytosine	4.46	[100]	4.3	100	4.1	[100]
Thymine................	9.9	[48]	9.8	98	9.68	[101]
Guanine	—	—	9.33	102	9.22	[102]
Hypoxanthine...........	8.94	[103]	—	—	8.75	[103]
Xanthine...............	7.53	[104]	—	—	5.50	[104]
Adenine	4.1	[102]	3.8	102	3.6	[102]

Note: Determinations were made spectrophotometrically. For derivatives of guanine and adenine in solution of ionic strength $\mu = 1$; in other cases in 0.1 M glycine or 0.025 M acetate buffers.

TABLE 3.12. Effect of Structure of Carbohydrate Residue on Value of pK$_a'$ of Derivatives of Nucleic Acid Bases

Compound	pK$_a'$	Literature cited
1-(β-D-Ribofuranosyl)-thymine..................	9.68	[101]
1-(β-D-Deoxyribofuranosyl)-thymine..............	9.8	[101]
1-(β-D-Xylofuranosyl)-thymine	9.75	[101]
1-(β-D-Arabinofuranosyl)-thymine	9.8	[101]
1-(β-D-Lyxofuranosyl)-thymine	9.92	[101]
1-(β-D-Ribofuranosyl)-5-fluorouracil..............	7.57	[100]
1-(β-D-Arabinofuranosyl)-5-fluorouracil.	7.63	[105]
1-(β-D-Ribofuranosyl)-5-fluorocytosine	2.26	[100]
1-(β-D-Arabinofuranosyl)-5-fluorocytosine	2.33	[106]
1-(β-D-2'-Deoxy-2'-fluororibofuranosyl)-cytosine	3.9	[107]

Role of substituents in the heterocyclic ring. The effect of substituents in different positions of the heterocyclic bases has received less study; the available data relate principally to pyrimidine derivatives. The effect of substituents in position 5 of the pyrimidine ring has been investigated most fully. Values of pK$_a'$ for some 5-substituted derivatives of pyrimidine nucleosides are given in Table 3.13.

It will be obvious that electron-donating substituents increase the value of pK$_a$ to some degree, whereas electron-acceptors reduce the basicity of the compounds. This type of effect was to be expected, having regard to the possibility of delocalization of the charge in the ionized bases.

By contrast, if electron-donating alkyl substituents are introduced into exocyclic amino groups of cytidine they lower the value of pK$_a$, and dialkyl

TABLE 3.13. Effect of Different Substituents in Position 5 of the Pyrimidine Ring on pK'$_a$ of Nucleosides

Nucleoside	pK'$_a$	Literature cited	Nucleoside	pK'$_a$	Literature cited
Cytidine..........	4.1	[100]	5-Fluoro-2'-deoxycyti-		
5-Methylcytidine.....	4.28	[108]	dine............	2.39	[100]
5-Fluorocytidine	2.26	[100]	Uridine	9.5	[100]
2'-Deoxycytidine.....	4.3	[100]	Ribothymidine	9.68	[101]
5-Methyl-2'-deoxycyti-			5-Fluorouridine	7.57	[100]
dine	4.40	[100]			

TABLE 3.14. Effect of Substituents in Exocyclic Amino Group on pK'$_a$ Values of Cytosine Derivatives [100]

Compound	pK'$_a$
Cytosine................................	4.61
4-exo-N-Methylcytosine	4.55
5-Fluorocytosine	2.90
5-Fluoro-4-exo-N-methylcytosine.................	2.66
Cytidine.............................	4.1
4-exo-N-Methylcytidine	3.92
4-exo-N,N-Dimethylcytidine	3.62
5-Fluorocytidine	2.26
5-Fluoro-4-exo-N-methylcytidine................	2.05
2'-Deoxycytidine.........................	4.3
4-exo-N-Methyldeoxycytidine	4.01
4-exo-N,N-Dimethyl-2'-deoxycytidine	3.79
5-Fluoro-2'-deoxycytidine....................	2.39
5-Fluoro-4-exo-N-methyl-2'-deoxycytidine	2.14
5-Fluoro-4-exo-N-dimethyl-2'-deoxycytidine	1.89
1-Methylcytosine.........................	4.55
1,4-exo-N-Dimethylcytosine	4.47
1,4-exo-N,N-Trimethylcytosine	4.20
5-Methyl-2'-deoxycytidine	4.40
4-exo-N,5-Dimethyl-2'-deoxycytidine............	4.04
4-exo-N-(β-Phenylethyl)-5-methyl-2'-deoxycytidine ...	3.83

Note: Measurements made in 0.025 M acetate buffer.

derivatives have appreciably lower pK$_a$ values than monoalkyl derivatives, as will be clear from the results shown in Table 3.14.

There are two possible explanations of this effect. First, the alkyl substituent may hinder the formation of a solvation sheath around the centres of localization of the positive charge in the protonated molecule, and second, it may reduce the ability of the protonated molecule to form hydrogen bonds with the solvent. For cytosine derivatives, for example:

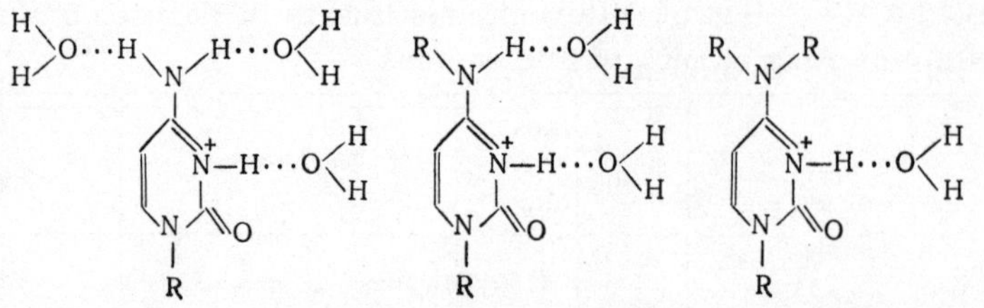

Both these effects evidently play an important role in the stabilization or destabilization of the cation.

Values of pK'_a for a number of cytosine derivatives with an alkyl substituent in the exocyclic amino group (in 0.025 M acetate buffer) are given below [100]:

	pK'_a
Cytosine.	4.61
4-exo-N-Methylcytosine	4.55
4-exo-N-Ethylcytosine	4.58
4-exo-N-(n-Butyl)-cytosine	4.69
5-Fluoro-2'-deoxycytidine.	2.39
5-Fluoro-4-exo-N-methyl-2'-deoxycytidine	2.14
5-Fluoro-4-exo-N-ethyl-2'-deoxycytidine	2.21
5-Fluoro-4-exo-N-(n-propyl)-2'-deoxycytidine	2.19
5-Fluoro-4-exo-N-(n-butyl)-2'-deoxycytidine	2.21

It is difficult to account for the slight increase in pK'_a with an increase in length of the alkyl substituent. Possibly the inductive effect of the alkyl radical (+I), which increases with an increase in length of the alkyl chain, may play a role in this case.

The great difference between the values of pK'_a as a result of the transition from 5-methyl-2'-deoxycytidine to 4-exo-N,5-dimethyl-2'-deoxycytidine, namely 0.36 pK unit, merits attention. Possibly in this case, besides steric hindrance of solvation, and the decrease in number of hydrogen bonds with the solvent, steric interaction may also take place between the 5-methyl and 4-exo-N-methyl groups. As a result of this interaction the amino group is displaced from the plane of the ring, and this in turn causes a disturbance of conjugation. Since the amino group is [109] a substituent of the (+M, −I)* type, the disturbance of conjugation must reduce the magnitude of the mesomeric (+M) effect, whereas the inductive effect (−I) remains unchanged. This increases the electron-acceptor properties of the amino group and reduces pK_a through relative destabilization of the protonated form. A similar

---

* Substituents of the (+M, −I) type are electronegative substituents such as nitrogen, oxygen, or halogens contributing two π electrons to the conjugated system (see page 122). These substituents attract electrons from the system of σ bonds to themselves (−I effect), but they are electron donors in the system of π bonds (+M effect).

effect is observed in the case of adenine derivatives also [110, 111]:

<table>
<tr><td></td><td>pK'_a</td></tr>
<tr><td>Adenine. . . . .</td><td>4.22</td></tr>
<tr><td>6-exo-N-Methyladenine . . . .</td><td>4.18</td></tr>
<tr><td>6-exo-N,N-Dimethyladenine . . . .</td><td>3.87</td></tr>
<tr><td>Adenosine . . . .</td><td>3.8</td></tr>
<tr><td>6-exo-N-(β-Hydroxyethyl)-adenosine. .</td><td>3.1</td></tr>
</table>

Clearly, therefore, introduction of a substituent at various positions of the nucleic acid bases appreciably modifies their acid – base properties. The systematic study of the effect of substituents on the value of pK_a could shed considerable light on the chemical properties and biological role of the heterocyclic bases of nucleic acids, including minor components. From this point of view it would be interesting to discover whether a correlation exists between the action of a given substituent on different bases, i.e., whether in this case the rule of the linear dependence of free energies [112-114] well-known in theoretical organic chemistry is also observed (for heterocyclic compounds, see the survey by Jaffe [115]). In fact, the relationship between pK'_a of uracil derivatives and pK'_a of cytosine derivatives is very close to linear (Fig. 3.12; Table 3.15). This means that in the case of bases of the nucleic acids, these correlations apply to the full, and it can be assumed that they will be useful for studying the mechanisms of chemical and, possibly, enzymic conversions of nucleic acids.

Values of pK'_a for compounds with a 5-methyl substituent are exceptions to this general linear rule. This effect can be assumed to be due to steric hindrances from solvation of the ions. In the case of cytosine derivatives, the stability of the cation must thereby be reduced and, consequently, in accordance with Equations (3) and (4), the value of pK'_a must be reduced. In the case of uracil the decrease in stability of the anion must lead to an increase in the value of pK'_a in accordance with equations (5) and (6), in full agreement with the picture observed (see Fig. 3.12).

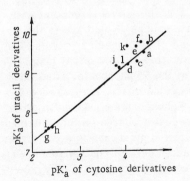

Fig. 3.12. Correlation between values of pK'_a for uracil derivatives and cytosine derivatives (for an explanation of the letters identifying the points, see column 1 in Table 3.15).

The role of phosphate groups in nucleotides. Phosphate groups in nucleotides exert an appreciable influence on the pK_a value of the base. Values of pK_a of ionization of several nucleosides and nucleotides are given in Table 3.16 [102].

The reason for the observed increase in pK_a in the series: nucleoside $\leq$ nucleoside-2',3'-cyclic phosphate < nucleoside-2'- and -3'-phosphates < nucleoside-5'-phosphate

TABLE 3.15. Correlation between pK'_a Values for Some Derivatives of Uracil and Cytosine

Identification letters in Fig. 3.12	Substituent in bases	pK'_a for uracil series	Literature cited	pK'_a for cytosine series	Literature cited
a	–	9.5	[100]	4.45	[100]
b	1-Methyl-.....................	9.75	[98]	4.55	[98]
c	1-(β-D-2'-Deoxyribofuranosyl)-......	9.3	[100]	4.3	[100]
d	1-(β-D-Ribofuranosyl)-............	9.25	[100]	4.1	[100]
e	1-(β-D-Ribofuranosyl)-5-methyl-.....	9.68	[116]	4.28	[108]
f	1-(β-D-2'-Deoxyribofuranosyl)-5-methyl-	9.8	[98]	4.40	[109]
g	1-(β-D-Ribofuranosyl)-5-fluoro-	7.57	[100]	2.26	[100]
h	1-(β-D-2'-Deoxyribofuranosyl)-5-fluoro-	7.66	[100]	2.39	[100]
i	1-(β-D-Arabinofuranosyl)-5-fluoro-	7.63	[105]	2.33	[106]
j	1-(Pyranosyl)-*.................	9.2	[98]	3.85	[98]
k	1-(Pyranosyl)-5-methyl-*..........	9.7	[98]	4.1	[98]
l	1-(β-D-2'-fluororibofuranosyl)-......	9.14	[117]	3.9	[107]

Note: Value determined spectrophotometrically in 0.025 M acetate buffer for derivatives of cytosine and in 0.1 M glycine buffer for derivatives of uracil.

* The following pyranosyl derivatives were used: 1-(β-D-glucopyranosyl)-, 1-(β-D-galactopyranosyl)-, 1-(β-D-arabinopyranosyl)-, and 1-(β-D-xylopyranosyl)-, for which the values of pK'_a are the same.

is evidently interaction of the ionized base with the phosphate group of the nucleotide.

It can be concluded from the conformations of ribose and deoxyribose (see Chapter 2) that the possibility of such interaction is particularly great in the case of nucleoside-5'-phosphates, the phosphate groups of which may be in spatial proximity to the ring of the base. Much less close contact with the base is expected for the 2'- and 3'-phosphate groups, which are in the trans-position relative to the base*. In the case of cytosine derivatives, for which zwitterions are formed as a result of ionization of the central ring, their stability must thus be higher in the case of the 5'-phosphates, and this must evidently lead to a greater increase in the value of pK'_a compared with the nucleoside than in the case of the 2'- and 3'-phosphates. On dissociation of uracil derivatives, interaction between the negative charge on the ionized base and the negative charges of the phosphate group must lead to a decrease in the stability of the ionized form of the base and a shift of equilibrium toward the neutral form, i.e., pK'_a must increase in this case also, in agreement with the experimental data (see Table 3.16).

---

* A greater degree of interaction between 5'-phosphate groups than of 2'- and 3'-phosphate groups is shown, in particular, in their influence on the chemical shift of protons in the NMR spectra of the corresponding nucleotides compared with the nucleosides [92, 118].

TABLE 3.16. Effect of Phosphate Group of Nucleotide on Ionization Constant of Base [102]

Derivatives	pK$_a'$					
	ribonucle-oside	2'-phos-phate	3'-phos-phate	2'(3')-phosphate	5'-phos-phate	2',3'-cyclic phos-phate
Cytosine	4.17	—	—	4.43	4.54	4.12
Adenine	3.52	3.81	3.70	—	3.88	—
Uracil	9.38	—	—	9.96	10.06	9.47
Thymine	9.93	—	—	—	10.47	—
Guanine	9.42*	9.87	9.84	—	10.00	—
Hypoxanthine	9.02	—	—	—	9.62	—

Note: Values determined spectrophotometrically at 20°C and at zero ionic strength.
* pK$_a$ for deoxyguanosine 9.52; pK$_a$ for deoxyguanosine-5'-phosphate 10.00.

TABLE 3.17. Values of pK$_a$ for Secondary Ionization of Nucleoside-5'-mono-, -di-, and -triphosphates [122]

Derivatives	pK$_a$			Derivatives	pK$_a$		
	5'-mono-phosphate	5'-diphos-phate	5'-tri-phosphate		5'-mono-phosphate	5'-diphos-phate	5'-tri-phosphate
Adenosine	6.57	7.20	7.68	Cytidine	6.62	7.18	7.65
Guanosine	6.66	7.19	7.65	Uridine	6.63	7.16	7.58
Inosine	6.66	7.18	7.68				

Note: Values determined at 25°C with inclusion of a correction for ionic strength, but with no allowance for statistical correction.

Ionization of phosphate groups in nucleotides. In mononucleotides (except cyclic phosphates) the phosphate residues possess the properties of dibasic acids and accordingly they have two ionization constants. The first ionization constant has the value pK$_a$ ~1, whereas the second lies within the range pK$_a$ 6-7. Cyclic phosphates are monobasic acids with pK$_a$ ~ 1. The nature of the base has only a slight effect on the value of the secondary ionization constant of the phosphate groups in nucleoside phosphates (Table 3.17).

Dihydropyrimidines. On saturation of the double bond in the pyrimidine ring, a sharp increase (Table 3.18) is observed in the corresponding values of pK$_a'$, and in the case of cytosine derivatives this may be associated with an increase in the contribution of other tautomeric forms to the tautomeric equilibrium, whereas in the case of uracil derivatives the conditions of delocalization of the negative charge in the anion, being more favourable in uracil than dihydrouracil, evidently play the principal role.

External Factors

Effect of ionic strength of the medium. If expressions for coefficients of activity determined in accordance with the Debye – Hückel law (see [123, 124]) are substituted in Equations (4) and (6) (see page 148), the relationship

TABLE 3.18. Values of pK'_a for Dihydropyrimidine Derivatives

Compound	pK'_a	Literature cited	Compound	pK'_a	Literature cited
1-Methyluracil.	9.75	[119]	Cytidine.	4.1	[100]
1-Methyl-5,6-dihydro-			5,6-Dihydrocytidine. . .	6.1	[120]
uracil	~12	[119]	1-Methyl-5,6-dihydro-		
Cytosine.	4.45	[100]	cytosine.	6.62	[73]
5,6-Dihydrocytosine . .	6.3	[119]	6-Hydroxy-5,6-dihydro-		
			cytidine.	5.56	[121]

between the apparent and thermodynamic equilibrium constants will be determined by the following equation:

$$pK'_a = pK^0_a + \frac{0.5 \sqrt{\mu} \, (2m - 1)}{1 + b \sqrt{\mu}} \tag{9}$$

where μ is the ionic strength of the solution; m the charge of the conjugate acid; and b a constant depending, in particular, on the effective radius of the ion in solution.

It must be expected on the basis of Equation (9) that with an increase in the ionic strength of the medium, compounds dissociating in accordance with Equation (5) (page 148), and not possessing charges in the undissociated form, in the alkaline region (uridine, inosine, guanosine, but not their nucleotides), will reduce the observed value of pK_a. Meanwhile, during dissociation in the acid region, cytidine, guanosine, inosine, and adenosine (but not their nucleotides), the conjugate acids of which possess a charge m equal to +1 [Equation (3) on page 148], must increase the values of pK'_a with an increase in ionic strength. As a first approximation, Equation (9) can be used to estimate the effect of the ionic strength of the medium on nucleotides by determining the total charge of the molecule of the conjugate acid [102]. From considerations such as these a table can be compiled for the qualitative prediction of changes in the values of pK'_a with an increase in ionic strength of the medium (Table 3.19) [102].

These conclusions are fully confirmed experimentally (Table 3.20).

With an increase in the ionic strength of the medium the ionization constants of nucleotides approach those of nucleosides because of screening of the phosphate group. At sufficiently low values of ionic strength, when in Equation (9) the expression $b\sqrt{\mu} \ll 1$, we obtain:

$$pK'_a = pK^0_a + 0.5 \sqrt{\mu} \, (2m - 1) \tag{10}$$

Equation (10) shows that at low values of ionic strength linear relationships can be obtained between pK'_a and $\sqrt{\mu}$ (see [125]).

In investigations in which the ionic strength of the medium is modified, allowance must also be made for the effect of structure and charge of the cation and anion. The following principles can be noted here [102]:

TABLE 3.19. Changes in pK_a' with an Increase in Ionic Strength Predicted by Equation (9)

Compound	pH Region in which pK_a is determined	Total charge m of conjugate acid	Changes in pK_a' with increase in ionic strength of medium, $\Delta pK_a'$
Adenine, adenosine.	2.0–7.0	+1	>0
Cytosine, cytidine	2.0–7.0	+1	>0
Cytidine and guanosine phosphates	2.0–7.0	0	<0
Uracil, uridine	7.0–12.0	0	<0
Thymine, thymidine.	7.0–12.0	0	<0
Guanosine.	7.0–12.0	0	<0
Uridine-2',3'cyclic phosphate	7.0–12.0	−1	<0
Uridine, guanosine, and inosine phosphates	7.0–12.0	−2	<0

1. Bivalent ions reduce the coefficient of activity of bases of the nucleic acids more effectively than monovalent.

2. Ions with a smaller ionic radius have a stronger influence on the value of pK_a; monovalent ions can be arranged in the magnitude of their effect in the following order:

$$Li^+ > Na^+ > K^+ > Cs^+$$

Effect of temperature.   One of the most important factors influencing the equilibrium constant in general and, naturally, the acid–base equilibrium constant in particular, is temperature.   Changes in pK_a' for cytidine [126] and uridine-5'-phosphate [125] in acid and alkaline media at different temperatures are shown in Table 3.21.   In both cases appreciable decrease in pK_a with an increase in temperature is observed.   This change in the ionization constant is described by the equation:

$$pK_a = \frac{1}{2.3} \cdot \frac{\Delta F}{RT} = \frac{1}{2.3} \left(\frac{\Delta H}{RT} \cdot \frac{\Delta S}{R} \right) \tag{11}$$

where ΔF, ΔH, and ΔS represent the free energy, enthalpy, and entropy of the ionization reaction, T the absolute temperature, and R is the universal gas constant.

Recently determined values of ΔF, ΔH, and ΔS for some bases, nucleosides, and nucleotides are given in Table 3.22.

Action of the solvent and concentration effect.   It has recently been shown that purine bases, nucleosides, and nucleotides form complexes in solution, the concentration of which increases, in accordance with the law of mass action, with an increase in the concentration of the components (see Chapter 4).   It will be expected that their dissociation constants will alter with a change in the concentration of the components in solution.   In fact, in

TABLE 3.20 Experimentally Determined Values of pK_a for Bases, Nucleosides, and Nucleotides obtained at different Ionic Strengths [102]

Compound	pK_a'		$\Delta pK_a'$
	at $\mu = 0$	at $\mu = 1$	
Cytosine	4.68	4.78	+0.10
Cytidine	4.17	4.19	+0.02
Cytidine-3'(2')-phosphate	4.43	4.32	−0.11
Cytidine-5'-phosphate	4.54	4.43	−0.11
Cytidine-2',3'-cyclic phosphate	4.19	4.10	−0.09
Adenine	4.18	4.35	+0.17
Adenosine	3.52	3.70	+0.18
Adenosine-2'-phosphate	3.81	3.72	−0.09
Adenosine-3'-phosphate	3.70	3.68	−0.02
Adenosine-5'-phosphate	3.88	3.80	−0.08
2'-Deoxyadenosine	—	3.79	—
2'-Deoxyadenosine-5'-phosphate	—	3.79	—
Uracil	9.52	9.26	−0.26
Uridine	9.38	9.11	−0.27
Uridine-3'(2')-phosphate	9.96	9.12	−0.84
Uridine-5'-phosphate	10.96	9.24	−0.82
2'-Deoxyuridine	9.50	9.22	−0.28
Thymine	10.00	9.74	−0.26
Thymidine-5'-phosphate	10.47	9.65	−0.82
Thymidine	9.93	9.65	−0.28
Uridine-2',3'-cyclic phosphate	9.47	8.95	−0.52
Guanosine	9.42	9.22	−0.20
Guanosine-2'-phosphate	9.87	9.20	−0.67
Guanosine-3'-phosphate	9.84	9.19	−0.65
Guanosine-5'-phosphate	10.00	9.33	−0.67
2'-Deoxyguanosine	9.52	9.33	−0.19
2'-Deoxyguanosine-5'-phosphate	10.00	9.33	−0.67
Inosine	9.02	8.81	−0.21
Inosine-5'-phosphate	9.62	8.92	−0.70

Note: Values determined spectrophotometrically at 20°C.

TABLE 3.21. Changes in pK_a' of Cytidine and Uridine-5'-phosphate with Temperature

Temperature, °C	pK_a'		Temperature, °C	pK_a'	
	cytidine [126]	uridine-5'-phosphate [125]		cytidine [126]	uridine-5'-phosphate [125]
20	4.32	9.34	40	4.15	9.01
30	4.22	9.18	50	4.06	8.85

Note: Values determined potentiometrically at ionic strength $\mu = 0.2$.

TABLE 3.22 Thermodynamic Parameters of Ionization of Bases, Nucleosides, and Nucleotides [125, 127, 132]

Compound	Dissociation in acid medium				Dissociation in alkaline medium			
	pK_a	ΔF, kcal /mole	ΔH, kcal /mole	ΔS, cal /mole ·deg	pK_a	ΔF, kcal /mole	ΔH, kcal /mole	ΔS, cal /mole ·deg
Adenine.........	4.1	5.6	4.2	4.7	9.7	13.2	9.5	−12.4
	4.1	−	3.99	−				
Adenosine........	3.6	4.8	3.8	−3.4	−	−	−	−
	3.6	−	3.81	−	−	−	−	−
Deoxyadenosine.....	3.8	−	3.87	−	−	−	−	−
Cytosine........	4.5	6.1	5.0	−3.7	11.8	16.1	11.0	−17.1
	4.5	−	4.47	−	−	−	−	−
Cytidine........	4.2	5.7	4.4	−5.0	−	−	−	−
Deoxycytidine......	4.3	−	4.3	−	−	−	−	−
Guanine.........	−	−	−	−	9.4	12.8	10.1	−9.1
Guanosine........	1.6	2.2	1.00	−4.0	9.2	12.6	8.6	13.0
	1.6	−	0.99	−	−	−	−	−
Deoxyguanosine.....	2.5	−	1.91	−	−	−	−	−
Deoxyguanosine-5'-phosphate........	2.9	−	0.14	−	−	−	−	−
Hypoxanthine......	−	−	−	−	8.8	12.0	7.2	−16.1
Inosine..........	−	−	−	−	8.9	12.1	7.2	−16.4
Uridine..........	−	−	−	−	9.5	13.0	8.0	−16.8
Uridine-5'-phosphate .	−	−	−	−	9.4	12.6	6.6	−19.1

Note: Values of pK_a determined spectrophotometrically at $\mu = 0.1$ and 25°C, or potentiometrically at $\mu = 0.1$ and 20°C.

the case of guanosine this relationship has been discovered [131], and as is clear from Fig. 3.13, with an increase in the concentration of nucleoside the value of its pK'_a falls. This may perhaps explain the different values obtained when pK'_a is determined spectrophotometrically (when the concentration of the compound varies from 10^{-4} to 10^{-5} mole/litre) or potentiometrically (when much higher concentrations of the substances, possibly leading to association, are used). Wherever possible, it is preferable to use the spectrophotometric method*.

* Discrepancies are observed in the values of thermodynamic ionization constants of guanosine and its pK'_a determined by different authors. Sukhorukov [127], for instance, gives pK'_a 1.6 for guanosine (spectrophotometric method of determination; values of free energy and heat and entropy of ionization 2.2 kcal/mole, 1 kcal/mole, and 4 cal/mole·deg respectively). The same value for pK'_a of guanosine is given in [128, 129], and it has also been found by the potentiometric method [130]. Meanwhile, in other investigations [131], pK'_a of guanosine was found spectrophotometrically to be 2.2. Corresponding values of the free energy and heat and entropy of ionization were 2.62 kcal/mole, 2.22 kcal/mole, and 2.5 cal/mole·deg respectively. Bunville and Schwalbe explain the difference between their findings [131] and results obtained potentiometrically [130] by concentration effects. The differences between results obtained by different workers using the same method are completely inexplicable.

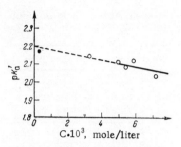

Fig. 3.13. pK'_a of guanosine as a function of its concentration in solution [131]: filled circles indicate results of spectrophotometric determinations, empty circles results of potentiometric determinations; 25°C; $\mu = 0.1$.

A very important factor affecting the value of the acid – base equilibrium constant is the nature of the solvent. The effect of ionic strength on the equilibrium has already been examined above, but in addition, it must be pointed out that dissolved substances of a nonionic character can also lead to changes in the equilibrium constant. The mechanism of this effect may be either nonspecific (a change in the degree of solvation, for example) and different for the different forms participating in the equilibrium, or it may be specific, involving the formation of chemical bonds between molecules of the dissolved substance and the solvent. According to one report [133], pK'_a of guanylic acid is slightly reduced in 48% sucrose solution; this must evidently be attributed to nonspecific effects. However, in 8 M urea solution the value of pK'_a is much higher (compared with aqueous solutions) for all four bases of nucleotides. The difference between the values of pK'_a in the acid region is 0.47 unit for cytidylic acid and 0.5 unit for adenylic acid; in the alkaline region the difference is 0.58 unit for guanylic acid and 0.53 unit for uridylic acid [133]. The magnitude of the shift increases in a linear manner with an increase in urea concentration; appreciable changes appear in the UV spectra, probably as a result of the formation of hydrogen bonds between the bases and urea.

The ability of the bases, nucleosides, and nucleotides to ionize determines many of the physical and chemical properties of these compounds, notably their reactivity, and also such important criteria as their spectral, chromatographic, or electrophoretic characteristics. In all these cases, it is usually essential to know the relative content of the ionized form in the solution of a compound, and for a given pH value this can be determined by the formulae:

$$\frac{C_i}{C_0} = \frac{1}{1 + 10^{pH - pK'_a}} \quad \text{ionization in accordance with Equation (3)}$$

$$\frac{C_i}{C_0} = \frac{1}{1 + 10^{pK'_a - pH}} \quad \text{ionization in accordance with Equation (5)}$$

where C_i represents the concentration of the ionized form and C_0 the total concentration of the investigated substance.

The ratio C_i/C_0 is sometimes called the charge of the molecule and it is widely used for preliminary assessment of the chromatographic or electrophoretic mobility of a substance. In ion-exchange chromatography and in electrophoresis, in order to obtain the highest possible degree of separation of the various monomers, it is naturally essential to use pH values at which the difference between charges on the molecules is maximal. Knowledge of the pK_a values under the experimental conditions must greatly simplify the task of selecting optimal conditions of fractionation.

VI. General matters concerned with reactivity of the bases of nucleic acids

On the basis of the electronic characteristics, tautomerism, and acid – base properties of the bases of the nucleic acids it should be possible, on a priori grounds, to predict some of the more general principles governing their reactivity. It is natural to expect that different tautomeric forms will differ in their reactivity. Unfortunately, however, it is very difficult to determine experimentally which form in fact participates in any given reaction, because equilibrium between the tautomeric forms is established very rapidly. Neutral and ionized forms of the bases also must differ in their reactivity; in this case it is much easier experimentally to decide which is participating in the reaction. Cationic and neutral (but not anionic) forms of the bases must be expected to react more readily with nucleophilic agents, whereas anionic and neutral forms (but not cationic) must react more readily with electrophilic agents.

It was mentioned above that at present in the overwhelming majority of cases we do not know which of the tautomeric forms participates in a particular reaction, and in the subsequent account all arguments will be based on the supposition that the principal tautomeric form (under the particular conditions concerned) is reacting although, generally speaking, this need not be the case. This assumption (regarding the participation of the principal form only) is to some extent justified because many of the predictions based on it are in agreement with the experimental findings.

In this theoretical discussion of reactivity we shall consider neutral forms of the bases, assuming that the relative distribution of the electron density of the molecule, while changing on ionization, nevertheless remains qualitatively close to the distribution of a neutral molecule. This assumption is also confirmed experimentally: if charged molecules take part in a reaction the direction of the reaction is correctly predicted on the basis of examination of neutral molecules. Many approaches can be used to estimate the relative reactivity of different atoms (groups) in one molecule or of identical atoms (groups) in different molecules. Some of these approaches are based on quantum-chemical calculations; others on empirically determined rules and correlations. In nucleic acid chemistry approaches of the first type are more commonly used at the present time. This is convenient from the point of view that predictions can be made without preliminary experiments on the compounds concerned or on their close analogues, frequently by using principles obtained for completely different classes of compounds. The approach using correlation equations requires investigation of the close analogues of the compound concerned in order to predict the property of interest to the experimenter. However, analogues of bases, nucleosides, and nucleotides are not easily accessible, and although this method is very widely used in organic chemistry, its application to the chemistry of nucleic acid bases has been relatively limited. These two approaches will next be examined.

1. The use of quantum chemical calculations

Most reactions of the bases of nucleic acids take place without disturbance of their cyclic structure, and for this reason during the study of their reactions it is often sufficient to consider simply those properties of the molecules which are determined by their π electron distribution. Furthermore, this approximation is often sufficient, even in cases when the σ bonds are ruptured. Later, therefore, we shall in general examine the problem of reactivity solely from the standpoint of the properties of π electron orbitals.

There are two possible approaches to the prediction of the chemical properties of the molecule [134-136], the first of which examines the characteristics of the isolated molecule, while the second takes into consideration the changes taking place in the molecule during its interaction with a reagent. The first approach is known as the isolated-molecule approximation, the second as the reacting-molecule approximation.

The isolated-molecule approximation method. At the beginning of this chapter a number of quantum-chemical characteristics of the bases of nucleic acids were examined: the electron densities on the atoms, the bond orders, and the energies of the filled and empty electronic levels. All these characteristics can be used to predict the reactivity of the molecule, and in organic chemistry the most usual course is to use electron densities on atoms as a measure of their reactivity. It is considered that in reactions with electrophilic reagents, the atom with the highest electron density will possess the greatest reactivity, whereas in reactions with nucleophilic reagents, the highest reactivity will be possessed by the atom with the lowest electron density. From this point of view the results of calculations shown on the molecular diagrams on pages 124-129 predict that the C2, C4, and C6 atoms, carrying a net positive charge in pyrimidine bases and the C2, C4, C6, and C8 atoms in purine bases must be attacked by nucleophilic reagents. On the other hand, the C5 atoms in pyrimidine bases, the oxygen atoms of carbonyl groups and the nitrogen atom of pyridine type, i.e., atoms carrying a net negative charge, must favour electrophilic attack. As will be shown below, these conclusions are largely confirmed by experimental data, and in this way the simplest approximation is justified as a means of qualitative prediction of reactivity of atoms forming the bases of nucleic acids.

Another characteristic of reactivity is the bond order. It is usually accepted that the higher the bond order, the easier it must be for electrophilic attack at that bond to take place. It follows from the molecular

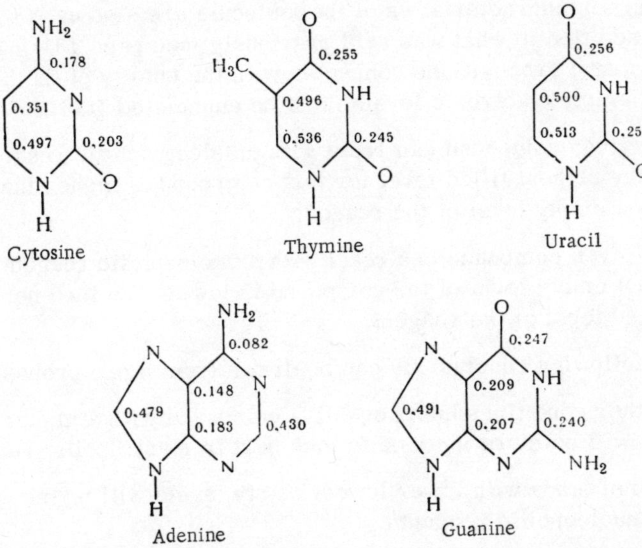

Fig. 3.14. Free valence indices for carbon atoms of bases of nucleic acids.

diagrams (see page 124 et seq.) that in pyrimidine bases the C5 – C6 double bond has a much higher order than the C4 – C5 bond in the purines. In the first case, therefore, addition reactions must take place much more readily, a conclusion in agreement with experiments (although other factors perhaps also play an important role in this situation).

The reactivity of molecules in radical reactions is usually characterized by the free valence index. Calculated free valence indices for carbon atoms of the bases of the nucleic acids are shown in Fig. 3.14. The general conclusion drawn from all methods of calculation is as follows: the most reactive bond relative to homolytic reagents in pyrimidine bases is the C5 – C6 double bond. Unfortunately, because no detailed investigations of radical reactions have been made in the field of the nucleic acid bases and their components, it is impossible at present to conclude if this is really so. The concept of "free valence index" is also used to characterize nucleophilic and electrophilic reactions, where the general position remains the same as with radical processes: the higher the free valence index, the more readily the reaction proceeds. On this basis it can be concluded that the C5 – C6 double bond in pyrimidines is the most reactive position in the pyrimidine molecule and that uracil is more reactive than cytosine. These conclusions are in general in agreement with the experimental findings, although certain exceptions do exist. For example, both the free valence index and the charges on the atoms predict that the C2 and C4 atoms in the uracil molecule must be about equal in reactivity, and that in the cytosine molecule the C2 atom must be much more active toward nucleophilic reagents than the C4. However, in both cases the most reactive atom in the bases and nucleosides and nucleotides derived from them is the C4 atom of the pyrimidine ring.

Energetic characteristics of the molecule are also used to predict reactivity. In addition to what was said previously (see page 131), at this juncture two general propositions connected with the energy of the highest filled and lowest empty electronic levels must be enunciated [136]:

1. A given compound can react with an electrophilic reagent if the energy of the highest filled level for this compound is higher than the energy of the lowest empty level of the reagent.

2. A given compound can react with a nucleophilic reagent if the energy of the lowest empty level of the compound is lower than the energy of the highest filled level of the reagent.

The following conclusions can be drawn from these propositions:

1. Other conditions being equal, a compound with a higher highest filled level will be more inclined to take part in electrophilic reactions.

2. Compounds with lower lowest empty levels will be more inclined to react with nucleophilic reagents.

On this basis, the purine with the greatest tendency to take part in electrophilic reactions must be guanine, as is confirmed by all methods of calculation, and cytosine must be the most reactive of the pyrimidines in this respect, if it is assumed that more sophisticated methods of calculation yield the most reliable results. These conclusions are also in agreement with the experimental facts. In particular, when halogens act on purine nucleosides and nucleotides, guanine derivatives react very readily whereas the corresponding adenine derivatives react only under vigorous conditions (see Chapter 5). Unfortunately, the great differences between the results for the calculated energies of the lowest empty orbitals (see page 133) do not permit definite conclusions to be drawn regarding the tendency for the bases to take part in nucleophilic reactions.

Predictions based on the isolated-molecule method are valid if the structure of the reaction complex is close to the structure of the original molecule. On the other hand, if the original molecule undergoes considerable modifications in the reaction complex, these changes must be taken into account, and to some extent this can be done by the "reacting-molecule approximation" method.

The reacting-molecule approximation method. One of the most widely used quantum-chemical approaches to the study of the reacting molecule is that based on Wheland's concept of the reaction complex, in which the electronic structure of an atom under attack corresponds to an sp^3 hybrid and its orbitals are excluded from conjugation. The energy required to remove an atom from conjugation is called the localization energy which is the difference between the energies of the π electrons of the initial structure and of the structure with the atom excluded from conjugation (Fig. 3.15). Depending on the type of reaction, at the energy levels calculated for the conjugated part

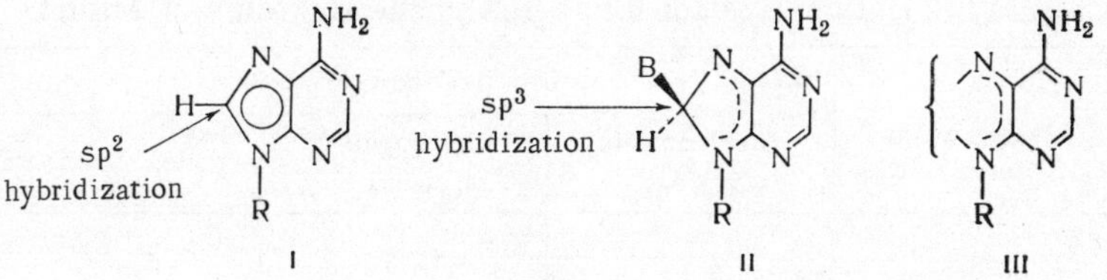

R denotes hydrogen atom or various radicals

Fig. 3.15. Illustration of calculation of the localization energy exemplified by an adenine derivative: I) original molecule with energy of π electrons equal to E_0; II) Wheland reaction complex with total energy of π electrons in conjugated system III and of electrons on localized atom equal to E_1; III) conjugated system of electrons in reaction complex. The localization energy $E_L = E_0 - E_1$ differs depending on the number of electrons located at energy levels for which calculations are made.

of the structure there is a certain number of electrons, equal to the number of electrons in the initial molecule (nucleophilic reactions), 2 less than the number of electrons in the initial molecule (electrophilic reactions), or 1 less than the number of electrons in the initial molecule (homolytic reactions). The number of electrons remaining on the localized atoms is 0, 2, and 1 respectively. The corresponding localization energies are called nucleophilic, electrophilic, and radical. If the reaction is one of addition at a double bond, the two atoms linked by this bond are simultaneously excluded from conjugation.

Comparison of the localization energy with experimentally determined values of the free activation energy ΔF, the activation enthalpy ΔH (or with the activation energy E which differs from it by RT), and the activation entropy ΔS leads to the conclusion that for a given class of reactions the localization energy is a measure of the relative reactivity of the various atoms in a given molecule or of the same atoms in different molecules. Two conditions must be satisfied: first, for all such reactions the energies of modification of the σ bonds in the molecule during formation of the activated complex must be identical, and second, the activation entropies must be constant or proportional to the changes in energy of the π bonds. Few calculated data for localization energies are as yet available.

The Pullmans, in their monograph [134], give in particular calculated values of the localization energies of the C2, C6, and C8 atoms of purine and uric acid. For both compounds the order of the change in localization energies during the transition from atom to atom is the same. The table with the calculations for purine taken from this book is reproduced below (Table 3.23).

These results indicate that in nucleophilic, electrophilic, and homolytic reactions the C8 atom of the purine skeleton has the highest reactivity. The difference in uric acid, from this same point of view, is that the calculations predict higher reactivity in nucleophilic reactions for its C6 atom.

TABLE 3.23. Localization Energies of the C Atoms of Purine

Purine atom	Localization energy, β		
	nucleophilic reactions	electrophilic reactions	radical reactions
C2	2.32	2.57	2.45
C6	2.18	2.48	2.33
C8	2.18	2.39	2.29

If these results are now applied to other purines, considerable reactivity of the C8 atom can be predicted in reactions of different types, higher than for C2 and C6 in electrophilic reactions, possibly a little lower than for C6 in nucleophilic reactions, but much higher than C2. This conclusion agrees completely with the experimental data and, in particular, with the ease of proton exchange at C8 of purines for deuterium and tritium, and with the absence of exchange of the proton bound to the C2 atom (see page 290). It was stated above that halogenation of the purines take place with substitution of hydrogen at C8. This conversion is an example of the electrophilic reaction.

Unfortunately few examples of nucleophilic reactions involving purines are known, so that it is difficult to compare theoretical deductions with experimental conclusions. Localization energies have been calculated for nucleophilic addition at the double bond and substitution of the amino group in cytidine and its 5-methyl derivative [137], and satisfactory agreement has been obtained between the calculated data and the relative velocities of the reactions of these compounds with hydroxylamine. Comparison of the localization energies of addition at the double bond in uracil and cytosine also agree closely with the relative reactivity of these compounds (see Chapter 5). The lower value of the localization energy of addition at the double bond in uracil (compared with cytosine) is also in agreement with the known fact that the photohydrate of uridine [1-(β-D-ribofuranosyl)-6-hydroxy-5,6-dihydrouracil] is more stable than the photohydrate of cytidine [1-(β-D-ribofuranosyl)-6-hydroxy-5,6-dihydrocytosine] (see Chapter 12).

By way of summing up, the following general principles governing the determination of the most reactive positions in molecules on the basis of quantum-mechanical calculations can be enunciated.

1. The most active atoms relative to electrophilic reactions are those with the least electrophilic localization energy, the highest electron density, and the highest free valence index. These atoms are: C8 in purines (as regards localization energy and free valence index) and C5 in pyrimidines (as regards the free valence index and electron density).

2. The most reactive atoms relative to nucleophilic reactions are those with minimal nucleophilic localization energy, with the highest free valence index and the lowest electron density. These atoms are: in pyrimidines, C6 (as regards electron density and free valence index) and C2 and C4 (as regards

electron density); in purines, C8 (as regards localization energy and free valence index) and C6 (as regards localization energy and electron density).

3. The most reactive atoms toward radical reactions are those with the least radical localization energy and the highest free valence index, namely: C8 in purines (as regards both characteristics), and C5 and C6 in pyrimidines.

4. So far as addition at the double bond is concerned, compounds with minimal localization energy and maximal order of the double bond are the most reactive.

A limitation of the quantum-chemical approaches examined above is that they do not take into account many extremely important factors partly determining the direction and velocity of the reaction. These include, for example, the effect of the solvent, spatial effects of substituents, and so on. Spatial effects of substituents have already been discussed in the examination of the acid – base properties of the bases. A few more examples of this effect will be considered. It follows from the facts relating to conformation of nucleosides and nucleotides (see Chapter 2), in particular, that under ordinary conditions the anti-conformation is preferable for these compounds, in which the ribose residue and carbonyl group in pyrimidine nucleosides and nucleotides are on different sides of the N-glycoside bond. If this conformation remains preferable in solution, it can be expected that the sugar residue will give rise to steric hindrance to nucleophilic addition at the double bond, so that reactions of this type with bulky reagents may actually become impossible. For example, the possibility is not ruled out that reagents such as semicarbazide or Girard's reagent (see page 302) are not added to the double bond because of these steric hindrances. The fact that photochemical hydration of the double bond in uridine-5'-phosphate is more difficult than the corresponding process with uridine-3'-phosphate is perhaps also explained by steric effects (see Chapter 12). Despite these numerous facts, no detailed investigation of the steric effect of the sugar residue on reactivity of the bases has yet been undertaken.

Substituents in position 5 of the pyrimidine ring likewise exert definite steric effects on the course of some reactions. Their effect on pK_a was mentioned above, and difficulty of addition to the double bond in nucleophilic reactions of 5-substituted pyrimidines (evidently due to inductive and steric effects) must now be mentioned. For example, in the reaction between thymidine and hydrazine or hydroxylamine derivatives in an alkaline medium, the velocity of the modification is much lower than that of the corresponding reactions with uridine (see pages 401 et seq. and 407 et seq.).

Yet another factor which is disregarded by quantum-chemical calculations is the possible formation of intramolecular bonds of various types, such as hydrogen bonding of the sugar residue with the base (see page 116). It is natural to expect that the formation of such bonds must also affect reactivity, increasing it if these additional bonds in the reaction complex become stronger than in the initial molecule, or reducing it if rupture of these intramolecular bonds is necessary for the formation of the intermediate complex.

Finally, the solvent also influences the velocity and direction of the reaction. It solvates the initial molecules and intermediate complexes non-specifically or it can form specific chemical bonds (hydrogen, for example) with the initial molecules and intermediate complexes. If solvation or the ability to form these bonds is stronger in the intermediate complex than in the initial molecules, that particular solvent will accelerate the reaction; if, on the other hand, the formation of these bonds is energetically more advantageous in the initial molecules than in the reaction complex, the solvent will delay the reaction. These effects may be so powerful that the direction of the reaction may also be changed; for this reason, results obtained in one solvent must not be directly transferred to other reagent. In addition, the solvent may compete with a reagent in the course of the reaction. Cases when reactions take place in different directions in different solvents are well known. For example, during halogenation of pyrimidine bases in an anhydrous solvent, the hydrogen atom in position 5 of the heterocyclic ring is substituted; in the presence of water this reaction takes place by addition of bromine and a hydroxy group at C5 and C6 of the double bond (for further details, see page 284 et seq.). The effect of all these factors must be taken into consideration when attempts are made to assess the reactivity of compounds.

If different potentially reactive positions in the same molecule are compared, in a nucleoside, for example, they can be seen to be nonequivalent, if only from the standpoint of the steric effect of the sugar residue. For a nucleoside with the anti-configuration, the C8 atom in purine derivatives must be more effectively screened from steric effects than the C2 and C6 atoms. Correspondingly, reactions at C8 will take place more slowly than would be postulated on the basis of a simple quantum-chemical examination.

Even for reactions with the same reagents, different parts of the molecule can be expected to form reaction complexes with different degrees of solvation, also causing a discrepancy between the experimental data and those predicted theoretically. From this point of view it would seem more correct to use quantum-chemical calculations to compare equivalent reaction centres in different molecules, thereby neutralizing to some extent the other factors influencing reactivity. This type of approach is used in organic chemistry to study the effect of conditions on reaction velocity and, in particular, to study correlations between reactivity and the electronic characteristics of substituents.

2. The use of correlation equations [112–114]

In order to study the chemical features of the bases themselves and their behaviour as components of nucleosides and nucleotides, it is essential to determine the position of these compounds in a series of very close analogues differing from each other in some uniformly changing characteristic, such as

in the presence of different substituents possessing different inductive effects
or a different steric effect. For instance, when studying a particular reaction
of a series of 1-alkyl-substituted pyrimidines with a variety of alkyl substituents, it might be better to assess the role of the ribose (or deoxyribose)
moiety in determining the chemical properties of the base as a component of
a nucleoside or nucleotide. A similar approach is widely used in organic
chemistry to study reaction mechanisms, when it is found that the free activation energy of many reactions is a linear function of a certain characteristic
which varies from one substituent to another, but is constant for a given substituent in different compounds. This principle is sufficiently familiar and it
is expressed as the rule of the linear dependence of free energies. Special
cases of this rule are well known: the equations of Hammett or of Taft. They
express the reactivity of a series of related compounds toward the same reagent as a function of the electronic characteristics of the substituents in
these compounds in the form of equations of the type:

$$\log \frac{k_i}{k_0} = \rho \sigma_i$$

In this equation k_i is the velocity constant of the reaction of combination
with substituent i; k_0 the velocity constant of the reaction with a standard substituent; ρ the coefficient of proportionality characteristic of that reaction;
and σ_i the characteristic of the electronic properties of the substituent which
is constant for a given substituent in different compounds and reactions *.

Relationships of this type are sufficiently universal, and correlations
with values of σ exist not only for the free energies of reactions, but also
for various characteristics of a compound, notably for frequencies in IR-
spectra, values of chemical shifts in NMR spectra, and so on. Although
examples of correlations such as this are not numerous, they do exist in a
series of derivatives of bases of the nucleic acids, thus demonstrating that,
in principle, the rule of the linear dependence of free energies is applicable
to this class of compounds also. We have already examined above (see page
157) the correlation of pK_a in the series of uracil and cytosine. The fact
that this correlation exists means that in both these cases the same substituent can be characterized by a definite value, constant for it alone; this is
another way of expressing the rule of the linear dependence of free energies.

Cases are also known when correlations have been drawn directly with
one or several σ constants simultaneously. Coburn and his collaborators
[138, 139] obtained a near-linear relationship (Fig. 3.16) between the chemical
shift of C2 and C8 proton signals in the NMR spectrum of 6-substituted
purines and Brown's constant [114] of the substituent in position 6. A similar
picture was obtained for the relationship between the chemical shift of the C8
proton and the sum of the σ^+ values for substituents in positions 2 and 6 in a

---

* In some cases the same substituent may have several values of σ depending, for example, on
whether it participates in conjugation or not.

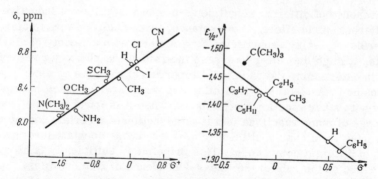

Fig. 3.16. Chemical shift of the C8 proton in 6-substituted purines as a function of the σ^+ constants of the substituents [139].

Fig. 3.17. Half-wave potential of polarographic reduction (relative to saturated calomel electrode) of 5-substituted 6-azauracils as a function of σ^+ constants of the 5-substituent (in aqueous solution at pH 8.3) [140].

series of 2,6-disubstituted purines [139]. The same workers attempted to draw a correlation between chemical shifts of protons and another series of σ constants, namely the inductive constant σ_I and the resonance constant σ_R of Taft [138, 139]. In this case, the following equations were obtained for 6-substituted purines:

$$\Delta\delta = 0.4\sigma_I + 0.7\sigma_R \quad \text{(for proton at C8)}$$
$$\Delta\delta = 0\sigma_I + 0.9\sigma_R \quad \text{(for proton at C2)}$$

where $\Delta\delta$ represents the difference between chemical shifts of the corresponding proton in substituted and unsubstituted purines.

It follows from these equations that the chemical shift of the C2 proton is determined entirely by mesomeric effects, whereas the chemical shift of the C8 proton is determined by both mesomeric and inductive effects (see page 156). It can be concluded from an examination of the mesomeric formulae that only very slight mesomeric interaction can exist between positions 2 and 6 in the 6-substituted purine (the meta-arrangement). The observed dependence may mean that the chemical shift is largely determined by the electron density on the neighbouring nitrogen atoms, one of which is in the ortho-, the other in the para-position relative to the 6-substituent. In the case of the chemical shift of the C8 proton, the relationship obtained is not unexpected, since the C8 atom can interact with the 6-substituent both by inductive and by mesomeric mechanisms. This example clearly shows that this type of analysis can yield much information concerning the fine structure of the electronic influence of substituents on the properties of molecules.

Another example [140] is the use of Taft's constants to correlate the relationship between the half-wave potential of polarographic reduction on a mercury dropping electrode of 5-substituted 6-azauracils and the σ constant of the substituent. The linear relationship obtained for a series of substituents

(Fig. 3.17) demonstrates the common mechanism of reduction of the corresponding compounds and its dependence on polar effects of the substituents. The fact that the half-wave reduction potential in the case of the 5-tert-butyl-derivative falls outside the general rule probably means that steric effects play an important role in the case of bulky substituents. The negative slope of the straight line is perfectly natural: it means that with an increase in electronegativity of the substituent, reduction (i.e., the transmission of an electron to the compound) is facilitated.

There are therefore no fundamental limitations to the application of the rule of linear dependence of free energies to the investigation of the properties of derivatives of nucleic acid bases. The use of these rules can provide much valuable information concerning the properties of components of the nucleic acids.

Effect of ionization on reactivity of the bases. It was pointed out above that many reactions in the series of compounds with which we are concerned involve the protonation or deprotonation of the bases. When an electrophilic or nucleophilic reagent is itself incapable of ionization, and the reaction is a simple one-stage irreversible reaction, the reaction velocity can be expected to be a simple function of the hydrogen ion concentration. In the case of a nucleophilic reaction of a nucleophilic agent, which is not protonated under the particular conditions, with a protonated molecule of the base, the observed velocity constant is determined by the following equation:

$$k_{obs} = k_0 \frac{[H^+]}{K'_a + [H^+]} \tag{12}$$

where k_{obs} is the observed velocity constant, k_0 the true velocity constant of the reaction between the cation and nucleophilic agent, K'_a the apparent ionization constant of the base derived from Equation (3), and $[H^+]$ the hydrogen ion concentration.

Relationships of this type can be expected in certain hydrolytic conversions of the nucleic acid bases. Deamination of dihydrocytosine derivatives can serve as an example (see Chapter 5).

The relationship between the velocity of deamination of the photohydrate of cytidine (6-hydroxy-5,6-dihydrocytidine) [121] and pH is characterized by a steady increase in the value of the observed velocity constant with a decrease in pH. This also follows from Equation (12). Analysis shows, however, that the neutral form of the base (as a component of the nucleoside) also plays an important role in the deamination process. At higher pH values, therefore, the velocity constant is not zero, as might be expected if only the protonated form reacted, but it has a definite value independent of pH.

When an electrophilic reaction takes place with an anion formed by dissociation of a neutral molecule of the uridine or guanosine type, and when the electrophilic agent itself cannot be deprotonated, the relationship between the observed velocity constant of the reaction and hydrogen ion concentration can be expressed in the form:

$$k_{obs} = k_0 \frac{K'_a}{[H^+] + K'_a} \tag{13}$$

where k_0 is the true velocity constant of the reaction between the anion and electrophilic agent, and K'_a the apparent ionization constant of the base derived from Equation (5).

Certain alkylation reactions, such as those of guanosine, uridine, inosine, and pseudouridine with acrylonitrile (for further details, see Chapter 5), are evidently reactions of this type. In the reaction of acrylonitrile with uridine, a study of the dependence of reaction velocity on pH shows that values of k_0 calculated by Equation (15) from experimental values of k_{obs} remain constant within the limits of experimental error.

pH	k_{obs}, min^{-1}	k_0, min^{-1}
9.15	4.57	17.5
9.70	8.97	16.1
10.05	15.40	20.9

The constancy of k_0 confirms the validity of the suggested mechanism for the reaction between the electrophilic agent and anion.

However, these very simple cases when only one of the reacting substances is ionized are rarely seen in reactions of the nucleic acid bases. Processes in which both reacting compounds are capable of ionization are much more common. Some of the most common types of conversions can now usefully be examined.

1. Reactions of the cationic form of a base with a nucleophilic reagent capable of protonation. Reactions of this type include most of the nucleophilic reactions of cytidine, adenosine, and guanosine. If the reaction is a one-stage process of interaction between cation and nucleophilic agent, the observed velocity constant of the reaction will be determined by the following equation:

$$k_{obs} = k_0 \frac{(K'_a)_N \cdot [H^+]}{\{(K'_a)_B + [H^+]\} \cdot \{(K'_a)_N + [H^+]\}} \tag{14}$$

where k_0 is the true velocity constant of the reaction between the base cation and the nucleophilic reagent; $[H^+]$ is the hydrogen ion concentration, and $(K'_a)_B$ and $(K'_a)_N$ the constants of protonation of the base and nucleophilic reagent calculated by Equations (3) and (7).

It can be concluded from analysis of this expression that the curve of observed velocity constant as a function of pH has a maximum at the value:

$$pH = \frac{(pK'_a)_B + (pK'_a)_N}{2}$$

and that at pH values equal to $(pK'_a)_B$ and $(pK'_a)_N$ the observed velocity constants are equal. The reaction of substitution of a semicarbazide group for the amino group in cytidine [142] is an example of conversions of this type (see also page 301).

2. The analogous equation is also obtained for the relationship between k_{obs} and hydrogen ion concentration during a reaction of a neutral form of the base, capable of releasing a proton, with an anionic nucleophilic agent. In this case, in Equation (14) the terms $(K'_a)_B$ and $(K'_a)_N$ are the ionization constants of the base and nucleophilic reagent as in Equation (5), and their values are determined by Equation (8).

3. An equation of this same type describes the relationship between the observed velocity constant and hydrogen ion concentration in the case when a neutral base, capable of deprotonation in accordance with Equation (5), reacts with a neutral nucleophilic reagent capable of protonation in accordance with Equation (3). In formula (14) for this case, $(K'_a)_N$ is the ionization constant of the nucleophilic reagent from Equation (3), the value of which is determined by expression (8).

4. The velocity constant k_{obs} for the reaction between the neutral form of a base capable of protonation with the neutral form of a nucleophilic reagent, also capable of protonation, is described by the formula [the symbols have the same meaning as in Equation (14)]:

$$k_{obs} = k_0 \frac{(K'_a)_B \cdot (K'_a)_N}{\{(K'_a)_B + [H^+]\} \cdot \{(K'_a)_N + [H^+]\}} \tag{15}$$

Although reactions of neutral bases with neutral reagents proceed more slowly than reactions of ionized bases with neutral reagents, nevertheless they frequently distort the relationship given above. Moreover, the true mechanisms of reactions which actually take place are very often much more complex than simple one-stage conversions. Although in practice all reactions of the bases of nucleic acids exhibit pH-dependence, nevertheless they cannot always be satisfactorily described by the simple formulae given above, and in each concrete case detailed investigation of the mechanisms of the conversions is necessary in order to understand the relationship between reaction velocity and acidity of the medium. Insofar as this is possible at the present time, concrete reactions will be analysed in this manner in the second half of this book.

Effect of ionic strength on reaction velocity. The apparent ionization constant K'_a figures in all the equations for the relationship between the observed velocity constant of a reaction and the hydrogen ion concentration given above [143]. As has been mentioned already (see page 159), this constant depends on the ionic strength. It must therefore be expected that the velocity constant of a reaction will depend on the ionic strength in cases when preliminary protonation or deprotonation of the reagents is necessary for the reaction. For the reaction of a charged base with a charged reagent, the dependence of the reaction velocity on the ionic strength is associated not only with a change in the ionization constants of the compounds, but also with a more general effect, which follows from the assumption that the velocity of any reaction

$$A + B \underset{\rightleftarrows}{} X_{act} \rightarrow C$$

is proportional to the concentration of the activated complex X_{act} in equilibrium with the original products. As a rule the constant of this equilibrium is expressed by the ratio between the activities of the components, and the concentration X_{act} depends on the coefficients of activity and, consequently, on the ionic strength.

In any concrete case investigation of the effect of ionic strength on the reaction velocity can yield considerable information on the mechanism of the reaction. Unfortunately, effects associated with changes in ionic strength in the reactions of nucleic acid bases and their derivatives have virtually never been studied.

In this chapter we have therefore examined some of the problems connected with reactivity of the bases themselves and of the bases as components of nucleosides and mononucleotides. The concepts described above can be used to study conversions of the precursors of nucleic acids, nucleotide coenzymes and allied compounds in which the heterocyclic base does not react or reacts only weakly with the other bases or non-nucleotide substances.

Bibliography

1. A. Streitwieser (Jr.), Molecular Orbital Theory for Organic Chemists, Wiley, New York (1961).
2. B. Pullman and A. Pullman, Quantum Biochemistry, Wiley, New York (1963).
3. K. Higasi, H. Baba, and A. Rembaum, Quantum Organic Chemistry, New York (1965).
4. J.-P. Mathieu and A. Allais, Principes de Synthèse Organique, Masson, Paris (1957).
5. T. A. Hofman and J. Ladic, Adv. Chem. Phys., 7:84 (1964).
6. J. I. Fernandez-Alonzo, Adv. Chem. Phys., 7:5 (1964).
7. C. Giessner-Prettre and A. Pullman, Theoret. Chim. Acta, 9:279 (1968).
8. J. S. Kwiatkowski, Theoret. Chim. Acta, 10:47 (1968).
9. A. Pullman, E. Kochanski, M. Gilbert, and A. Denis, Theoret. Chim. Acta, 10, 231 (1968).
10. V. A. Kuprievich, V. I. Danilov, and O. V. Shramko, Teor. i Éksper. Khim., 2:734 (1966).
11. A. Veillard and B. Pullman, J. Theoret. Biol., 4:37 (1963).
12. C. Nagata, A. Imamura, Y. Tagashira, and M. Kodama, Bull. Chem. Soc. Japan, 38:1638 (1965).
13. B. Pullman and A. Pullman, Quantum Biochemistry, Wiley, New York (1963).
14. M. Sundaralingam and L. H. Jensen, J. Mol. Biol., 13:930 (1965).
15. B. Pullman and A. Pullman, Quantum Biochemistry, Wiley, New York (1963).
16. A. Imamura, H. Fujta, and C. Nagata, Bull. Chem. Soc. Japan, 40:522 (1967).
17. G. Fraenkel, R. E. Carter, A. McLachlan, and J. H. Richards, J. Am. Chem. Soc., 83:5846 (1960).
18. H. Spieseche and W. G. Schneider, Tetrahedron Letters, 468 (1961).
19. T. Shaefer and W. G. Schneider, Canad. J. Chem., 41:966 (1963).
20. C. MacLean and E. L. Mackor, Mol. Phys., 4:241 (1961).
21. B. Pullman, Tetrahedron Letters, 231 (1963).
22. P. G. Lykos and R. L. Miller, Tetrahedron Letters, 1743 (1963).
23. A. Veillard, J. Chim. Phys., 59:1056 (1962).
24. A. Veillard and B. Pullman, C. r., 253:2418 (1961).
25. M. P. Schweizer, S. J. Chan, G. H. Helmkamp, and P. O. P. Ts'o, J. Am. Chem. Soc., 86:696 (1964).
26. A. Streitwieser (Jr.), Molecular Orbital Theory for Organic Chemists, Wiley, New York (1961).
27. H. Berthod and A. Pullman, J. Chim. Phys., 62:942 (1965).
28. J. Delre, J. Chem. Soc., 4031 (1958).
29. C. P. Smith, Dielectric Behaviour and Structure, McGraw-Hill, New York (1955).
30. H. DeVoe and I. Tinico (Jr.), Mol. Biol., 4:500 (1962).
31. H. Berthod, G. Giessner-Prettre, and A. Pullman, Theoret. Chim. Acta, 5:53 (1966).
32. H. Berthod, G. Giessner-Prettre, and A. Pullman, Theoret. Chim. Acta, 8:212 (1967).
33. A. Streitwieser (Jr), Molecular Orbital Theory for Organic Chemists, Wiley, New York (1961).
34. B. Pullman and A. Pullman, Quantum Biochemistry, Wiley, New York (1963).
35. C. A. Coulson, Valence, Oxford Univ. Press, Oxford (1961).
36. A. Streitwieser (Jr.), Molecular Orbital Theory for Organic Chemists, Wiley, New York (1961).
37. C. A. Coulson, Valence, Oxford Univ. Press, Oxford (1961).
38. A. Streitwieser (Jr.), Molecular Orbital Theory for Organic Chemists, Wiley, New York (1961).

39.   B. Pullman and A. Pullman, Quantum Biochemistry, Wiley, New York (1963).
40.   C. Lifschitz, E. D. Bergwan, and B. Pullman, Tetrahedron Letters, 4583
      (1967).
41.   C. Janion and D. Shugar, Acta Biochim. Polon., 15:261 (1968).
42.   C. Janion and D. Shugar, Biochem. Biophys. Res. Comm., 18:617 (1965).
43.   A. R. Katritzky and J. M. Lagowski, in: Advances in Heterocyclic Chemistry,
      Vol. 1, Academic Press, New York – London (1963), p. 312.
44.   S. F. Mason, The Chemistry and Biology of Purines. A Ciba Foundation Symposi-
      um, London (1957), p. 60.
45.   M. Kh. Karapet'yants, Chemical Thermodynamics [in Russian], Goskhimizdat
      (1953), p. 46.
46.   V. I. Danilov, Biofizika, 12:540 (1967).
47.   K. N. Trueblood, P. Horn, and V. Luzzati, Acta Cryst., 14:965 (1961); E.
      Shefter and K. N. Trueblood, Acta Cryst., 18:1067 (1965).
48.   D. Shugar and J. J. Fox, Biochim. Biophys. Acta, 9:199 (1952).
49.   H. T. Miles, Biochim. Biophys. Acta, 22:247 (1956).
50.   H. T. Miles, Biochim. Biophys. Acta, 27:46 (1958).
51.   C. L. Angell, J. Chem. Soc., 504 (1961).
52.   B. I. Sukhorukov, V. Ts. Aikazyan, and Yu. A. Ershov, Biofizika, 11:753 (1966).
53.   J. P. Kokko, J. H. Goldstein, and L. Mandell, J. Am. Chem. Soc., 83:2909
      (1961).
54.   A. R. Katritzky and A. Waring, J. Chem. Soc., 1540 (1962).
55.   M. A. Paul and F. A. Long, Chem. Rev., 57:1 (1957).
56.   M. Spencer, Acta Cryst., 12:59 (1959).
57.   D. J. Brown and J. M. Lyall, Austr. J. Chem., 15:851 (1962).
58.   C. Hebne and J. L. Rivail, C. r., D264:861 (1967).
59.   T. L. V. Ulbricht, Tetrahedron Letters, 1027 (1963).
60.   H. T. Miles, J. Am. Chem. Soc., 85:1007 (1963).
61.   L. Gatlin and J. C. Davis, J. Am. Chem. Soc., 84:4464 (1962).
62.   H. T. Miles, B. R. Bradley, and E. D. Becker, Science, 142:1569 (1963).
63.   J. Pitha and J. Beranek, Coll. Czech. Chem. Comm., 28:1507 (1963).
64.   T. L. V. Ulbricht, J. Chem. Soc., 6134 (1965).
65.   P. Pithova, A. Piskala, J. Pitha, and F. Šorm, Coll. Czech. Chem. Comm.,
      30:1626 (1965).
66.   G. W. Kenner, C. B. Reese, and A. R. Todd, J. Chem. Soc., 855 (1955).
67.   J. Wempen, R. Duschinsky, L. Kaplan, and J. J. Fox, J. Am. Chem. Soc.,
      83:4755 (1961).
68.   P. Brookes and P. D. Lawley, J. Chem. Soc., 1348 (1962).
69.   C. Helène, A. Hang, M. Delbrück, and P. Douzou, C. r., 259:3385 (1964).
70.   C. Helène and P. Douzou, C. r., 259:4853 (1964).
71.   D. M. Brown, M. J. E. Hewlins, and P. Schell, J. Chem. Soc., C, 1925 (1968).
72.   S. Furberg and L. H. Jensen, J. Am. Chem. Soc., 90:470 (1968).
73.   D. M. Brown and M. J. E. Hewlins, J. Chem. Soc., C, 2050 (1968).
74.   N. Dupuy-Mamelle and B. Pullman, J. Chim. Phys., 64:708 (1967).
75.   D. J. Brown and S. F. Mason, J. Chem. Soc., 682 (1957).
76.   H. T. Miles, F. B. Howard, and J. Frazier, Science, 142:1458 (1963).
77.   F. B. Howard and H. T. Miles, J. Biol. Chem., 240:801 (1965).
78.   H. T. Miles, Biochim. Biophys. Acta, 35:275 (1959).
79a. D. G. Watson, D. J. Sutor, and P. Tollin, Acta Cryst., 19:111 (1965);
79b. R. F. Steward and L. H. Jensen, J. Chem. Phys., 40:2071 (1964).
80.   B. Pullman and A. Pullman, Quantum Biochemistry, Wiley, New York (1963).
81.   A. Albert, Heterocyclic Chemistry, London (1959), pp. 336–346.

82. W. Cochran, Acta Cryst., 4:81 (1951).
83. M. Sundaralingam, Acta Cryst., 21:495 (1966).
84. J. Kraut and L. H. Jensen, Acta Cryst., 16:79 (1963).
85. J. M. Broomhead, Acta Cryst., 1:324 (1948); Acta Cryst., 4:92 (1951).
86. C. E. Bugg and R. E. Marsh, J. Mol. Biol., 25:67 (1967).
87. M. Sundaralingam and L. H. Jensen, J. Mol. Biol., 13:914, 930 (1965).
88. R. E. Marsh, R. Bierstedt, and E. L. Eichhorn, Acta Cryst., 15:310 (1962).
89. C. R. Dekker, Ann. Rev. Biochem., 29:453 (1960).
90. M. Tsuboi, Y. Kyogoku, and T. Shimanouchi, Biochim. Biophys. Acta, 55:1 (1962).
91. C. D. Jardetzky and O. Jardetzky, J. Am. Chem. Soc., 82:222 (1960).
92. S. Danyluk and F. E. Hruska, Biochemistry, 7:1038 (1968).
93. F. J. Bullock and O. Jardetzky, J. Org. Chem., 29:1988 (1964).
94. A. R. Katritzky and A. J. Waring, J. Chem. Soc., 3046 (1963).
95. O. Jardetzky, P. Pappas, and N. C. Wade, J. Am. Chem. Soc., 85:1657 (1963).
96. H. C. Börresen, Acta Chem. Scand., 21:2463 (1967).
97. L. B. Clark and J. Tinoco, J. Am. Chem. Soc., 87:11 (1965).
98. J. J. Fox and D. Shugar, Biochim. Biophys. Acta, 9:369 (1952).
99. R. J. Sinsheimer, R. L. Nutter, and G. R. Hopkins, Biochim. Biophys. Acta, 18:13 (1955).
100. J. Wempen, R. Duschinsky, L. Kaplan, and J. J. Fox, J. Am. Chem. Soc., 83:4755 (1961).
101. J. J. Fox, J. F. Codington, N. C. Yung, L. Kaplan, and J. O. Lampen, J. Am. Chem. Soc., 80:5155 (1958).
102. J. Clauwaert and J. Stockx, Z. Naturforsch., 23b:25 (1968).
103. J. J. Fox, J. Wempen, A. Hampton, and I. L. Doerr, J. Am. Chem. Soc., 80:1669 (1958).
104. L. F. Cavalieri, J. J. Fox, A. Stone, and N. Chang, J. Am. Chem. Soc., 76:1119 (1954).
105. N. C. Yung, J. H. Burchenol, R. Fecher, R. Duschinsky, and J. J. Fox, J. Am. Chem. Soc., 83:4060 (1961).
106. J. J. Fox, N. Miller, and J. Wempen, J. Med. Chem., 9:101 (1966).
107. I. L. Doerr and J. J. Fox, J. Org. Chem., 32:1462 (1967).
108. J. J. Fox, D. Van Praag, J. Wempen, I. L. Doerr, L. Cheong, J. E. Knoll, M. L. Eidinof, A. Bendich, and G. B. Brown, J. Am. Chem. Soc., 81:178 (1959).
109. J.-P. Mathieu and A. Allais, Principes de Synthèse Organique, Masson, Paris (1957).
110. F. Windmueller and N. O. Kaplan, J. Biol. Chem., 236:2716 (1961).
111. S. F. Mason, The Chemistry and Biology of Purines. A Ciba Foundation Symposium, London (1957), p. 66; J. Chem. Soc., 2071 (1954).
112. R. W. Taft, in: Steric Effects in Organic Chemistry, M. S. Newman (editor), Wiley, New York (1956).
113. V. A. Pal'm, Bases of the Quantitative Theory of Organic Reactions [in Russian], Khimiya (1967).
114. P. R. Wells, Chem. Rev., 63:171 (1963).
115. H. H. Jaffe and H. L. Jones, in: Advances in Heterocyclic Chemistry, A. R. Katritzky (editor), Academic Press, New York – London (1964), p. 209.
116. J. J. Fox, N. Yung, J. Davall, and G. B. Brown, J. Am. Chem. Soc., 78:2117 (1956).
117. J. F. Codington, J. J. Doerr, and J. J. Fox, J. Org. Chem., 29:558 (1964).
118. P. O. P. Ts'o, S. A. Rappaport, and F. J. Bollum, Biochemistry, 5:4153 (1966).
119. C. Janion and D. Shugar, Acta Biochim. Polon., 7:309 (1960).

120. J. Ono, R. G. Wilson, and L. Grossman, J. Mol. Biol., 11:600 (1965).

121. H. E. Johns, J. C. LeBlanc, and K. B. Freeman, J. Mol. Biol., 13:849 (1965).

122. R. Philips, P. Eisenberg, P. George, and R. Rutman, J. Biol. Chem., 240:4393 (1965).

123. J. T. Edsall and J. Wyman, Biophysical Chemistry, Vol. 1, Academic Press, New York (1958), p. 441.

124. A. R. Katritzky (editor), Physical Methods in Heterocyclic Chemistry, Academic Press, New York (1963).

125. N. N. Aylward, J. Chem. Soc., B, 401 (1967).

126. S. Lewin and D. A. Humphreys, J. Chem. Soc., B, 210 (1966).

127. B. I. Sukhorukov, V. I. Poltev, and L. A. Blyumenfel'd, Dokl. Akad. Nauk SSSR, 149:1380 (1963).

128. T. L. V. Ulbricht, Comprehensive Biochemistry, 8:199 (1963).

129. D. O. Jordan, The Chemistry of the Nucleic Acids, Butterworth, London (1960), p. 134.

130. P. A. Levene and H. S. Simms, J. Biol. Chem., 65:519 (1925).

131. L. G. Bunvill and S. J. Schwalbe, Biochemistry, 5:3521 (1966).

132. M. Rawitscher and J. M. Sturtevant, J. Am. Chem. Soc., 82:3739 (1960).

133. J. Stockx and L. Vandendriessche, Biochim. Biophys. Acta, 72:137 (1963).

134. B. Pullman and A. Pullman, Quantum Biochemistry, Wiley, New York (1963).

135. A. Streitwieser (Jr.), Molecular Orbital Theory for Organic Chemists, Wiley, New York (1961).

136. T. Fueno, Ann. Rev. Phys. Chem., 12:303 (1961).

137. É. I. Budovskii, E. D. Sverdlov, R. P. Shibaeva, G. S. Monastyrskaya, and N. K. Kochetkov, Molekul. Biol., 2:329 (1968).

138. W. C. Coburn, M. C. Thorpe, J. A. Montgomery, and K. Hewson, J. Org. Chem., 30:1110 (1965).

139. W. C. Coburn, M. C. Thorpe, J. A. Montgomery, and K. Hewson, J. Org. Chem., 30:1114 (1965).

140. J. Krupička and J. Gut, Coll. Czech. Chem. Comm., 27:546 (1962).

141. R. W. Chambers, Biochemistry, 4:219 (1965).

142. H. Hayatsu, K.-I. Takeishi, and T. Ukita, Biochim. Biophys. Acta, 123:445 (1966).

143. R. V. Wolfenden, J. Mol. Biol., 40:307 (1969).

Chapter 4
The Secondary Structure of
Nucleic Acids

I. Introduction

The properties of macromolecules of nucleic acids are largely determined by interaction between their components: phosphate residues, carbohydrates, and heterocyclic bases. The properties of the molecule which are determined by the presence of a large number of negatively charged phosphate residues are essentially indistinguishable from the properties of other polymer anions, and as far as this is possible at the present time they can be described in terms of electrostatic concepts. The specific properties of oligonucleotides and polynucleotides are determined principally by interaction between the bases. This interaction may be of two types:

1) interaction between the planes of rings of neighbouring bases in the same chain (interplanar, vertical, and longitudinal interaction); this takes place in both single-stranded and double-stranded molecules;

2) interaction between bases located in the same plane but in different chains (base pairing); this includes, in particular, hydrogen bonding.

In this chapter we shall examine the various types of interactions, briefly consider the theories put forward to explain them, and review the experimental facts confirming the theories. Interactions between bases will be considered separately; effects associated with interaction between charges on phosphate groups will be considered in the sections devoted to specific features of the secondary structure of nucleic acids.

---

* See note on page 4.

II. General aspects of interaction between the bases of nucleic acids

1. Pairing of complementary bases

The Watson – Crick hypothesis of the double-helical structure of DNA, according to which the bases are paired (adenine with thymine and guanine with cytosine) in two complementary chains, was put forward in 1953. Since that time considerable effort has been made, in both theoretical and experimental directions, to prove that base pairing in fact takes place in this way and not otherwise, to verify its specificity, and to establish the structure of the base pairs or, in other words, to determine how the atoms in the pairs are linked together.

Watson and Crick put forward their concept of specific pairing on the basis of the then available data for the nucleotide composition of different forms of DNA. These data indicated that the adenine:thymine and guanine:cytosine ratios were near to unity. Watson and Crick suggested that the most probable scheme of base pairing was as follows:

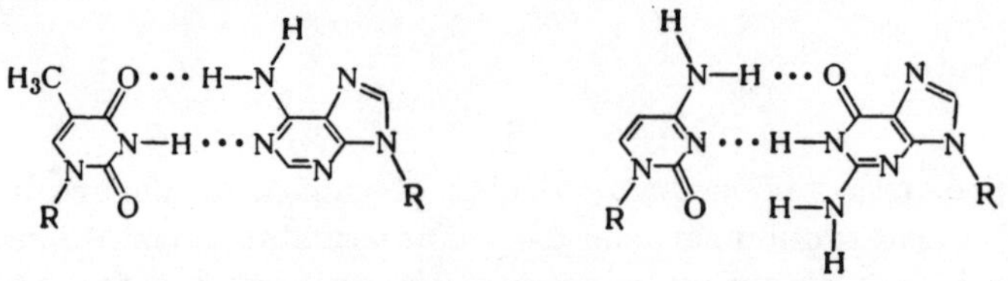

R denotes deoxyribose residue

The specificity of base pairing as suggested by Watson and Crick was subsequently verified experimentally. This experimental evidence will be briefly reviewed in this chapter insofar as it relates to nucleotides, nucleosides, and bases. The additional evidence of the specificity of interaction obtained at the polynucleotide level will be discussed separately.

The pairing of complementary bases is a weak type of interaction and the bond is easily broken by the action of heat or other agents which are themselves capable of hydrogen bonding, notably water. Investigations of the specificity of interaction between bases are therefore carried out as a rule in organic solvents such as carbon tetrachloride, chloroform, or dimethyl sulphoxide, or in crystalline samples obtained from a solution of the test mixture. The principal methods of investigation are IR and NMR spectroscopy and x-ray structural analysis.

Experimental Studies of the Specificity of Base Pairing

IR Spectra. A band at 3392 cm^{-1}, corresponding to N – H valence vibrations, is observed in the IR-spectrum of a dilute (0.022 M) solution of 1-cyclohexyluracil in chloroform [2], but the presence of hydrogen bonds can be

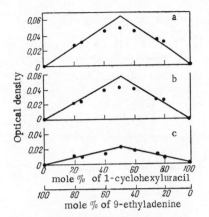

Fig. 4.1. Optical density of bands at 3330 cm^{-1} (a), 3490 cm^{-1} (b), and 3260 cm^{-1} (c), formed during mixing of 1-cyclohexyluracil and 9-ethyladenine in deuterochloroform, as a function of the composition of the mixture [2]. Total concentration of components 0.022 M. Length of optical path 1 mm; 20°C.

demonstrated only in much higher concentrations *; this follows from the appearance of bands at 3210, 3110, and 3050 cm^{-1}. In the same way, strong maxima corresponding to symmetrical (3416 cm^{-1}) and antisymmetrical (3527 cm^{-1}) valence vibrations of the NH$_2$ group, but only weak bands corresponding to the formation of hydrogen-bonded self-associates (3482 and 3312 cm^{-1}), are observed in the spectrum of a 0.022 M solution of 9-ethyladenine. The intensity of the last two bands is increased in a saturated solution.

In a mixture of 1-cyclohexyluracil and 9-ethyladenine, with a total concentration of 0.022 M, relatively strong bands are observed at 3490 and 3330 cm^{-1}, together with a weak band at 3260 cm^{-1}. This is evidence of hydrogen bonding to a much higher degree than in solutions of the individual compounds. Investigation of the intensity of these new bands as a function of the composition of the mixture, the total molar concentration of the components being constant, leads to the conclusion that the composition of the resulting complex corresponds to the ratio 1:1 (Fig. 4.1).

Similar results have been obtained [4] by the study of interaction between 2',3'-O-benzylidene-5'-O-trityl derivatives of guanosine and cytidine in CDCl$_3$. In this case, however, a considerable degree of self-association of the guanosine derivative is found; nevertheless, complex formation with the cytidine derivative takes place to a more marked degree. Investigation of the composition of the resulting complex under the same conditions as above (characteristic frequencies 3488 and 3300 cm^{-1}) gives a stoichiometric ratio of 1:1 between the components.

The formation of these complexes is an equilibrium reaction, and with a decrease in concentration of the mixture equilibrium is shifted toward the original components. If the total concentration of components is 1.5·10^{-4} mole/liter, no complex formation takes place between 9-ethyladenine and 2',3',5'-O-triacetyluridine or between 9-ethyladenine and 3',5'-O- diacetylthymidine in CDCl$_3$. However, within this same concentration range, 2',3', 5'-O-triacetylguanosine forms 1:1 complexes with 1-methylcytosine, dimethyl-5-azacytosine, and 2',3',5'-O-tribenzoyl-6-azacytidine [5]. The reason for this difference is the higher energy of interaction between the guanine·cytosine pair than between the adenine·uracil pair (see below).

_____
* Hydrogen bonding between uracil molecules is also observed in the crystalline state, as shown by the results of x-ray structural analysis of 1-methylthymine crystals [3].

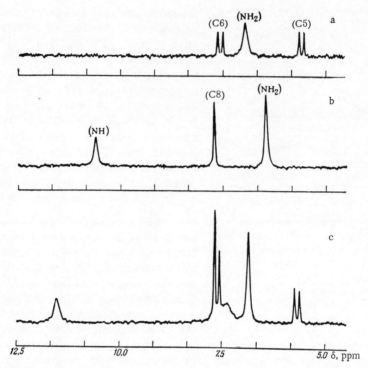

Fig. 4.2. Proton magnetic resonance spectra of 1-methylcytosine (a), 9-ethyl-guanine (b), and an equimolar mixture of the two (c) in completely deuterated dimethyl sulphoxide. Total concentration of components in all cases 0.2 mole/litre; 20°C [7].

The formation of hydrogen-bonded complexes was not found in solutions of mixtures of 9-ethyladenine with 2',3'-O-benzylidene-5'-O-tritylcytidine or guanosine in deuterochloroform when the total concentrations of the components were 0.0016 and 0.0080 M, nor in mixtures of 2',3'-O-ben-zylidene-5'-O-tritylguanosine with cyclohexyluracil under the same conditions and with the same total concentrations [4].

Derivatives of the bases of nucleic acids thus interact specifically with each other to form hypothetically hydrogen-bonded adenine·uracil and guanine·cytosine pairs*.

NMR Spectra. In NMR spectra of different compounds, the signal from protons taking part in hydrogen bonding is shifted toward weaker fields by comparison with nonbonded protons [6]. The use of this rule to study the specificity of hydrogen-bonded pairing between the bases of nucleic acids [7-9] confirms the conclusions drawn from the results of IR-spectroscopy.

As an example, the NMR spectra of 1-methylcytosine, 9-ethylguanine, and an equimolar mixture of them in dimethyl sulphoxide are compared in

---

* Specific hydrogen bonding was also observed between 2',3'-O-benzylidene-5'-O-tritylinosine and 2',3'-O-benzylidene-5'-O-tritylcytidine in chloroform solutions [317].

TABLE 4.1. Specificity of Hydrogen-Bonded Base Pairing as Shown by NMR Data

Base pairs	Concentration of bases, mole/litre	Temperature, °C	Chemical shift of protons of groups compared with single base, ppm				
			NH	NH$_2$	NH$_2$ of 1-methyl-cytosine	NH of 1-methyl-thymine	NH$_2$ of 9-ethyl-adenine
			of 9-ethyl-guanine				
9-Ethylguanine + 1-methyl-cytosine	0.1 + 0.1	40	0.5	0.24	0.29	—	—
	0.1 + 0.1	20	0.71	0.34	0.36	—	—
	0.2 + 0.2	40	0.7	0.34	0.46	—	—
	0.2 + 0.2	20	0.94	0.41	0.43	—	—
9-Ethyladenine + 1-methyl-thymine	0.1 + 0.1	20	—	—	—	0.05	0.02
	0.2 + 0.2	20	—	—	—	0.08	0.02
9-Ethylguanine + 1-methyl-thymine	0.2 + 0.2	20	0.00	0.00	—	0.00	—
9-Ethylguanine + 9-ethyl-adenine	0.2 + 0.2	20	0.02	0.00	—	—	0.00

Fig. 4.2.   The data represented in this figure confirm that protons of amino groups of cytosine and guanine and of the imino group of guanine participate in hydrogen bonding. Similar changes in the NMR spectra compared with spectra of the individual bases take place if 1-methylthymine and 9-ethyladenine, and 1-methylthymine and 2-aminoadenosine are mixed in this same solvent, whereas if nonspecific pairs are mixed, no changes take place in the spectra, and the spectrum of the mixture corresponds to the combined spectra of the individual bases in corresponding concentrations [7, 8] (Table 4.1).

Examination of Table 4.1 shows that the change in the chemical shift during mixing of solutions of the corresponding bases is much less in the case of pairing of 9-ethyladenine + 1-methylthymine than in the case of pairing of 9-ethylguanine + 1-methylcytosine. This indicates weaker interaction between the bases in the first pair.

The same conclusion can be drawn, but even more clearly, from NMR spectra of mixtures of nucleosides and their analogues in different solvents [8]. For instance, whereas guanosine and cytidine interact strongly in dimethyl sulphoxide, interaction between 9-ethyladenine and 1-cyclohexyluracil is observed only in nonpolar solvents such as deuterochloroform [8]. Nevertheless, even in this case specificity of interaction is preserved. In high concentrations, very slight self-association of adenine and guanine derivatives also is observed.

It is also clear from Table 4.1 that, with an increase in the concentration of the components and a decrease of temperature [7, 8], interaction between the bases is intensified.

Hence, it can be concluded from the results of IR- and NMR spectroscopy that strict specificity is observed during base pairing in full agreement with the Watson – Crick hypothesis. However, conclusions regarding the structure of the resulting pairs can be drawn from these results only in the case of the guanine·cytosine pair, for which both methods show [5, 7, 8] that the amino group of guanine and cytosine and the imino group of guanine take part in hydrogen bonding. This can happen only if the structure of the pair corresponds to that given below, as suggested by Pauling [10].

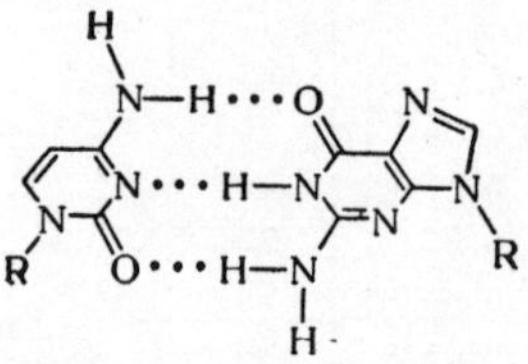

R denotes various radicals

The structure of base pairs can be determined more exactly by x-ray structural analysis, but in this case it must be remembered that the effect of packing forces in the crystal may cause changes from the structure of the same pair in solution.

X-Ray Structural Analysis [11].   Although no unequivocal conclusions regarding the structure of adenine·thymine and adenine·uracil pairs (or pairs of their derivatives) can be deduced from the results of IR- and NMR spectroscopy, this can be done with the aid of x-ray structural analysis.

After Hoogsteen

After Haschemeyer ("reverse" of Hoogsteen's structure)

After Watson and Crick

"Reverse" of Watson and Crick's structure

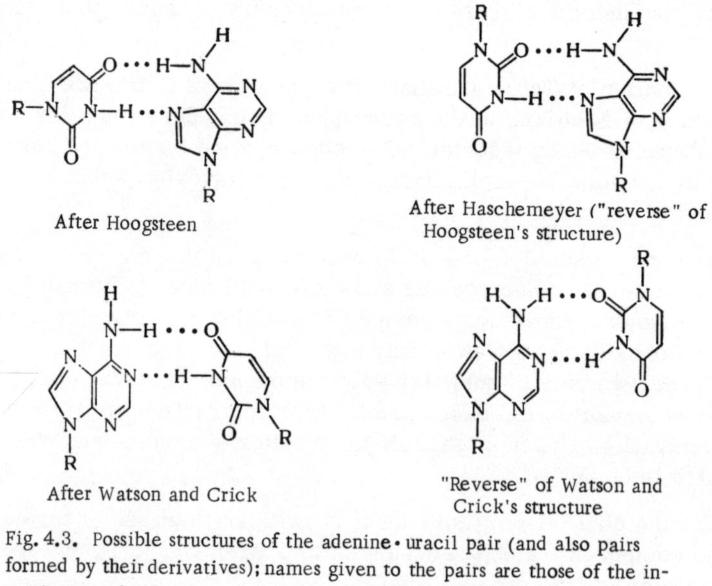

Fig. 4.3. Possible structures of the adenine·uracil pair (and also pairs formed by their derivatives); names given to the pairs are those of the investigators who discovered their structure: see Table 4.2.

TABLE 4.2. Formation of Hydrogen-Bonded Complexes between Base Pairs as Shown by x-Ray Structural Analysis

Components of complex	Atomic groups participating in hydrogen bonding		Length of hydrogen bond, Å	Type of complex formed	Literature cited
	in purine component	in pyrimidine component			
9-Methyladenine + 1-methylthymine	6-exo-NH_2 N7	4-exo-O H at N3	2.85 2.92	After Hoogsteen	[12]
9-Ethyladenine + 1-methyluracil	6-exo-NH_2 N7	4-exo-O H at N3	2.98 2.83	After Hoogsteen	[13]
9-Methyladenine + 1-methyl-5-bromo-uracil	6-exo-NH_2 N7	4-exo-O H at N3	2.98 2.86	After Hoogsteen	[14]
9-Ethyladenine + 1-methyl-5-fluorouracil	6-exo-NH_2 N7	4-exo-O H at N3	2.96 2.78	After Hoogsteen	[15]
Adenosine + 5-bromo-uridine	6-exo-NH_2 N7	2-exo-O H at N3	3.10 2.80	After Hasche-meyer	[16]
9-Ethyladenine + 1-methyl-5-bromo-uracil	6-exo-NH_2 N7	2-exo-O H at N3	3.04 2.80	After Hasche-meyer	[17]
9-Ethylguanine + 1-methylcytosine	6-exo-O H at N1 2-exo-NH_2	4-exo-NH_2 N3 2-exo-O	2.93 2.91 2.82	After Watson and Crick	[18, 19]
9-Ethylguanine + 1-methyl-5-bromo-cytosine	6-exo-O H at N1 2-exo-NH_2	4-exo-NH_2 N3 2-exo-O	2.86 2.95 2.91	After Watson and Crick	[20]
2'-Deoxyguanosine + 5-bromodeoxycyti-dine	6-exo-O H at N1 2-exo-NH_2	4-exo-NH_2 N3 2-exo-O	2.82 2.91 2.78	After Watson and Crick	[21]
9-Ethylguanine + 1-methyl-5-fluoro-cytosine	6-exo-O H at N1 2-exo-NH_2	4-exo-NH_2 N3 2-exo-O	2.96 2.94 2.82	After Watson and Crick	[19, 22]

In principle, four "isomeric" adenine·thymine or adenine·uracil pairs can exist (Fig. 4.3).

The x-ray structural data given in Table 4.2 show that by interaction between derivatives of guanine and cytosine in crystals, pair formation takes place in accordance with the structure suggested by Pauling [10]. The structure of the adenine·uracil pair (or, more precisely, the pair formed from derivatives of these bases) never corresponds to the Watson and Crick structure, but corresponds either to Hoogsteen's model or to Haschemeyer's model (depending on the substituents in the nuclei of the interacting bases).

It is still not quite clear whether the reasons for this scheme of pairing are concerned with the energetics of interaction between the bases or with the demands of crystalline packing.

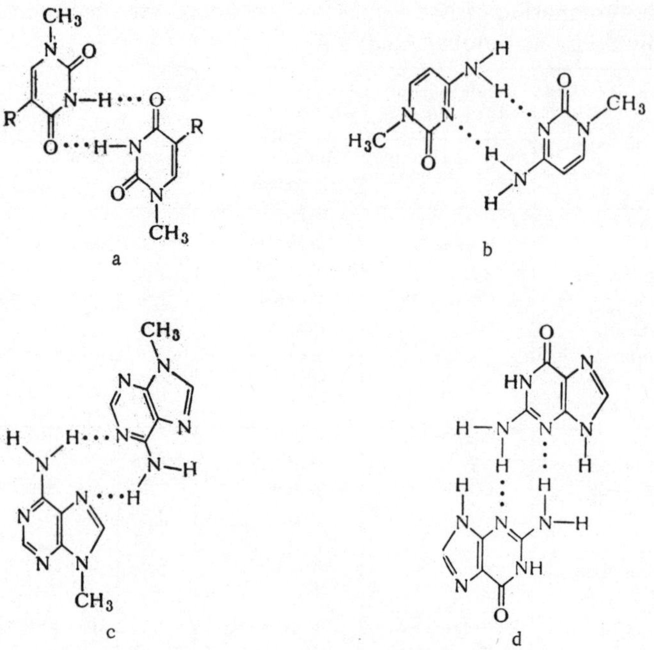

Fig. 4.4. Structures of self-associative dimers of derivatives of bases in a crystalline state: a) self-associative dimers of 1-methylthymine (R = CH₃) and 1-methyluracil (R = H) [23]; b) of 1-methylcytosine [24]; c) of 9-methyladenine [25]; d) dimer of guanine (from results of x-ray analysis of guaninium chloride crystals) [26].

Considerable light on the nature of the forces stabilizing base pairs is also shed by the structure of their self-associates, because base pairing is a reaction in competition with self-association. The structures of hydrogen-bonded dimers formed by crystallization of some derivatives of the bases of nucleic acids, deduced from x-ray structural data [23-27], are given in Fig. 4.4 (for a discussion of this question, see [27]). In every case, the resulting dimers are linked by two hydrogen bonds. In the case of complexes of uracil and adenine, this state of affairs evidently still holds good for solutions in deuterochloroform also, as is apparent from the thermodynamic characteristics obtained by IR-spectroscopy [28] (although the structure of the complexes is not proved in the work cited).

It thus follows from the qualitative data that in all cases investigated the formation of base pairs is more advantageous than the formation of self-associates. This conclusion is confirmed by quantitative investigations of the association constants.

Quantitative Characteristics of Interaction between Bases during the Formation of Hydrogen-Bonded Pairs. Values of the self-association constants for several compounds are given in Tables 4.3 and 4.4. It will readily be seen that the base pairing constants are much higher than the self-association constants.

TABLE 4.3. Self-Association Constants of Several Derivatives of 1-Cyclohexyluracil and 9-Ethyladenine [29] (from IR-spectra in deuterochloroform at 25°C)

Derivative of 1-cyclohexyluracil	K_{self}, litres /mole	Derivative of 9-ethyladenine	K_{self}, litres /mole
3-Methyl-.	0	6-exo-N,N-Dimethyl-.	0
5,6-Dihydro-	2.9	6-exo-N-Methyl-.	1.5
4-Thio-	2.7	Unsubstituted.	3.1
Unsubstituted	6.1	6-Deamino-2-amino-(2-aminopurine).	2.0
5-Methyl-.	3.2	2-Amino-.	11
5-Bromo-	4.1	8-Bromo-	120
5-Iodo-.	5.7		

Calorimetric determinations give a value of 92 litres/mole for the association constant between 1-cyclohexyluracil and 9-ethyladenine, whereas the self-association constants of 1-cyclohexyluracil and of 9-ethyladenine are 19 and 13 litres/mole respectively [30]. Lower values of the self-association constants compared with the constants of base pairing were also obtained [31] by investigation of base pairing between 2',3'-O-isopropylidene-5'-O-trityl derivatives of adenosine and uridine in CCl_4. In this case, however, the values of the constants are approximately ten times higher than in $CDCl_3$, evidently because of the lower polarity of CCl_4 and its lower ability to form hydrogen bonds with the substances in solution.

Interaction during the formation of the guanine·cytosine pair is considerably stronger (as is also clear from the qualitative data). The approximate value of the complex formation constant between 1-methylcytosine and 2',3',5'-O-triacetylguanosine [5] is 10^5 litres/mole in $CDCl_3$, which is about a thousand times higher than the corresponding constant for the adenine·uracil pair.

Strong association is also observed between derivatives of hypoxanthine and cytosine. The association constant at 25°C in chloroform for 2',3'-benzylidene-5'-O-tritylinosine and 2',3'-benzylidene-5'-O-tritylcytidine, according to the results of IR-spectroscopy [317], is $2·10^3$ litres/mole.

The following rules for the effect of substituents on the base pairing constant can be formulated.

1. Introduction of methyl substituents at N3 in the uracil molecule or at the exocyclic amino group of adenine prevents pairing; this means that in unsubstituted compounds the protons at N3 participate in hydrogen bonding.

2. Saturation of the C5 − C6 double bond in the pyrimidine ring sharply reduces the stability of the complex, possibly because of a decrease in the positive character of the hydrogen at N3 (as is shown by the higher value of pK_a for dihydrouracil than uracil; see page 159) and with a corresponding decrease in the tendency toward hydrogen bonding.

TABLE 4.4. Complex Formation Constants between Derivatives of 1-Cyclo-hexyluracil and 9-Ethyladenine [29] (from IR-spectra in deuterochloroform at 25°C)

Pyrimidine base	K_{comp}, litres/mole					
	with 6-exo-N,N-dimethyl-9-ethyl-adenine	with 6-exo-N-methyl-9-ethyl-adenine	with 9-ethyl-adenine	with 2-amino-9-ethylpurine	with 2-amino-9-ethyladenine	with 8-bromo-9-ethyladenine
3-Methyl-1-cyclohexyl-uracil.............	—	—	<1	—	—	—
1-Cyclohexyl-5,6-dihydro-uracil............	—	15	30	10	100	—
1-Cyclohexyluracil-4-thiouracil..........	—	—	90	—	—	—
1-Cyclohexyluracil.....	1.5	50	100 (150 ± 50 [24])	45	170 (300 ± 100 [24])	140
1-Cyclohexylthymine ...	—	70	130	56	210	—
5-Bromo-1-cyclohexyluracil	—	100	240 (350 ± 75 [24])	75	550 (600 ± 150 [24])	—
5-Iodo-1-cyclohexyluracil	—	—	220	—	—	—

3. No definite rules can be observed regarding the effect on stability of the complex of substitution in position 5 of the uracil residue. Both electron-donor (CH$_3$) and electron-acceptor substituents (Br, I), for instance, increase the stability of complexes of substituted uracils compared with that of unsubstituted.

It must be specially emphasized that no direct relationship exists between the number of hydrogen bonds and the stability of the complex. Despite the fact that the number of hydrogen bonds in dimeric self-associates of uracil and adenine and in the adenine·uracil complex is the same, the base pair is much more stable than the self-associates. In the same way, although the number of hydrogen bonds is the same in a complex of 2-aminopurine derivatives with uracil derivatives and in the corresponding complexes of adenine (6-aminopurine), the latter are twice as stable. Finally, despite the results of IR-spectroscopy [29, 32] and NMR spectroscopy [7], indicating that three hydrogen bonds are formed by interaction between 2,6-diaminopurine and uracil, the stability of the resulting complex is much lower than the stability of the guanine·cytosine complex. Hence, besides the number of hydrogen bonds, other forces are evidently concerned in the stabilization of the complexes. It is now generally accepted that these are Van der Waals – London interactions between bases located in the same plane.

Theoretical Examination of the Problem of Stability

of Hydrogen-Bonded Pairs

The question which naturally arises after examination of the experimental data is this: why is it that, although they have the same number of hydrogen bonds, some associates are more stable than others? A satisfactory answer can be obtained in terms of modern theoretical concepts of the role of Van der Waals — London forces in the stabilization of complexes formed by hydrogen bonds. At the present time the energy of interaction taking place through these forces between bases can be estimated in two ways. The first way [33] regards the interacting bases as two dipoles, and accordingly, the term Van der Waals — London forces implies the total effect produced by dipole — dipole interactions, interactions between the dipole of one base and the induced dipole of the other, and by dispersion (London) interactions, i. e., fluctuation-dipole — induced-dipole interactions. An additional contribution to the energy of interaction is made by hydrogen bonds which can be considered independently. This approach can give only a very rough assessment, because the bases in a complementary pair are too close together to be regarded as two independent dipoles.

The other approach — to regard interaction between bases in a pair as the combined effect of interaction between individual atoms of one base with the atoms of the other — is evidently more correct. In this case, by Van der Waals — London interactions are meant the combined effect of Coulomb interactions between the net charges on the atoms of the two bases (the corresponding forces are designated $F_{\rho\rho}$), interactions between net charges on the atoms of one base, and dipoles induced by them in the other base ($F_{\rho\alpha}$) and, finally, dispersion forces of interaction (London forces F_L) between the fluctuation-induced dipole on one base and dipoles induced by it on the other. The total energy of interaction between the bases E_m can thus be expressed as the sum:

$$E_m = E_{\rho\rho} + E_{\rho a} + E_L$$

Each of these terms in the energetic equation can be estimated independently. The greatest contribution to the total is made by the term $E_{\rho\rho}$, which is calculated from the net σ and π charges between the atoms of the bases, estimated by quantum-chemical methods (see Chapter 3).

Values of the energy of interaction between various bases, calculated for experimentally determined configurations of the corresponding self-associates or pairs, and also for some other probable (as regards the possibility of hydrogen bonding) configurations, are given in Table 4.5. Calculations of this type have correctly predicted that the stability of adenine·thymine pairs with a Hoogsteen or Haschemeyer structure (the second and third pairs in Table 4.5) is higher than that of the pair with a Watson and Crick structure (the first pair in that Table), and higher than the stability of any of the dimer self-associates of adenine and thymine. The much higher stability of the

TABLE 4.5. Energetic Characteristics of Interaction between Base Pairs and Self-Associates [27]

Interacting bases	Atomic groups participating in hydrogen bonding		$E_{\rho\rho}$	$E_{\rho\alpha}$	E_L	E_m
	in purine component	in pyrimidine component	kcal/mole			
Adenine + thymine	N1 6-exo-NH$_2$	H at N3 4-exo-O	−4.6	−0.2	−0.7	−5.5
	N7 6-exo-NH$_2$	H at N3 4-exo-O	−5.9	−0.2	−0.9	−7.0
	N7 6-exo-NH$_2$	H at N3 2-exo-O	−5.6	0.15	−0.9	−6.65
Guanine + cytosine	2-exo-NH$_2$ H at N1 6-exo-O	2-exo-O N3 4-exo-NH$_2$	−15.91	−2.02	−1.25	−19.18
Adenine + adenine	6-exo-NH$_2$ N1	N1 6-exo-NH$_2$	−5.23	−0.11	−0.45	−5.79
Guanine + guanine	H at N1 6-exo-O	6-exo-O H at N1	−13.37	−0.62	−0.53	−14.52
	2-exo-NH$_2$ N7	N7 2-exo-NH$_2$	−5.79	−0.73	−0.63	−7.15
Thymine + thymine	H at N3 4-exo-O	4-exo-O H at N3	−3.62	−0.38	−1.19	−5.19
	H at N3 2-exo-O	2-exo-O H at N3	−2.61	−0.15	−1.10	−3.86
Cytosine + cytosine	N3 4-exo-NH$_2$	4-exo-NH$_2$ N3	−10.65	−1.09	−1.23	−12.97
Adenine + cytosine	6-exo-NH$_2$ N7	N3 4-exo-NH$_2$	−6.20	−0.59	−0.96	−7.75
Guanine + thymine	H at N1 6-exo-O	2-exo-O H at N3	−4.41	−0.50	−0.58	−5.49
	H at N1 6-exo-O	4-exo-O H at N3	−6.24	−0.58	−0.58	−7.40

guanine·cytosine pair than of the adenine·thymine pair or of any of the self-associative pairs of guanine and cytosine, has also been correctly predicted. This means that adenine·thymine and guanine·cytosine pairs are formed from monomer units to a greater degree than self-associates.

The energy of guanine·thymine pairs (−7.40 kcal/mole) and adenine ·cytosine pairs (−7.75 kcal/mole) is lower than the energy of formation of self-associates of guanosine (−14.52 kcal/mole) and of cytosine (−12.97 kcal /mole), i.e., the formation of heterogeneous pairs in this case is less advantageous than the formation of self-associates. Calculations also have as a rule correctly predicted the structure of self-associates of bases in the crystalline state. Similar results, but allowing only for electrostatic interaction, have also been obtained in other investigations [34, 35].

This calculation applies to fixed configurations of base pairs and does not rule out the probability that, from the point of view of energy of Van der Waals interactions, there may be other, more advantageous configurations. It is therefore interesting to study the change in the energy of interaction during a change in the relative position of the bases [35]. This can be done by fixing the position of one of the bases, and then altering the position of the other as much as possible, stipulating only one limiting condition: the atoms of the bases must not be brought closer together than a distance representing the sum of their Van der Waals radii. The calculated energies of Coulomb interaction, expressed as functions of the mutual arrangement of the bases, have clearly defined maxima corresponding to the configurations of hydrogen-bonded pairs. It must be pointed out that in calculations of this type no allowance is made for the formation of hydrogen bonds between atoms, and electrostatic interactions only are taken into account. Nevertheless, the most stable configurations thus obtained correspond to the traditional configurations of hydrogen-bonded base pairs.

These calculations do not take into account effects due to the solvent. In addition, the combined σ and π electron densities, calculated quantum-chemically, are used in them as initial data, and as was shown in the previous chapter, they are not very accurate*. However, the close agreement between the experimental data suggests that the principal stabilizing forces in the interaction between complementary bases are in fact Van der Waals — London forces.

III. Characteristics of interaction between the bases of nucleic acids and their derivatives in aqueous solutions

Whereas bases, nucleosides, and nucleotides in nonaqueous solutions and in the crystalline state associate by hydrogen bonding, in aqueous solutions no interaction is observed between monomer constituents of nucleic acids (or their analogues). This is apparent from the study of spectra of Raman scattering of mixtures of complementary nucleosides at a concentration of 1.0 M [36]. Nevertheless, powerful association of a vertical type, in which bases located above each other in parallel planes form "stacks" (interplanar interaction), is observed for monomers in aqueous solutions. The experimental evidence will be examined below and the theoretical interpretation of this type of interaction will be briefly discussed.

---

* Nevertheless, calculations of interaction between bases employing different distributions of electron densities (obtained by different workers using different methods) give qualitatively similar pictures of the dependence of the energy of interaction on the mutual arrangement of the bases [35].

TABLE 4.6. Self-Association Constants of Bases and Nucleosides (from changes in vapour pressure of aqueous solutions at 25°C)

Compound	K_{self}, mole^{-1}	Literature cited
Inosine	3.0	[40]
	1.8*	[38]
1-Methylinosine	1.8-2.0	[38]
Purine	2.1	[37]
Ribosylpurine	1.9	[38]
	3.5†	[40]
Adenosine	4.5	[38]
2'-O-Methyladenosine	5.1	[38]
2'-Deoxyadenosine	4.7-7.5	[38]
	12	[40]
6-Methylpurine	6.7	[39]
6-exo-N-Methyladenosine	11.8-14.9	[38]
6-exo-N-Methyl-2'-deoxyadenosine	15.9	[38]
6-exo-N,N-Dimethyladenosine	22.2	[38]
Uridine	0.61	[38]
	0.70	[40]
Cytidine	0.87	[38]
5-Bromouridine	1.0	[39]
Thymidine	0.91‡	[40]
2'-Deoxycytidine	0.91	[40]
9-Methylpurine	1.8	[50]
9-Ethylpurine	2.06	[50]
9-Isopropylpurine	2.28	[50]
9-tert-Butylpurine	2.49	[50]
6-tert-Butylpurine	8.72	[50]
2-tert-Butylpurine	9.18	[50]
8-Isopropylpurine	4.37	[50]

* K_{self} has also been determined [40] from the change in sedimentation constant as a function of concentration during ultracentrifugation, giving a value of 2.0.

† K_{self}, determined [40] from the sedimentation coefficient, was 1.7.

‡ K_{self}, determined [40] from the sedimentation coefficient, was 1.2.

1. Association and self-association of bases, nucleosides, and nucleotides

One of the first investigations which demonstrated the existence of associates of bases and nucleotides was the study of the vapour pressure of a solution with changes in the concentration of purine, uridine, and cytidine [37]. Equilibrium association constants were determined from these findings, and the results showed that a series of associates containing different numbers of monomers (in accordance with the equilibrium series) can be formed. The equilibrium association constants obtained in that and in subsequent studies of the same type [38-40] are given in Table 4.6. The following conclusions can be drawn from a comparison of their values.

1. Association is not induced by hydrogen bonds, because the changes observed in the association constants in the series

2'-deoxyadenosine < 6-exo-N-methyl-2'-deoxyadenosine
adenosine < 6-exo-N-methyladenosine < 6-exo-N,N-dimethyladenosine
inosine < 1-methylinosine

are opposite to those which would be observed following substitution of protons capable of hydrogen bonding by an alkyl group.

2. The sugar residue has no significant effect on the magnitude of the interaction, as is shown by comparison of the equilibrium constants of purine, ribosylpurine, and also ribo-, deoxyribo-, and 2'-O-methylribo-derivatives, so that association is primarily a specific property of bases.

3. Introduction of a methyl substituent or halogen into the heterocyclic ring increases the association constant.

Association also takes place in solutions of mixtures of nucleosides, but the degree of interaction between different bases cannot be estimated from changes in vapour pressure [40]. Association constants between different bases can be determined from the change in solubility of some bases in the presence of others. Values obtained in this way (at 25.5°C) are given below:

Interacting pair	K_{assoc}, mole^{-1}	Interacting pair	K_{assoc}, mole^{-1}
Adenine + purine	9.3	Adenine + phenol	2.3−3,1
Adenine + cytosine	4.6−5.0	Thymine + purine	1.6−2,3
Adenine + uridine	4.3−4.9	Thymine + uridine	1.1−1,2
Adenine + pyrimidine	1.8−2.3	Thymine + pyrimidine	0.8−0,9

These results show that when the bases differ, hydrogen bonding is unlikely to play a significant role, since the interaction constant between the complementary pair adenine + uracil, for example, does not exceed the self-association constant of adenine, which can be estimated from the interaction constant of the pair adenine + purine.

Comparison of the data for association between different bases and the data for self-association (Table 4.6) given above indicates that the degree of association in aqueous solutions diminishes in the series:

$$\begin{array}{ccccc} \text{Purine} & & \text{Purine} & & \text{Pyrimidine} \\ \text{base} & & \text{base} & & \text{base} \\ + & > & + & > & + \\ \text{Purine} & & \text{Pyrimidine} & & \text{Pyrimidine} \\ \text{base} & & \text{base} & & \text{base} \end{array}$$

and, in particular

$$\begin{array}{ccccccc} \text{Adenine} & & \text{Thymine} & & \text{Cytosine} & & \text{Uracil} \\ + & > & + & \geqslant & + & > & + \\ \text{Adenine} & & \text{Thymine} & & \text{Cytosine} & & \text{Uracil} \end{array}$$

In the case of guanine, self-association is evidently stronger than for the other bases, as is shown by the marked differences between the properties even of dilute solutions of deoxyguanosine from the ideal [40]. Introduction

TABLE 4.7. Thermodynamic Constants of Self-Association of some Purine and Pyrimidine Derivatives (in water at 25°C)

Compound	ΔH, kcal/mole	Literature cited	ΔS, e.u.	Literature cited	ΔF, kcal /mole	Literature cited
Purine	-4.2 ± 0.2	[42]	-13	42	-0.44	[39]
6-Methylpurine	-6.0 ± 0.4	[42]	-16	42	-1.12	[39]
Purine riboside	-2.5 ± 0.1	[43]	-7	43	-0.380	[38]
Deoxyadenosine	-3.7 ± 0.6	[43]	-7	43	-1.5	[40]
Cytidine	-2.8 ± 0.1	[43]	-10	43	0.080	[37]
Uridine	-2.7 ± 0.1	[43]	-10	43	0.290	[37]

of a phosphate residue does not prevent self-association, as is shown, for example, by the results of a study of sedimentation equilibria of solutions of adenosine-5'-phosphate [41].

Taken as a whole, the results examined above thus provide information concerning the degree of interaction between bases, but they tell us nothing about the mutual arrangement of the interacting bases in the resulting complex. This type of information can be obtained by analysis of changes in the NMR spectra of solutions of different purine and pyrimidine derivatives in D_2O in relation to their concentration (see below).

2. Thermodynamic self-association constants of purine and pyrimidine derivatives

Enthalpies and entropies of self-association have been calculated for some purine and pyrimidine derivatives [42, 43]. These values are given in Table 4.7.

The absence of correlation between changes in the free energy of interaction and enthalpy changes will be obvious. All the compounds differ considerably in their self-association entropy.

3. Concentration changes in optical properties of solutions of monomer components of nucleic acids

Interaction between bases in aqueous solutions also causes deviation of the optical properties of these solutions from additivity. For example, if the concentration of solutions of deoxyadenosine is increased, a marked decrease in the molar extinction coefficient* (pH 6.7) is observed [40]:

---

* Association with a change in adsorption and dispersion of optical rotation has also been described for guanosine-3'- and guanosine-5'-phosphates [44, 45] and isoguanosine [46]. However, in these cases, besides pure interaction between the planes of the bases, high-polymer hydrogen-bonded complexes are evidently formed.

C, moles /litre	$\varepsilon_{259.5\,nm}$ $\cdot 10^{-3}$	$\varepsilon_{207\,nm}$ $\cdot 10^{-3}$	C, moles/litre	$\varepsilon_{259.5\,nm}$ $\cdot 10^{-3}$	$\varepsilon_{207\,nm}$ $\cdot 10^{-3}$
$2.75 \cdot 10^{-4}$	15.3	21.0	$2.36 \cdot 10^{-2}$	13.9	—
$7.64 \cdot 10^{-3}$	15.0	—	$3.04 \cdot 10^{-2}$	13.7	19.3
$1.02 \cdot 10^{-2}$	14.4	20.0	$4.25 \cdot 10^{-2}$	13.2	—
$2.06 \cdot 10^{-2}$	14.1	19.9	$4.50 \cdot 10^{-2}$	13.1	—

These results are evidence of the formation of self-associates.

4. Concentration changes in NMR spectra of solutions of bases and nucleosides

With an increase in the concentration of purine derivatives in aqueous solutions, appreciable shifts in the proton resonance signals are observed toward higher field, and the higher the association constant of a particular base, the greater the magnitude of this shift (Table 4.8). These effects are reduced by a rise in temperature [47, 48, 50], by replacement of water by organic solvents [47], and by protonation [47].

These phenomena can be explained by the formation of complexes in which the bases are stacked so that the plane of one base is parallel to the plane of the other [47-51]. In complexes with this conformation, the protons of each base must be additionally screened because of the magnetic anisotropy induced by ring currents of the other base. In the case of solutions of pyrimidine derivatives this effect is not observed [49], but in the presence of purine derivatives, such shifts of proton signals of pyrimidines

TABLE 4.8. Relationship between Magnitude of Proton Chemical Shift and Concentration of Compounds in D_2O [38] (change in concentration from 0 to 0.2 M; generator frequency 60 MHz)

Compound	Temperature, °C	$\Delta \delta$, Hz					K_{self}, mole^{-1}
		protons at C2	protons at C8	protons at C6	protons at C1'	protons of CH_3	
Inosine................	32	6.4	5.3	-	7.1	-	-
1-Methylinosine..........	33	8.9	6.4	-	6.8	5.3	1.8-2
Purine	25-27	12.6	9.6	14.2	-	-	2.1
6-Methylpurine............	25-27	19.4	13.3	-	-	17.0	6.7
Ribosylpurine............	30	10.7	6.4	13.1	8.8	-	1.9
Adenosine*..............	32	14.8	8.3	-	6.9	-	4.5
2'-Deoxyadenosine.........	30	19.8	13.0	-	13.6	-	4.7-7.5
6-exo-N-Methyl-2'-deoxyadenosine................	32	26.0	15.8	-	14.0	15.2	15.9
6-exo-N,N-Dimethyladenosine	28	27.2	14.5	-	14.4	25.5	22.2
6-exo-N-Methyladenosine.....	26	32.8	17.5	-	12.6	18.1	11.8-14.9
2'-Deoxyadenosine*........	30	14.8	10.0	-	9.8	-	4.7-7.5
2'-O-Methyladenosine.......	31	13.7	7.5	-	8.8	-	5.1
3'-Deoxyadenosine*..........	35	15.8	9.0	-	9.6	-	-

* Change in concentration from 0 to 0.1 M.

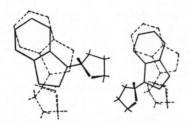

Fig. 4.5. Scheme showing probable "mean" conformations during interaction between purine nucleosides in aqueous solutions (broken lines represent nucleoside further away from observer) [38].

toward higher field are observed. In turn, pyrimidine derivatives reduce the shifts induced by interaction between purine bases [49].

Some information concerning the conformation of these complexes can be obtained from NMR data [38, 48, 50] (the mean conformation is implied, because the association energies are of the order of magnitude kT and, consequently, the complexes forming different conformations are rapidly converted from one into the other). It is clear from Table 4.8 that $\Delta\delta$ of the protons at C2 is substantially greater than for the protons at C8 or C1' in all the compounds investigated. This is evidently because of the existence of some preferential type of interaction between bases, when they occupy a certain relative position in the stack. The most probable conformations to account for the observed differences in proton chemical shifts in nucleosides are shown in Fig. 4.5. However, other methods of interaction between bases must evidently also exist (see [48, 50]).

5. Changes in the properties of bases when incorporated in oligonucleotides by comparison with monomers

Interplanar interactions between bases are even stronger if the bases are constituents of oligonucleotides and polynucleotides. The chief criterion of these interactions in oligonucleotides is the deviation of their optical properties from additivity, i.e., from the picture obtained by addition of the corresponding characteristics of the constituent monomers. This deviation can be clearly detected in the case of the simplest oligonucleotides: the dinucleoside monophosphates (the presence of a second phosphate group has a negligible effect on the optical properties). The formation of an intramolecular hydrogen bond between the bases is impossible in these compounds from steric considerations, and since the observed effects are independent of concentration*, there is no intermolecular association either. Consequently, the change in optical properties must be due to intramolecular interaction between the bases, i.e., to stacking.

Hypochromic effect. In quantum mechanics, interaction of light with matter is connected with a vector μ, known as the transition moment, which has a definite value for every substance. The magnitude and direction of

* Independence of concentration is also observed in the concentration interval 10^{-4}-10^{-5} mole /litre used for the optical determinations. In higher concentrations, of the order of 10^{-3}-10^{-1} mole /litre, which are used for NMR investigations, association of oligonucleotides is observed [318, 319].

this vector depend on the wave function of the state in which the electron was before interaction (the ground state) and on the wave function of the state into which the electron changes under the influence of the light quantum (excited state).

Only that part of the light whose electrical vector is parallel to the transition moment is absorbed. The intensity of absorption depends on the absolute value of $[\mu]$. In the classical formulation, a molecule interacting with light can be represented as an oscillator, oscillating in the direction of the vector of the transition moment, and the energy of the absorbed light can be taken as utilized in increasing the energy of oscillation of the oscillator. If two molecules are very close together and their positions are fixed, then because of interaction between their transition moments (or in the classical view, between their oscillators), the optical properties of each of these molecules must be modified [52-54]. This is manifested primarily as a change in the intensity of absorption and displacement of the absorption maximum, both of which depend on the mutual position of the oscillators. In particular, if the oscillators are arranged one beneath the other, and if there is a known angle φ between them, the intensity of absorption in the long-wave region must be reduced*.

The decrease in the intensity of absorption can be characterized by the "percentage hypochromicity" h, determined by the formula [56]:

$$h = \left(1 - \frac{\varepsilon_D(\lambda)}{\varepsilon_M(\lambda)}\right) 100 \tag{1}$$

where $\varepsilon_D(\lambda)$ and $\varepsilon_M(\lambda)$ represent molar coefficients of extinction of the dinucleoside monophosphate and an equimolar mixture of its constituent monomers.

The wavelength λ is usually chosen to correspond to the absorption maximum of the dinucleoside monophosphate. Another value which can be used to characterize the decrease in intensity of absorption is the percentage hypochromism H, determined by the formula [56]:

$$H = \left(1 - \frac{f_D}{f_M}\right) 100 \tag{2}$$

In this case f_D and f_M are known as the forces of the oscillators, directly connected with the value of the transition moment μ. They can be calculated theoretically or determined experimentally from the area of the absorption peak corresponding to the particular electron transition in the spectrum, by means of the formula:

$$f = 4.32 \cdot 10^{-2} \int_{\lambda_1}^{\lambda_2} \frac{\varepsilon(\lambda)}{\lambda^2} \, d\lambda \tag{3}$$

---

* For further details, see Yu. S. Lazurkin (Editor): Physical Methods of Investigation of Proteins and Nucleic Acids [in Russian], Nauka (1967), p. 113.

where λ_1 is the wavelength of the absorption minimum in the short-wave region, and λ_2 the wavelength in the long-wave region at which $\varepsilon(\lambda)$ becomes equal to zero.

Hypochromic effects in dinucleoside monophosphates of various compositions are well known [55-62]. As an example, the percentage hypochromicity and percentage hypochromism of all 16 possible ordinary dinucleoside monophosphates at three pH values are given in Table 4.9 [56]. In all cases, a marked decrease in absorption of UV-light is observed by comparison with the constituent monomers. If both bases in a dinucleoside phosphate at a given pH are charged equally, it can be expected that there will be no interaction between them because of repulsion of the like charges. In fact, in most cases a decrease in hypochromic effect is observed following protonation or deprotonation of both bases in a dinucleoside monophosphate, although there are some not completely understood exceptions in which the hypochromic effect is increased (UpU and GpU at pH 7 and 11.5 respectively).

With elevation of the temperature the optical density is increased and absorption of dinucleoside diphosphates becomes equal to or, at least, very close to the total absorption of a mixture of components of the same concentration [55, 57, 60, 61]. This also means that definite interaction must exist between the bases in dinucleoside phosphates.

The presence of a hypochromic effect is evidence of interaction between bases, but its absence does not necessarily mean that there is no interaction, and attempts by several workers to distinguish bases, on the basis of the hypochromic effect, into those which are capable of interplanar interaction and those which are not [56, 59] do not rest on a firm basis. This applies also to results obtained by measurement of optical rotatory dispersion and circular dichroism (see below).

The different values of hypochromic effects for isomeric dinucleoside diphosphates deserve attention. In this case the mutual arrangement of the bases in the stack is probably different for the two isomers.

Optical rotatory dispersion and circular dichroism. Changes in the properties of bases when incorporated in oligonucleotides (compared with the monomer components) are even more apparent when curves of optical rotatory dispersion and circular dichroism in the ultraviolet region are compared.

Most of the data at present available relate to dinucleoside monophosphates and trinucleoside diphosphates. The presence of a terminal phosphate group in the 5' position has been shown to have very little effect on the spectrum of circular dichroism and on the stability of association in dinucleoside monophosphates [327, 340], whereas a 3'-terminal phosphate group has a marked effect on the circular dichroism [327, 319] and on the optical rotatory dispersion [329] of dinucleoside monophosphates, appreciably weakening interaction between the bases in them [327].

TABLE 4.9. Percentage Hypochromism H and Percentage Hypochromicity h of Dinucleoside Monophosphates* at Different pH Values (25°C, ionic strength 0.1) [56]

Compound	$H,$ % at pH 1	$h,$ % at pH 1	$H,$ % at pH 7	$h,$ % at pH 7	$H,$ % at pH 11,5	$h,$ % at pH 11,5
GpC	(6.8)	(10.7)	7.2	8.7	4.4	6.0
CpG	(6.8)	(9.8)	6.2	3.2	5.4	9.2
UpG	6.5	5.6	7.6	5.3	(3.0)	(4.2)
GpU	4.9	6.2	—1.2	2.4	(0)	(4.3)
UpA	3.0	3.3	1.4	3.0	1.7	1.5
ApU	2.7	3.0	1.6	5.0	0	3.8
GpA	(2.7)	(3.7)	6.0	7.6	2.9	5.8
ApG	(1.4)	(3.0)	2.6	5.8	1.3	4.0
CpU	1.5	2.7	4.2	6.3	4.2	4.0
UpC	—0.3	3.2	0.7	2.4	1.2	1.8
ApC	(0)	(1.8)	7.3	7.6	9.7	9.0
CpA	(—2.3)	(—1.0)	5.2	7.8	6.3	6.5
ApA	(—0.5)	(0.3)	6.8	9.4 .	7.4	9.4
CpC	(—5.5)	(0.1)	4.9	7.2	5.7	7.2
GpG	(—1.0)	(0.5)	9.1	6.9	(4.4)	(—0.8)
UpU	—2.0	3.0	—3.6	1.7	(0.8)	(2.5)

* Values for compounds in which both bases are charged at that pH are given in parentheses.

Curves of optical density and optical rotatory dispersion for several dinucleoside monophosphates are compared in Fig. 4.6 with the sum of the optical rotatory dispersions of the individual components [56]. It is easy to see that these curves differ both in intensity of rotation and in shape. Isomers with different base sequences also differ in the shape of their curves (Fig. 4.7). These effects are independent of concentration. With a rise in temperature, the curve of optical rotatory dispersion of the oligonucleotide approaches the curve of optical rotatory dispersion [56, 61] of the sum of the monomers. Similar effects are also observed for trinucleoside diphosphates, in which intramolecular hydrogen bonding is hardly possible [63-66]. Just as in the case of the hypochromic effect, it can be expected that the optical rotatory dispersion will change with a change in pH, coming closer to the combined dispersion of the individual monomers, when both bases in the dinucleoside phosphate are charged.

A quantity which can be used as a measure of deviation of the optical rotatory dispersion of a dinucleoside phosphate from additivity is the absolute value of the maximum difference $[\Phi_D - \Sigma \Phi_M]$ between the rotation of the dinucleoside phosphate and the sum of the rotations of the monomers, which is determined by plotting $[\Phi_D - \Sigma \Psi_M]$ against wavelength λ. It can be concluded from the data in Table 4.10, giving the values of this difference for various dinucleoside monophosphates at three different pH values, that such an effect

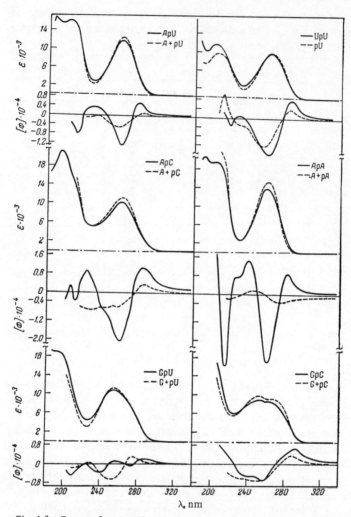

Fig. 4.6. Curves of optical density and optical rotatory dispersion of a series of dinucleoside monophosphates and of their monomer constituents at neutral pH values [236].

is in fact observed. The optical rotatory dispersion thus also shows that definite interaction exists between the bases in dinucleotides.

Certain conclusions regarding the mutual orientation of the bases in dinucleoside phosphates can be deduced from the optical rotatory dispersion data. Theoretical calculations of the optical rotatory dispersion effect [67], made on the assumption that the bases in a dinucleoside phosphate form a right-handed helix with angle of rotation $\gamma \approx 36°$, as illustrated in Fig. 4.8, give a qualitatively true picture of the dispersion, whereas the left-handed helix gives a Cotton effect of the opposite sign.

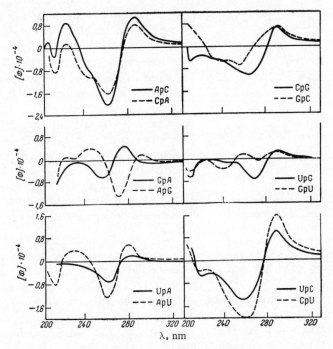

Fig. 4.7. Curve of optical rotatory dispersion of isomeric dinucleoside monophosphates at neutral pH values [56].

TABLE 4.10. Values of $[\Phi_D - \Sigma\Phi_M]$ of Dinucleoside Monophosphates at Different pH Values (25°C, ionic strength 0.1) [56]

Compound	$[\Phi_D - \Sigma\Phi_M] \cdot 10^{-4}$ at pH 1	λ, nm	$[\Phi_D - \Sigma\Phi_M] \cdot 10^{-4}$ at pH 7	λ, nm	$[\Phi_D - \Sigma\Phi_M] \cdot 10^{-4}$ at pH 11.5	λ, nm
GpC	(1.1) *	265	0.38	293	0.31	276
CpG	(0.30)	265	0.68	273	0.59	273
UpG	0.32	257	0.90	273	0.16	270; 250
GpU	1.62	257	0.74	255	0.17	251
UpA	0.23	263	0.32	267	0.03	262
ApU	0.18	262	0.80	261	0.37	262
GpA	(0.31)	275	0.76	277	0.16	265
ApG	(0.45)	273	1.08	270	0.75	258
CpU	0.70	272	1.22	265	0.78	267
UpC	0.49	275	0.57	267	0.22	270
ApC	(0.28)	270	1.52	265	1.33	264
CpA	(0.11)	270	1.11	263	1.11	265
ApA	(0.36)	257	2.86	260	2.72	260
CpC	(0.43)	280	1.61	272	1.57	271
GpG	(0.58)	253	0.66	249	(0)	—
UpU	0.67	263	0.67	264	(0.19)	240

* Values for compounds in which both bases at that pH are charged are given in
parentheses.

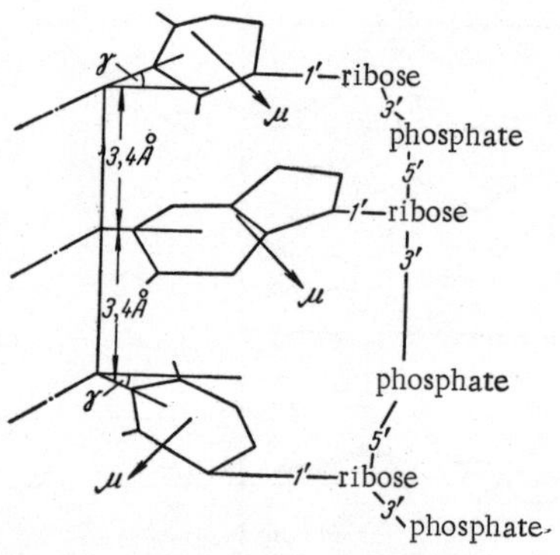

Fig. 4.8. Diagrams showing possible arrangement of bases in an oligonucleotide (a trinucleotide fragment is illustrated), deduced from theoretical and experimental optical rotatory dispersion data. Right-handed helix ($\gamma > 0$) [67].

Similar information is given by the circular dichroism spectrum [68-71]. The circular dichroism spectra of some dinucleoside monophosphates are compared in Fig. 4.9 with the spectra of the corresponding monomer units (nucleosides or nucleotides). The difference observed virtually disappears when the temperature is raised. It also follows from comparison of the circular dichroism data with theoretical calculations that the angle of rotation γ must be positive, and must have a value [71] between 30 and 45°.

Finally, NMR spectra of several oligonucleotides have recently been obtained [61, 72-75, 318-322]. It also follows from these data that interplanar interaction exists between the bases in dinucleoside phosphates. Interaction of this type is clearly seen, for example, in the NMR spectrum of adenylyladenosine cyclic phosphate (Fig. 4.10). At 5°C, four signals from protons of the dinucleotide are observed, indicating an irregular distribution of the bases in the complex. In addition, the signals are shifted toward higher field compared with those of adenylic acid, as must be observed if an interplanar complex is formed. If the temperature is raised to 67°C, two signals still remain in the spectrum and are shifted toward lower field, indicating a disturbance of the interaction between the bases.

6. Thermodynamic characteristics of interaction between bases in dinucleoside phosphates

The NMR data, as well as the optical rotatory dispersion and circular dichroism data (see page 202), show that if interaction between the bases takes place in an oligonucleotide, this oligonucleotide exists chiefly in the conformation of a right-handed helix, the nucleoside constituents of which are in the anti-configuration [318-322]. The mutual arrangement and, consequently, the energy of interaction between the constituent oligonucleotide bases are determined by several factors.

Optical rotatory dispersion and circular dichroism curves of dideoxyribonucleoside phosphates and diarabinonucleoside phosphates differ to a lesser degree from the calculated values for the sum of the components and they vary less with temperature [68, 323-326] than in the case of the ribo-derivatives. The circular dichroism spectra of mixed dinucleoside phosphates containing ribosyl and deoxyribosyl or arabinosyl constituents vary depending on the sequence of these constituents. If the deoxyribose or arabinose residue

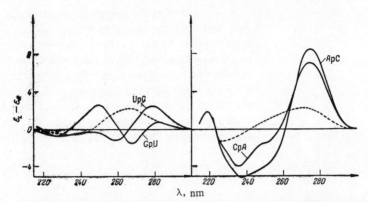

Fig. 4.9. Spectra of circular dichroism of a number of dinucleoside monophosphates (continuous curves) and of their constituent monomer units
(broken curves); 20°C, pH 7.0 [69].

participates in the formation of the phosphodiester bond through its hydroxyl
group at C3', the circular dichroism spectrum of this dinucleoside phosphate
will not differ from the spectrum of a mixture of its constituents [324, 325,
327]. If, however, the deoxyribose or arabinose residue participates in the
phosphodiester bond through the hydroxyl group at C5', the circular dichroism
spectrum of this dinucleoside phosphate will practically coincide with the
spectrum of the analogous diribonucleoside phosphate [324, 325, 327].

The circular dichroism spectra of oligonucleotides with a 2',5'-phospho-
diester bond differ very little from the spectra of the sum of their monomer
constituents, and they vary only very slightly with temperature, as has been
shown [327] in the case of cytidylyl-(2' → 5')-cytidine.

These findings may be evidence either of absence of interaction between
bases in dinucleoside phosphates or of the specific mutual arrangement of their
transition moments. However, the NMR spectra of adenylyl-(2'→5')-cytidine
[318] indicate considerable overlapping of their base planes, even more
marked than in the case of cytidylyl-(3'→5')-adenosine. It can accordingly
be postulated that absence of an interaction effect in the circular dichroism
spectra is the result of the specific mutual arrangement of the constituent
bases, and not of the absence of interaction [318].

It was pointed out above (see page 203) that isomeric dinucleoside phosphates differ in their optical rotatory dispersion and circular dichroism. The
NMR spectra suggest [318] that the uracil residue in ApU is overlapped by the
6-membered ring of the adenine residue, whereas in UpA this overlapping
takes place by the 5-membered ring of the adenine residue. This conclusion
is in agreement with the results of titration of these isomers with acids and
alkalies [78]. Association constants determined in this case for UpA and
ApU were 0.05 and 0.41 respectively.

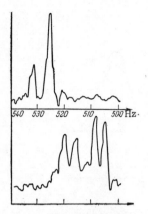

Fig. 4.10. NMR Spectra of ApA >p in D_2O at 5°C (lower curve) and 67°C (upper curve). Ionic strength 0.1, pD 7.1 [74]. Chemical shifts given in Hz relative to TMC signal.

It is difficult to determine from the existing data which of the bases is more capable and which less capable of association. Nevertheless, it has been concluded [69] from the optical rotatory dispersion and circular dichroism data that uracil shows a smaller tendency toward association than the other bases. This is confirmed by the values of the association constants obtained by the titration method (at 20°C; pH 7.0; 0.1 M NaCl) [78]:

UpU 0.00
UpA 0.05
ApU 0.41
ApA 5.38

This conclusion also agrees, apparently, with NMR data [318] showing that the uracil and guanine residues in UpG overlap to a lesser degree than the cytosine and guanine residues in CpG.

It has also been concluded from the circular dichroism spectra that guanine possesses low powers of association [328]. However, the NMR data suggest that overlapping of the base residues is greater in ApG and GpC than in ApA and ApC respectively.

Attempts at quantitative determination of the thermodynamic parameters of association between bases in oligonucleotides have been made. If it is accepted that there are only two possible states for dinucleoside phosphates (with interacting and noninteracting bases), the thermodynamic parameters of the transition:

$$\text{Interacting bases} \rightleftarrows \text{Noninteracting bases}$$

can be determined on the basis of the temperature dependence of any characteristics of the interaction, provided that these characteristics are known for the limiting states.

These determinations have been made for several dinucleoside monophosphates on the basis of changes in optical rotatory dispersion [61, 62, 77], circular dichroism [68, 69, 70, 71, 327, 328], and optical density [57, 60], and finally, of changes in NMR signals [74] with temperature.

Values of some thermodynamic parameters for the transition of dinucleoside monophosphates, obtained by the optical rotatory dispersion and circular dichroism methods, are given in Table 4.11. These results show that the two methods give decidedly different values for the thermodynamic parameters; values obtained for different pairs within each method are relatively close. This result is seen more clearly still when the constants obtained by these two and other methods are compared, for example, for ApA (Table 4.12).

TABLE 4.11. Thermodynamic Parameters of the Transition: Interacting Bases ⇌ Noninteracting Bases from Results Obtained by Various Methods

Compound	From optical rotatory dispersion data		From circular dichroism data		T_m *, °C
	$\Delta H°$, kcal/mole	$\Delta S°$, e.u.	$\Delta H°$, kcal/mole	$\Delta S°$, e.u.	
ApA	5.3	20	8	2.8	25
ApG	4.8	18	—	—	—.
ApC	6.2	22	6.1	21	25
ApU	8.4	32	6.7	24	11
GpA	5.6	20	6.1	22	9
GpC	7.8	28	—	—	—
GpU	6.8	25	—	—	—
CpA	7.3	27	7.0	24	15
CpC	6.9	25	7.5	25	24
CpU	7.8	28	6.8	24	6
UpA	5.1	21	—	—	—
UpG	6.0	23	—	—	—
UpC	6.2	22	—	—	—
UpU	7.8	29	—	—	—

Note: Measurements of optical rotatory dispersion made at pH 7, 25.2% LiCl [61]; determinations of circular dichroism at pH 7.5, 4.7 M KF, 0.01 M Tris buffer [69]. * T_m denotes the melting temperature for the transition at which the concentrations of compounds with interacting and noninteracting bases are equal (determined from circular dichroism data).

TABLE 4.12. Values of Enthalpy of Denaturation of ApA Obtained by Different Methods

Method	ΔH, kcal/mole	Literature cited
Optical rotatory dispersion.	5.3 6.5	[61] [77]
UV-Spectroscopy.	8.5 9.4 10	[61] [60] [57]
Circular dichroism.	8	[69]
NMR (for ApA >p)	8.2	[74]

These considerable discrepancies between the values of the thermodynamic constants of transition obtained by different methods may be due to two causes. First, the limiting values for completely interacting and completely noninteracting molecules may have been incorrectly found. Second, the hypothesis regarding the existence of only two states may be incorrect, and several intermediate states between completely associated and completely dissociated forms may exist.

There is absolutely no doubt that the limiting values are obtained extremely inaccurately by existing methods; nevertheless, since the differences between the values characterizing association are so great, there is

H

reason to suppose that the second cause is the more likely one [61]. Attempts have been made to develop a model simulating many states [79, 61], but the results so far obtained do not adequately explain the experimentally observed values.

The following are the principal conclusions which can be drawn from the experimental material on interaction between bases in an aqueous medium examined above:

1. In aqueous solutions, free bases, nucleosides and nucleotides, and also bases incorporated in oligonucleotides, are capable of association with the formation of complexes in which the plane of one base is parallel to the plane of the other.

2. Interaction of this type is stronger in the case of oligonucleotides.

3. The data for monomer components indicate that purine bases associate more strongly than pyrimidine; this is confirmed by data for dinucleoside monophosphates. The uracil ring interacts less strongly than the others.

4. The associates thus formed are asymmetrical, and in the case of oligonucleotides the mutual arrangement of the planes of neighbouring bases depends on their sequence.

5. These associates are destroyed by organic solvents, by a rise of temperature, and by a change in the acidity of the medium.

7. Nature of the forces stabilizing associations of bases
in aqueous solution

The stability of associates in solution is evidently determined by two factors: by interaction forces between bases and by effects connected with interaction between the solvent and the free and associated bases.

Interaction of one base with another is evidently attributable mainly to Van der Waals – London forces (see page 192).

Comparison of the calculated [80] data with the experimental data for ΔH^0 of the bases (see Table 4.7) shows that the results obtained by calculation are much too high, although the increase in association of 6-methylpurine compared with purine is correctly predicted. Unfortunately, only these two compounds can be compared, for the remaining determinations of enthalpy have been carried out on nucleosides, whereas the calculations have been made for bases. These calculations give free energy values without taking the entropy factor into consideration; they cannot be compared with values of the association constants given previously because entropy changes during self-association of bases may vary from one base to another (see Table 4.7).

Nevertheless, the great difference between the energies of interaction of uracil and cytosine during self-association and association with purine deserves attention. This may, perhaps, explain the experimentally observed difference between the properties of the corresponding dinucleoside mono-phosphates. However, the experimentally determined values of the enthalpy of self-association of uridine and cytidine are very similar (Table 4.7).

The theory of the influence of the solvent on association between the bases of nucleic acids has not yet been worked out in all its details. The fullest investigations in this direction have been made by Sinanoglu and Abdulnur [81, 82]. Similar calculations, although differing in details, have also been undertaken by Pullman [80]. The discussion given below follows the lines adopted by Pullman and collaborators. Two effects connected with the influence of the solvent are examined. The first of these is the effect of electrostatic interaction, which differs for associated and free molecules, leading to a change ΔF_{solv} in the free energy during association. The value ΔF_{solv} will subsequently be called the electrostatic change in free energy. Second, the change in the surface of contact between the dissolved substance and the solvent is examined during the transition from dissociated to asso-ciated molecules, leading to a change ΔF_{surf} in free energy; this value will subsequently be called the change in surface free energy.

The factor ΔF_{solv}. Through electrostatic interactions with the solvent, the state characterized by a higher dipole moment and a smaller size is stabil-ized. This leads to greater stabilization of the separate molecules compared with the complex, since the complex is larger in size and has a smaller dipole moment, because of compensation of dipole moments of the constituents in the position corresponding to the energy minimum. Therefore $\Delta F_{solv} > 0$, i.e., interaction of this type is disadvantageous to complex formation. Under these circumstances, ΔF_{solv} of the cytosine + cytosine pair $> \Delta F_{solv}$ of the uracil + uracil pair $> \Delta F_{solv}$ of the purine + purine pair. The value of ΔF_{solv} is relatively low compared with ΔF_{surf}.

The factor ΔF_{surf}. The change in free energy connected with a change in the surface of interaction between the dissolved substance and the solvent can, in turn, be regarded as consisting of two parts. First, a contribution is made by the change in energy of surface tension:

$$\Delta F'_{surf} = \gamma \, \Delta A$$

where γ represents the surface tension, and ΔA the change in the surface with the transition from the complex to dissociated molecules.

Second, the expression ΔF_{surf} also includes the component $\Delta F''_{surf}$, characterizing the change in number of ordered molecules of solvent around the dissolved substance during association:

$$\Delta F''_{surf} = - T \, \Delta S''_{surf}$$

where T is the absolute temperature, and $\Delta S''_{surf}$ the change in entropy of the solvent during the transition from its body to the surface layer.

The value of $\Delta F''_{surf}$ is small compared with $\Delta F'_{surf}$; both these changes in energy lead to stabilization of the complex compared with the dissociated molecules.

The values of the change in free energy during the formation of Pu + Pu and Py + Py associates can be estimated on the assumption that unassociated bases form a cylindrical cavity in the solvent, with the bases having areas of 30 Å^2 for Py and 50 Å^2 for Pu. The difference between the surface area of the free bases and complex in that case will be determined by the difference between the surfaces, amounting to 2×30 Å^2 in the case of pyrimidines and 2×50 Å^2 in the case of purines. The value of ΔF_{surf} (equal to the sum of the factors $\Delta F'_{surf}$ and $\Delta F''_{surf}$) for purines is about 14 kcal/mole, and for pyrimidines about 8.4 kcal/mole. This means that purines must have a greater tendency toward association than pyrimidines. This approach to the solvent effect answers the question why organic solvents, whose surface energy is lower than that of water, have a destabilizing effect on interplanar interactions, and it also reveals the principles governing the solvent effect on certain chemical reactions, notably on the photodimerization of thymine [83] (see Chapter 12).

The results of theoretical calculations undertaken to estimate Van der Waals – London forces and the solvent effect during association of bases can be regarded as only purely qualitative. Nevertheless, they are useful in order to understand the relative role of the various factors concerned in the secondary structure of nucleic acids.

The two types of interaction between bases which have been examined above – crosslinking (by hydrogen bonding with stabilization through Van der Waals–London forces) and interplanar interactions – determine the stability and specificity of structure of the nucleic acids. In certain cases both types of interaction occur (in DNA or RNA, for example), while in others only interplanar interactions exist (in single-helical homopolynucleotides). The concrete features of the secondary and tertiary structures of polynucleotides will be examined later.

IV. Investigation of the macrostructure of double-stranded DNA

A cornerstone in modern molecular biology is the hypothesis put forward by Watson and Crick [1] in 1953. This hypothesis generalized the data available at that time for the structure and functions of DNA and stimulated the development of qualitatively fresh approaches to the study of the chemistry, physics, and functional role of nucleic acids. In particular, the principle of complementarity, emerging from the Watson and Crick hypothesis [1], has been used to explain the mechanisms of transmission of genetic information both during reproduction of genes and in protein biosynthesis. These mechanisms were subsequently confirmed experimentally by the work of

Kornberg on template synthesis of DNA, of Berg on the template synthesis of RNA and DNA, and of Nirenberg and Khorana on the code for protein synthesis.

1. The Watson and Crick hypothesis

At the time when Watson and Crick put forward their hypothesis [1, 84], several types of DNA structure had been described, but none of them could explain many of the fundamental problems. The most important of these problems was how bases which differ in size and are arranged in a definite but nonrepeating order along the polynucleotide chain, can give a highly symmetrical x-ray diffraction pattern.

Watson and Crick were able to put forward their hypothesis because of the accumulation of facts concerning the chemical structure and nucleotide composition of DNA. When this matter is discussed, it is usually considered that the principal factor which made the hypothesis possible was the rule discovered by Chargaff (see page 44) for the nucleotide composition of different DNAs, according to which the adenine : thymine and guanine : cytosine ratios were approximately 1 for all molecules then investigated. Another factor which is mentioned is the x-ray structural data then available, from which it could be concluded that DNA has a helical and highly symmetrical structure. The main approach used by Watson and Crick when explaining the structure of DNA was to build stereochemical models which did not conflict with the experimental data they possessed. They could not have built their models had not data just been obtained to indicate the 3',5'-character of the phosphodiester bond linking the individual nucleotide components [85, 86], the unbranched nature of the polynucleotide chain, the furanose form of deoxyribose [87], and the β-configuration of the N-gylcoside bond [88]. They also possessed information (from x-ray structural analysis) about the structure of the heterocyclic bases adenine, guanine [89-91], and cytosine [92], and also about the conformation of the furanose ring [92] and the stereochemistry of the N-glycoside bond [93].

By using all these facts they were able to build an adequate model of DNA.

The first model of DNA was built for one of the two forms of DNA then known (see below), the B form, because more information concerning the geometry of the molecule could be extracted from the x-ray structural data for this form. The following conclusions were drawn from the analysis of several DNA specimens:

1. The structure of DNA can be represented as a highly symmetrical helix.

2. The parameters of this helix are independent of the composition of the DNA; its radius is 10-12 Å, its pitch is 34 Å, and the distance between bases, whose planes are perpendicular to the axis of the helix, is 3.4 Å, so that the pitch in one chain includes ten bases.

3. In conjunction with data for the density of DNA specimens, it must be assumed that the DNA molecule contains two polynucleotide chains.

Consequently the model of DNA structure must be such that, regardless of the nucleotide composition and sequence of the nucleotide components along the chain, the same helix was obtained. This could happen only if the monomer units of the double helix were equal in size and symmetry. Leaving aside for a time the problem of how bases of different sizes can form monomer components of identical size in the double helix, the attempt had to be made to build a model of a double helix with identical monomer units in order to explain the possible arrangements of the phosphate – carbohydrate backbone of the molecule. Once this stereochemical model satisfying the x-ray structural data and the principle of observance of Van der Waals radii had been built, it could be deduced from it that the phosphate groups must lie outside the helix and the bases inside it.

The next problem concerned with DNA structure is: does the pitch correspond to a complete turn of the helix or to half a turn. The second alternative is possible if the DNA double helix has an axis of symmetry of the second order, parallel to the axis of the helix. In the first case (to give ten nucleotides to each pitch), the angle between the direction of the N-glycoside bonds of neighbouring bases in one chain must be 36°, and in the second case 18°. By building a stereochemical model of one of the chains, the verdict could be given in favour of the first alternative – the pitch corresponds to a complete turn of the helix – for in the second case it would be impossible to build such a model without infringing the Van der Waals radii.

Hence, although no concrete observations were made regarding the structure of the monomer units, two important problems were solved: the arrangement of the phosphate groups and the establishment of the fact that the pitch of the helix corresponded to one complete turn.

The next step in building the model is the hypothesis concerning forces holding the two polynucleotide chains of DNA together. Watson and Crick postulated that this is the result of hydrogen bonding between the bases which, as had already been discovered, lie inside the double helix. Two essential results followed immediately from this assumption:

1. Bonds must be formed between purine and pyrimidine bases, because such pairs would be expected to be of similar size only if the bases differed; Pu + Pu and Py + Py pairs would differ very considerably in size.

2. In accordance with Chargaff's rule, pairs can be formed only between guanine and cytosine and between adenine and uracil.

To satisfy the requirement of high symmetry of the helix, the pairs must conform to the following demands:

1. The distance between the glycoside bonds of the bases forming the pair must be similar for both pairs.

2. The angles formed by the glycoside bonds must be approximately equal in both pairs.

3. To ensure equivalence of all the carbohydrate and phosphate groups in the double helix, the polynucleotide chain must possess axial symmetry of the second order perpendicular to the axis of the helix. Correspondingly, the N-glycoside bonds must also have an axis of symmetry of the second order. On the basis of x-ray structural data for the size and conformation of the bases, and also of the hypothesis of keto-amino tautomeric forms of the bases, models of base pairs incorporated into a polynucleotide were constructed (Fig. 4.11), and they were found to satisfy all the requirements listed above.

The last question, whether the helix is right-handed or left-handed, was also solved by building a stereochemical model in which the bases formed complementary adenine·uracil and guanine·cytosine pairs. Data for the conformation of deoxyribose, the length and arrangement of the N-glycoside bond, and the length and angles of the $P-O$ bonds for the tetrahedral configuration of the phosphorus atoms were used when building the model. It was found that the helix must be right-handed, for difficulties arise in the construction of a left-handed model without infringing the Van der Waals radii (see page 217). It must be emphasized once again that, if the dyad axis is perpendicular to the axis of the helix, the direction of the two chains forming the molecule must be opposite. Consequently, at each end of the double-helical molecule there is the 3'-end of one and the 5'-end of the other single-helical component. In such a model the angle between neighbouring base pairs (i.e., the lines connecting the C1' atoms of the deoxyribose residues in the complementary pairs) will be 36°. The bases lie parallel to one another along the axis, and their planes form an angle of about 90° with the axis of the helix; the planes of the deoxyribofuranose ring are almost parallel to the axis of the helix. Bases forming a complementary pair lie in the same plane. The resulting model is shown schematically in Fig. 4.11a.

The model suggested by Watson and Crick thus agreed with the x-ray structural data and gave a DNA structure which not only explains its physico-chemical properties, but also suggests how replication of DNA can take place on the basis of the principle of complementarity [94]. However, the model required experimental verification.

To begin with, it was necessary to prove that the keto and amino groups of the bases in fact participate in hydrogen bonding. Evidence of the validity of this assumption is as follows.

1. Even before the model was built, results of potentiometric titration of DNA were available. These showed that titration of native DNA preparations ultimately leads to irreversible changes in the acid—base properties of DNA. The back-titration curve differs essentially from the direct titration curve: the equivalence points of native DNA are shifted much more to the acid side during acid titration, and much more to the alkaline side during alkaline titration than the corresponding points during back titration.

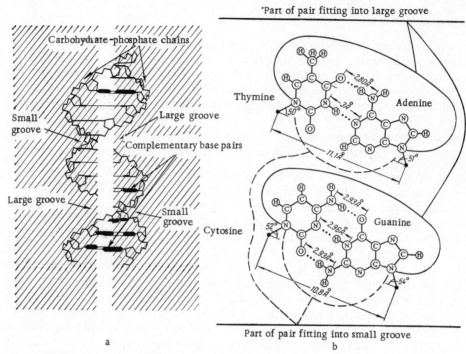

Fig. 4.11. Scheme of the double helix of DNA [84] (a) and of complementary base pairs (b).

This suggested that the titratable groups of the bases of nucleic acids participate in hydrogen bonding [95].

2. Shifts of the characteristic frequencies of the keto and amino groups of the bases, typical of hydrogen bonding, are observed in the IR-spectra of solutions of double-helical synthetic polyribonucleotides, which can act as models of DNA, in D_2O [96, 97].

3. The existence of hydrogen bonds in aqueous solutions of DNA is also confirmed by measurements of the velocity of isotopic exchange with deuterium and tritium [98, 99, 341, 342]. These measurements show that DNA contains a certain number of comparatively slowly exchanged protons n, which depends on the content of adenine·thymine and guanine·cytosine pairs in the DNA molecule [341, 342].

4. In the case of double-helical DNA no reaction between the amino groups of the bases and formaldehyde is observed (see Chapter 6). This indicates, if not their participation in hydrogen bonding, at least their strong steric screening (in agreement with the Watson and Crick model).

Strict specificity of interaction between the bases must also be proved. Evidence of this type was obtained both at the monomer level (see page 185), and also at the level of interaction between synthetic polynucleotides. Poly-meric derivatives of adenine (polynucleotides) were found to interact selec-tively with derivatives of uracil, bot not of cytosine, and in the same way guanine derivatives were found to interact with derivatives of cytosine, but not of uracil [100].

Next, the existence of tautomeric forms of the bases postulated by Watson and Crick had to be demonstrated. Evidence of the validity of this hypothesis also was obtained at the level of monomer units (see Chapter 3) and by investigation of synthetic polynucleotides [97]. Finally, the validity of the Watson and Crick scheme (see Fig. 4.11a) of hydrogen bonding was subsequently confirmed by x-ray structural analysis of the lithium salt of the B form of DNA [101].

In this way all the basic assumptions of the Watson and Crick hypothesis regarding DNA structure were confirmed. Later work by Kornberg and his collaborators also confirmed their second hypothesis, concerning the possi-bility of DNA replication in biological systems. Subsequent x-ray structural analyses of DNA [101-103] have led to only slight changes in the parameters proposed by Watson and Crick. A brief description of the three main crystal-line forms of DNA is given below. These forms are readily converted from one to another, depending on the relative humidity and the type of cation bound to the phosphate residue.

Form A [104] is the true crystalline form of a DNA salt formed when the relative humidity of the specimen is below 80%. In this form of DNA each turn of the helix contains 11 bases, inclined at an angle of 20° to a line per-pendicular to the axis of the helix. The distance between the bases is 2.56 Å and their angle of rotation is 32.7°. The two bases constituting a pair are not coplanar, and the dihedral angle between them is 16°. The molecule has a dyad axis perpendicular to the axis of the helix, so that the two chains composing the double helix are opposite in direction. Form A is a right-handed helix, and a left-handed helical structure would be impossible for it.

Form B exists at a relative humidity >80% and is a paracrystalline form. In the light of recent findings, the B form could be built as a left-handed helix [105]. However, since the A and B forms are easily converted from one to the other, because of the right-handed character of the A form, it follows that the helix in the B form must be in the same direction [105].

The B form plays a particularly important role because it is, evidently, the form in which DNA exists in aqueous solution. In recent investigations of the B form [101-103], more accurate measurements of the parameters given by Watson and Crick have been obtained. In particular, the diameter of the helix is found to be 18 Å, the bases are arranged closer to the axis of the helix, while the carbohydrate residues lie at a considerable angle to the axis of the helix. However, the basic propositions of the Watson and Crick hypo-thesis have been left unaltered.

H*

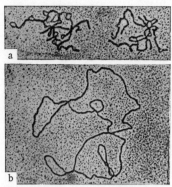

Fig. 4.12. Electron micrograph of covalent-bonded circular DNAs from mitochondria of <u>Xenopus</u> <u>laevis</u> (a) and of circular DNA of the same origin with one break in the chain (b) [118].

Maintenance of the B form in aqueous solution is evident from low-angle x-ray scattering data, which determine the distance between neighbouring base pairs in DNA solution as 3.2 Å, in agreement with the model of the B form [107, 333, 334]. In aqueous solution, however, DNA behaves as a linear molecule, even though its length is small (mol. wt. $< 0.3 \cdot 10^6$). As the length of a polynucleotide increases, its properties gradually approximate to those of a statistical coil, and with molecular weights exceeding $100 \cdot 10^6$, they can be completely described by such a model [108-110].

Form C [106], the third and least humid of all the three forms, is a helix whose turn corresponds to 9.3 nucleotide pairs, and the bases in it are inclined at 5° to the perpendicular drawn to the axis of the helix.

2. Secondary and higher structures of circular DNAs

The structure of the recently discovered circular DNAs is of great interest. The existence of two types of circular DNA molecules has already been mentioned above (see page 34): types in which the DNA ring is closed by noncovalent or covalent bonds. The secondary structure of these molecules is indistinguishable from the secondary structure of linear DNAs, i. e., these molecules are evidently formed by the ordinary double helix of the B form.

In the case of covalent-bonded DNAs, characteristic features appear, and these must be examined in more detail because of the important role of this type of DNA in biological systems [111, 335].

The evidence for the existence of these DNAs has been examined in Chapter 1 (see page 34). These molecules differ from linear and from non-covalent-bonded circular DNAs in their much higher sedimentation coefficient [111, 112]. For example, the replicative form of DNA of phage øX174, with a covalent-bonded structure [112], has a sedimentation coefficient in 2 M NaCl at neutral pH values of 21 S, whereas this same DNA, but with breaks in one of the chains, has a sedimentation coefficient of 17 S*. Electron micrographs reveal definite differences between noncovalent-bonded and covalent-bonded circular DNAs (Fig. 4.12). In the second case, the circular forms are clearly visible, whereas in the first case numerous intersections between the chains can be observed [111-119]. These results suggest that

---

* For the relationship between the hydrodynamic characteristics and molecular weight of circular DNA, see [330].

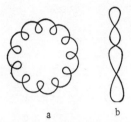

a   b

Fig. 4.13. Diagram showing
toroidal (a) and coiled (b)
forms of the superhelical
molecule of covalent-bonded
circular DNA.

covalent-bonded circular DNAs (from natural sources) have a special tertiary structure in which the double helix of Watson – Crick type is so arranged that it forms an additional superhelix, as illustrated schematically in Fig. 4.13. This superhelix may be either toroidal (Fig. 4.13a) or coiled on itself (Fig. 4.13b). The superhelix-forming effect is observed only for covalent-bonded circular DNAs isolated from natural sources or obtained with the aid of polynucleotide ligase (DNA-ligase [21, 215, 316]). Rupture of even one phosphodiester bond in the DNA superhelix (by the action of DNase, for example) leads to the formation of an uncoiled circular molecule.

The formation of superhelical structures can be examined from the topological standpoint [122-124]. Let it be assumed that under certain conditions the number of turns in the double helix of a linear, double-stranded, double-helical molecule is equal to α*, and that under these conditions it is closed into a circular structure so that the 5'-end of each chain is linked with the 3'-end of the same chain. Under these circumstances the axis of the double helix will form a flat ring, and each of the chains, as before, will make α turns around it, just as in the linear molecule. If, now, the external conditions are changed so that the number of turns in the double helix is altered and is no longer equal to α (let us call this new number of turns β), then because of closure of the ring the axis of the double helix will now itself become a helix, thus forming a superhelix and compensating in this way for the change in number of turns in the double helix.

The hypothetical case in which in the original linear double-stranded molecule α is equal to 0 (two independent, nonhelicized chains) is represented schematically in Fig. 4.14. Closure of such a molecule into a ring makes the two chains into a circular structure without superhelix-formation (Fig. 4.14a). If the chains of this last structure are twisted into a right-handed double helix, of the Watson and Crick type for example, the axis of the molecule must form either a right-handed self-coiled (Fig. 4.14b) or a left-handed toroidal (Fig. 4.14c) superhelix. The number of turns τ of the superhelices (both types) in this case is equal to the number of newly formed turns of the double helix. If the two chains formed a left-handed double helix, the direction of the turns of the superhelices would be opposite to that discussed above.

---

* In the Watson and Crick structure the number of turns of the double helix is 1/10 of the total number of base pairs in the molecule. However, with a change in the conditions (temperature, pH of the medium, etc.) this number may be reduced (see page 220). The number α is also equal to the number of turns of each chain around the axis of the helix or around the other chain.

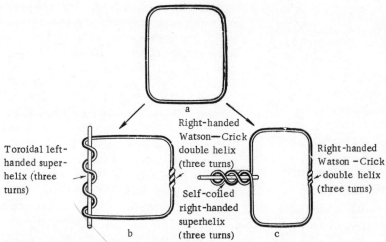

Fig. 4.14. Scheme to show the formation of superhelical structures in covalent-bonded circular DNA under the condition that α < β (see text).

If the original linear molecule was already a right-handed helix, and after closure into a ring under a change of the conditions the number of right turns was increased so that $[\beta] > [\alpha]$, the axis of the molecule must form either a left-handed toroidal or a right-handed self-coiled superhelix, with a number of turns equal to the number of newly formed turns of the double helix $[\tau] = [\alpha] - [\beta]$. If the number of turns of the double helix was reduced after cyclization, superhelices in the opposite direction would be obtained. If the values of α and β are conventionally regarded as positive for right-handed double helices, the number and direction of the turns of the superhelix are given by the simple formula $\tau = \alpha - \beta$. When $\beta > \alpha$, a left-handed toroidal or right-handed self-coiled superhelix is formed. In agreement with this formula, these superhelices possess a negative value of τ. A right-handed toroidal and left-handed self-coiled superhelix are formed when $\beta < \alpha$, and they are characterized by a positive value of τ.

All covalent-bonded circular DNAs known at the present time and iso-lated from natural sources possess superhelices of the right-handed self-coiled type ($\tau < 0$). Superhelices with $\tau > 0$ are formed under artificial con-ditions by untwisting of the double helix in covalent-bonded DNAs (from na-tural sources) by the action of external factors (see, for example, page 231).

The number and direction of the turns of the superhelix thus depend on the conditions under which closure of the covalent ring takes place (they de-termine the value of α), and on the conditions under which the molecule in question is examined (these determine the value of β). The study of super-helicization of DNAs closed in vitro by means of DNA-ligase [315, 316, 331, 332] confirms this statement. This shows [315] that the degree of super-helicization under each given set of conditions depends on the relationship between the temperature and ionic strength of the medium during formation

of the superhelix and during closure of the ring*. A very high degree of super-
helicization can be obtained [316] if the ring is closed in the presence of inter-
calating dyes (dyes producing partial uncoiling of the DNA double helix), and if
these are removed from the DNA molecule after cyclization. Investigation of
superhelicization of covalent-bonded DNAs isolated from different sources
may shed light on the conditions of their formation in vivo. As a measure of
the degree of superhelicization of DNA, the number of turns of the superhelix
per unit length of the molecule can be adopted. Sometimes the value σ, known
as the density of the superhelix, is used. This is the quotient obtained by
dividing the number of superhelical turns τ by the number of β^0, equal to
1/20 of the total number of nucleotides in the double-helical molecule concerned.

$$\sigma = \tau/\beta^\circ$$

The study of the tertiary structure of circular DNAs be means of inter-
calating dyes. It has recently been shown [128] that the phenanthridine dye
ethidium bromide

can interact with DNA. It is assumed [337] that during this reaction the aromatic ring of the dye is introduced (intercalated) between the two neighbouring base pairs†, reducing the angle of rotation between them by approximately 12°. As intercalation of ethidium bromide takes place, the double helix
must consequently be uncoiled, so that the value of β for the circular DNA is
reduced. With the initial condition $\alpha < \beta$, the action of the dye must lead to
a decrease in the number of superturns, and the DNA molecule as a whole
must tend toward a flat circular configuration in which $\alpha = \beta$. With further
intercalation of the dye, β must become smaller than α, and coiling of the ·
superhelix in the opposite direction must be expected.

In the case when in the initial molecule $\alpha > \beta$, the DNA molecule will
not be converted by the action of ethidium bromide into the original flat form,
but will be coiled more strongly as the quantity of dye intercalated into it
increases. Experiments have shown that with an increase in the quantity of
intercalated dye the sedimentation coefficient of circular DNAs at first decreases, passes through a minimum, and then starts to increase, ultimately
attaining a constant value [126, 127].

The minimal value of the sedimentation coefficient evidently corresponds to conversion of the molecule from the superhelicized to a flat circular configuration; the subsequent increase in the sedimentation coefficient

* It is assumed that a change in the ionic strength of the medium leads to a change in the mean
angle of rotation of neighbouring base pairs in the double helix relative to each other [331, 332].

† A similar effect is produced by antibiotics such as actinomycin [338].

corresponds to coiling in the opposite direction. These experiments show that in DNAs isolated from natural sources $\alpha < \beta$. The number of super-helical turns in circular DNA can be determined from data of this type. Knowing the number of moles of fixed dye per pair of nucleotides at the equivalence point, the molecular weight, and the angle of rotation produced by intercalation of the molecule of dye, it is easy to calculate the number of turns of the double helix and, consequently of the superhelix which are uncoiled by the action of this quantity of dye.

It has been concluded from experiments of this type that in many cases the densities of the superhelix (σ) for different DNAs are similar in value, and that for the DNA of papilloma viruses [127, 128], for mitochondrial DNA of chicken liver [126], DNA of virus SV 40 [124], and polyoma virus [127, 128], σ varies from -0.03 to -0.04. Similar values have also been obtained for the circular DNAs from Escherichia coli [332], phage λ [332], and the replicative form of DNA phage ϕX174 [338]. Considerably lower values of the density of the superhelix have been found for the DNAs of HeLa cells and sea urchin eggs [316].

The number and sign of the superhelical turns can be determined also by electron microscopy [114, 129, 336].

It may be considered that tertiary structures similar to those described above are observed in all covalent-bonded circular DNAs. This gives these molecules special properties different from those of ordinary linear DNAs.

V. Investigation of the secondary structure of double-stranded RNA

A considerable number of double-helical RNAs are now known [130–141], differing from ordinary single-helical RNAs in the following respects:

1. They obey Chargaff's rule (i.e., their adenine:uracil and guanine:cytosine ratios are approximately equal to 1).

2. They have a considerably lower molar extinction than ordinary RNAs.

3. On heating they undergo cooperative conversion from the helical into the denatured form (see below).

4. Under ordinary conditions they do not react with formaldehyde, a characteristic property of double-helical hydrogen-bonded molecules.

5. They have a much greater resistance to the action of ribonuclease than ordinary RNAs.

The structure of double-helical RNAs from various sources has recently been investigated by the x-ray structural method [130–133, 136, 140], but as yet the conclusions which have been obtained are nowhere near

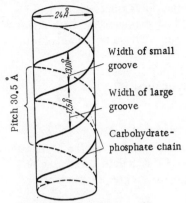

Fig. 4.15. One of the possible macrostructures of the double-helical RNA of rice dwarf virus [136].

so definite as those for DNA. For the RNA of rice dwarf virus [136], on the basis of x-ray structural analysis and circular dichroism, a structure has been accepted which differs from that of the A and B forms of DNA. If the phosphorus atoms of this RNA are taken to form the walls of a cylinder, the diameter of the cylinder will be 18 or 24 Å, and the number of nucleotides per turn of the helix will be 10 (Fig. 4.15). However, recent investigations of the structure of crystals of the double-helical RNAs of reovirus [131-133] show that the roentgenograms agree closely with two possible models of the double-helical RNA, one of which includes 10 nucleotide components per turn of the helix, the other 11. In the first model the bases must be inclined at an angle of 10° to the perpendicular drawn to the axis of the helix, and in the second case 14°. In the first model (10 nucleotides per turn) the bases are turned through 36° relative to each other, compared with 32.7° in the second model. The distance between the base pairs for the two models is 3 and 2.73 Å respectively, while the pitch of the turn in both cases is 30 Å. Despite the fact that the second model – with 11 nucleotides per turn of the helix – agrees slightly better with the experimental data, even now the possibility cannot be ruled out that the first model, with 10 nucleotides per turn of the helix, is correct*.

A distinguishing feature of double-helical RNA is the independence of its conformation on the water content in the crystal [142, 143], which is also observed with the synthetic double-helical polyribonucleotides. These distinguishing features are evidently not connected with the fact that RNA contains uracil instead of thymine (in DNA), since the DNA of phage PBS2, which contains uracil instead of thymine, possesses the ordinary DNA conformation [144]. It can therefore be postulated that either the type of base pairing in the molecule of double-helical RNA differs from that in double-helical DNA, or the sugar residue, which differs in these two types of polynucleotides, has some influence on the conformation of the polynucleotide. The first of these suggestions must be ruled out because x-ray structural analysis of double-helical molecules obtained by interaction between synthetic polyribonucleotides shows that a diffraction pattern similar to that observed for double-helical RNAs is given only by those double helices which are formed by complementary

---

* The possibility that three forms of RNA may exist has recently been discussed (on the basis of the results of x-ray structural analysis of double-helical synthetic polyribonucleotides). These forms differ in the number of base pairs per turn of the helix and in other quantitative characteristics, but they resemble each other much more closely than the A and B forms of DNA [339].

polynucleotides [143].   The difference between the conformations of double-helical RNA and double-helical DNA is thus evidently associated with differences in the structure of the carbohydrate residue in these two macromolecules.

VI. Destruction of the macromolecular structure

of double-helical molecules (denaturation)

The secondary structure of double-helical* double-stranded nucleic acids can be destroyed by various procedures, and depending on the strength of the agent used, the characteristic properties of the double helix may be completely or only partially lost, and ultimately two single polynucleotide chains may be formed.

If a solution of a double-helical polynucleotide is gradually heated, within a certain temperature range which is specific for a given polynucleotide, the properties of the solution suddenly change: the optical density increases, the optical rotation and viscosity decrease, the sedimentation coefficient rises, and so on.   These changes are associated with degradation of the double-helical molecule, leading ultimately to separation of the two complementary polynucleotide chains.   The method most widely used to detect the time of this transition is measurement of the optical density during denaturation.   As an example, the change in optical density of various types of DNA on heating is shown in Fig. 4.16.   Clearly an increase in density takes place over a narrow range of temperatures, and starting from a certain temperature the absorption again becomes constant.   If it is assumed that the optical density at a low temperature corresponds to the completely helicized molecule, and the optical density at the end of melting corresponds to a molecule with completely destroyed secondary structure, the proportion of helical segments (θ) and of nonhelical segments ($1 - \theta$) at any intermediate point of transition can be calculated easily by the formula:

$$\theta = \frac{D_\infty - D_t}{D_\infty - D_0}$$

where D_∞ is the optical density of a solution of the completely denatured molecule, D_t the optical density of a solution of the polynucleotide at a given intermediate temperature; and D_0 the optical density of a solution of the polynucleotide at a low temperature.

---

* Here and subsequently the term "single-helical" polynucleotide will be used to describe a single-stranded structure in which only interplanar (stacking) interactions exist between neighbouring bases of the chain. The term "double-helical" polynucleotide will be used to describe a structure in which, besides this stacking interaction, complementary pairs are formed from bases belonging to different parts of the same chain or to the different polynucleotide chains. These complementation interactions may involve all bases along the polynucleotide chains (double-helical DNA) or only some of them, in the partially double-helical structures (such as tRNA).

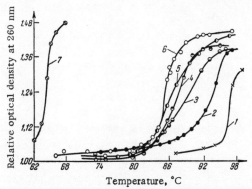

Fig. 4.16. Change in optical density at 260 nm during denaturation of double-helical polynucleotides [145]: 1) DNA of
Mycobacterium phlei; 2) DNA of Serratia marcescens; 3)
DNA of Escherichia coli; 4) DNA from calf thymus; 5) DNA
from salmon sperm; 6) DNA from Diplococcus pneumoniae;
7) Double-helical poly-dAdT complex; all in 0.1 M NaCl
solution with addition of 0.015 M sodium citrate.

The temperature at which the proportions of helical and nonhelical segments are equal ($\theta = 1 - \theta = 0.5$) is called the melting temperature of the polynucleotide and is designated T_m (Fig.4.17). For a given double-helical polynucleotide, under constant external conditions (pH, ionic strength of the
medium, pressure, etc.), the value of T_m is constant and it characterizes
the stability of the double-helical structure. Another characteristic feature
of the denaturation process is the width of the transition interval from the
native to the denatured state (ΔT_m), reflecting the cooperativeness of the
transition, i.e., the degree of simultaneity of destruction of all elements
of the helical structure as the temperature rises. If all the elements of the
helical structure of DNA were destroyed simultaneously at the same temperature, then ΔT_m would be equal to zero. However, DNA never "melts" in that way,
and the process passes through a continuous series of partially denatured states,
so that ΔT_m is not equal to zero (see page 224). The value of ΔT_m is
determined by the difference between the temperatures at which the tangent
at point T_m to the curve of $1 - \theta$ versus temperature intersects the straight
lines $1 - \theta = 1$ (total denaturation) and $1 - \theta = 0$ (absence of denaturation), as
illustrated in Fig. 4.17 [146]. Double-helical molecules are characterized by
a relatively low value of ΔT_m (3-7°C), and this distinguishes them essentially
from single-helical and partially double-helical molecules.

The melting temperature of double-helical molecules depends on many
factors, the more important of which will be considered below.

1. Factors influencing thermal denaturation

Length of the molecule. With a decrease in the molecular weight of a
polynucleotide, its T_m value also decreases [147-151]. This effect has been
investigated in natural and synthetic polynucleotides, and it is seen particularly

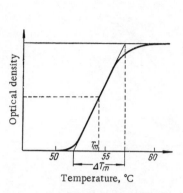

Fig. 4.17. Analysis of the denaturation curve of DNA (see text).

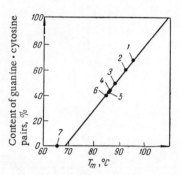

Fig. 4.18. T_m of double-helical polynucleotides as a function of content of guanine · cytosine pairs (for key to points on straight line, see Fig. 4.16).

clearly in the case of short oligomers. For example, by interaction between poly-A and poly-U in the presence of magnesium ions in D_2O at pD 7, the tetranucleotides form a complex with T_m = 17°C, while hexanucleotides form a complex with T_m = 27°C [147]. With an increase in length of the molecule, the differences between the melting temperatures of the double-helical poly-nucleotides are reduced.

Nucleotide composition. The melting temperature of double-helical polynucleotides also depends on their composition. With an increase in the content of guanine·cytosine pairs, T_m of the double-helical molecules in-creases in a straight line [121, 145, 212, 343-347], as shown in Fig. 4.18 (compare with the data for stability of the base pairs on page 191). As a model to establish how T_m of a polynucleotide depends on its base composi-tion, synthetic oligonucleotides can be used. If it is possible to form three hydrogen bonds per pair of bases, the melting temperature of the corre-sponding double-helical polynucleotides will be increased. Nevertheless, the number of hydrogen bonds per base pair by itself does not determine the value of T_m of a double-helical polynucleotide unequivocally. For example, T_m for the double-helical complex of poly-2-aminoadenylic acid with polyuridylic acid is much lower than for the double-helical complex of polycytidylic acid with polyguanylic acid, although they both contain the same number of hydro-gen bonds [147]. This effect has been mentioned previously in connection with the interaction between monomeric components of the nucleic acids (see page 192).

Nature of the carbohydrate residue. Double-helical polynucleotides can be arranged in the following order of stability [147]:

(poly-dI)·(poly-C) < (poly-dI)·(poly-dC) < (poly-I)·(poly-dC) < (poly-I)·(poly-C)

(poly-dG)·(poly-dC) < (poly-dG)·(poly-C) < (poly-G)·(poly-dC) < (poly-G)·(poly-C)

These series show that double-helical polyribonucleotides are more stable than double-helical polydeoxyribonucleotides. The same rule is also observed for natural polynucleotides. For example, T_m of double-helical RNA from rice dwarf virus is 15°C higher than T_m for DNA with the same base composition [135]. This effect is possibly connected with hydrogen bonding between the hydroxyl group of the ribose residue and the corresponding base in the polynucleotide, leading to strengthening of interplanar interactions as the result of a decrease in the freedom of rotation of the base around the glycoside bond [76] (see page 118). However, this explanation cannot be taken as the only one. The difference between the stabilities of the double-helical complexes may be due to interactions with the solvent or to a change in the ring conformation of the carbohydrate residue.

Dependence on external factors. The melting temperature of double-helical polynucleotide complexes rises with an increase in the ionic strength of the solution, as a result of screening of the charges on the phosphate groups. The practically straight line which is observed experimentally when T_m is plotted against the logarithm of the concentration of monovalent ions (up to a concentration of 0.3 mole/litre) [348], is well predicted theoretically on the basis of an examination of the difference between the values of the electrostatic free energy of the phosphate groups in the helical and denatured states [147] (see page 239). At high values of ionic strength, T_m for double-helical polynucleotides reaches a maximum, after which it remains unchanged or falls [152a, 347]. Attempts have been made to represent T_m of double-helical polynucleotides as a function of the content of guanine·cytosine pairs and also as a function of the logarithm of the ionic strength of the medium [212, 346]. In some cases the equations obtained have agreed reasonably well with the experimental data [346].

In a salt concentration of $\sim 10^{-4}$ M, DNA is denatured at room temperature. A definite specificity relative to the ions in the solution is observed. The presence of some bivalent ions (Mg^{++}, Ba^{++}, Co^{++}, Mn^{++}, Ni^{++}, Zn^{++}) increases the T_m of DNA, whereas others (Cu^{++}, Cd^{++}, Pb^{++}) have the opposite effect.

At present there is no general agreement why these ions differ in their action, and although cations are known to be strongly bound with DNA it is not clear whether this takes place through nonspecific electrostatic effects, or whether specific chemical bonding also takes place [147].

The temperature of transition from an orderly to a denatured state also depends strongly on the acid — base properties of the medium [153-155]; at pH values below 2.7 and above 12, denaturation of DNA is observed at room temperature. In the pH interval 5.5-8.5, however, T_m of DNA is relatively independent of pH [109, [154]. The dependence on pH which is observed can be explained by ionization of the bases [154]. It has been concluded from the changes in the UV-spectra [348-351], optical rotatory dispersion [350-353], and circular dichroism [354] during acid titration of double-helical polynucleotides that denaturation is preceded by marked protonation of the residues of one of the bases of the guanine-cytosine pair. However, it is not yet certain which of these two bases undergoes protonation to the greater degree. In alkaline denaturation the first stage is evidently ionization of the guanine residues [355].

Finally, the value of T_m of double-helical molecules is also influenced by the presence of certain substances such as urea, guanidine, etc., in the solution [109]. The presence of organic solvents has an influence on the T_m of DNA. This can be understood from the standpoint of Sinanoglu's theory (see page 211), for with a decrease in the surface energy of the solvent, the difference between the free energy of native DNA, with its small surface, and of denatured DNA, with its large surface, is reduced. The helix—coil transition for DNA is also influenced by pressure [156, 356], the DNA concentration [157, 357], and so on.

The end result of denaturation by the action of temperature, acid, or alkali may be separation of the chains of the double-helical molecule [109]. This is shown by the very slow recovery of the characteristic properties of the original (native DNA) in the course of renaturation (see below) after prolonged incubation of DNA under denaturation conditions. This conclusion is also confirmed by the second order of the reaction velocity of renaturation, the results of electron microscopy [358, 359], and also changes in the hydrodynamic characteristics of the molecule during denaturation [109, 360]. However, the most convincing evidence comes from the possibility of separating the complementary chains after denaturation [109, 185-195].

Denaturation by the action of organic solvents is less clearly understood. Although separation of the chains does take place by the action of formamide [198], cases of denaturation without separation of the chains probably may also occur [158].

Denaturation is not a process of the "all or nothing" type but passes through several intermediate states, characterized at each given temperature by a definite equilibrium distribution of locally denatured regions. This is shown, in particular, by comparison of changes in viscosity and optical density during denaturation [182]. Comparisons of this type also can be used to estimate the mean length of the locally denatured regions within the melting interval [182]. The formation of partially denatured molecules in the course of denaturation is the reason for the well-known discrepancy between the melting temperatures of DNA determined by measurement of changes in optical density actually during heating of the solution and after cooling of the solution heated to each given temperature. The value of T_m determined by the first method is several degrees lower, because the change in optical density actually during heating reflects a disturbance of the helical structure, not necessarily accompanied by separation of the complementary chains, whereas in the second method, separation of the complementary chains is an essential condition for the change in optical density observed [109]. These experiments led to the suggestion that DNA may contain thermostable "nuclei," segments rich in guanine·cytosine pairs which hold together the complementary chains of the partly denatured molecule.

The ability of the double-helical DNA molecule to exist as an entity despite considerable disturbances within the double helix is apparent from experiments to study denaturation of DNA by the action of heat in the presence of formaldehyde [179-181]. Electron micrographs of DNA treated in this way show locally denatured segments, in which the relative content of nucleotides increases with a rise of temperature or with an increase in the duration of formaldehyde treatment. The distribution of these segments is not random, as they always occupy definite places in the molecule. It can be postulated that these are segments rich in adenine·thymine pairs. As the temperature rises, the size of these segments increases, they join together, and the end result is separation of the complementary chains. Denaturation in the presence of formaldehyde, it will be noted, differs from the usual thermal denaturation because it is accompanied by chemical reaction with the formaldehyde, so that renaturation of the regions containing modified constituents is impossible. As a result, at any temperature, not only at the melting temperature, the end result of denaturation is separation of the chains [181]. No conclusions regarding the true size and distribution of these regions at a given temperature in the absence of formaldehyde can therefore be drawn from the results of experiments in which formaldehyde is used to obtain partially denatured regions in DNA molecules within the melting interval.

2. Special features of the denaturation of circular DNAs

Whereas separation of the chains can take place during denaturation of linear and noncovalent-bonded DNAs, this process is impossible in double-helical covalent-bonded circular DNA. In addition the presence of a superhelical structure influences denaturation processes in this case.

During determination of the buoyant density of covalent-bonded DNA of polyoma virus the so-called DNA I) in a caesium chloride density gradient with variable pH (the alkaline region), a two-stage transition is observed to a denatured state characterized by a higher buoyant density. Comparison of these results with those for the change in buoyant density of the same DNA, but with cleavage of one of its chains produced, in particular, by the action of a DNase (the so-called DNA II, Fig. 4.19), shows that the early transition for DNA I is observed in the region of lower pH values than for DNA II, whereas the later stage of denaturation for DNA I takes place at appreciably higher pH values [122]. Similar changes are observed if the pH dependence of the sedimentation coefficient is studied [117, 159, 161].

This evidently means that the first stage of denaturation of covalent-bonded DNA is the untwisting of part of the Watson – Crick double-helical structure, with a resulting decrease in the value of β (see page 219) and, correspondingly, since $\alpha < \beta$ (see page 220), with a decrease in the number of superturns τ, i.e., the molecule is uncoiled and the process ends with the formation of a flat, circular molecule of DNA with a partly untwisted region or regions (I' in Fig. 4.19). With further alkalification of the DNA I', uncoiling of the Watson – Crick double helix continues, but unlike with linear and noncovalent-bonded DNA, this process takes place without separation of the chains. The entropy of the denatured state is thereby lowered and, consequently, denaturation is less advantageous than in the case of the linear and noncovalent-bonded DNA. The end result of denaturation is the formation of a compact coil with high sedimentation coefficient (IV in Fig. 4.19). If one of the chains in such a denatured molecule is broken, a circular single-stranded DNA with sedimentation coefficient 18S (V in Fig. 4.19) and a linear single-stranded DNA with sedimentation coefficient 16S (VI) are formed. The forms V and VI are also obtained by alkaline denaturation of DNA II.

Another interesting feature distinguishing the circular DNAs and observed during alkaline denaturation is the very rapid renaturation when the alkaline solution of such a denatured DNA is neutralized, provided that the pH did not exceed 12.5. As a result of denaturation with higher concentrations of alkali, it is impossible to obtain renaturation (after neutralization) [159] under the conditions when DNA II is readily renatured*. Irreversibly denatured covalent-bonded DNA undergoes spontaneous renaturation after neutralization and treatment with pancreatic DNase [361], forming circular DNA II with single-helical cleavages; this process takes place much more rapidly than renaturation of DNA II. This evidently means that irreversible denaturation of DNA I takes place without cleavage of the DNA chains, although in the process of denaturation they are shifted relative to each other along the axis of the helix, after which the formation of a completely complementary structure on neutralization is hindered by the random formation of "nonnative" complementary pairs [159, 362]. Recreation of the original complementary structure may also be difficult because of steric hindrance through the formation of a superhelical structure. Rapid spontaneous renaturation after the formation of a single cleavage in one of the chains is evidently explained by the fact that under these circumstances there is rapid uncoiling of the superhelical denatured structure, and the two chains, which are in close proximity, can easily form a fully complementary double-helical structure, as occurs, for example, during renaturation of denatured DNA with cross-linkages between the complementary chains [160].

Thermal denaturation of circular DNA occurs at a much higher temperature and is characterized by a higher value of ΔT_m than in the case of linear or noncovalent-bonded DNAs [122]. This increase in T_m of the circular DNA is readily explainable in terms of enthalpy – entropy changes taking place during "melting." Since $T_m = \Delta H / \Delta S$, and since, as was mentioned above, ΔS for denaturation of circular DNA is smaller than ΔS for denaturation of linear DNAs (because it is impossible for the chains to separate), whereas ΔH in both cases is the same, T_m for circular DNA must be substantially higher. The hyperchromic effect corresponding to the first stage of the transition (uncoiling of the superhelical structure) cannot be observed during

---

* In alkaline solution at pH 12 and 50°C, however, renaturation of this DNA takes place relatively quickly.

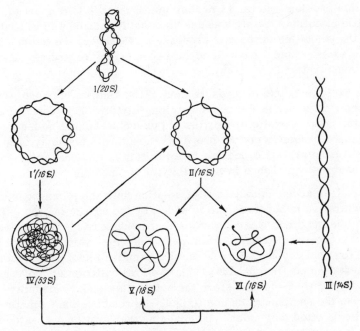

Fig. 4.19. Scheme of conversions of covalent-bonded circular DNA
(see text) [161].

thermal denaturation, because the change to form I is accompanied by untwisting of only about 1-3% of the helical structure.

The formation of form I is also observed during denaturation of covalent-
bonded circular DNAs in the presence of formaldehyde [162]. Heating DNA I
of polyoma virus in the presence of formaldehyde leads to the formation of a
molecule sedimenting at the same rate as DNA II, but having no cleavages.
Formaldehyde, by attaching itself to segments denatured by heating, prevents
their subsequent helicization on cooling [163]. Electron photomicrographs
of DNA of papilloma virus treated in this way [162] show that a mixture of
superhelical and flat circular DNAs is formed after this treatment, and
the relative proportion of the latter in the mixture increases with an increase
in the temperature at which the treatment is carried out. The number of flat
molecules formed reaches a maximum at 42°C; at this temperature only 8%
of superhelicized molecules can be found. However, with a rise of temperature their number increases again. This means that thermal denaturation
takes place by the same method as alkaline denaturation: form I is converted
into the flat form I', and this later forms the superhelical structure again,
but with the superhelix in the opposite direction.

Kinetics of denaturation. An understanding of the features distinguishing
the secondary structure of polymers is largely dependent on knowledge of the
kinetic characteristics of the processes of its breakdown and formation. Until

recently this problem, so far as nucleic acid research is concerned, has received little attention, mainly because no suitable method was available to measure the velocities of the fast processes. However, the information now available on this problem enables certain conclusions regarding the mechanism of denaturation to be drawn.

The basic principle of measurement of the kinetics of denaturation is to bring the DNA under denaturing conditions rapidly and to determine changes in any of the characteristic properties of the double-helical molecule in relation to time. Changes in optical properties are most commonly used to characterize the denaturation process. Denaturation is produced either by the action of heat or by a change in the acidity of the medium.

If alkali is added quickly to DNA solution until the pH changes to 12.5, rapid denaturation is observed, as is shown by the change in optical properties [164, 165] (the maximal hyperchromic effect is observed). If, however, the solution is neutralized after a certain period of time (which must not be too long), the original double-helical structure remains intact, as is shown by the sedimentation coefficient [166, 167]. Neutralization after more prolonged exposure to alkali does not restore the original secondary structure and, judging from the sedimentation coefficient, the result is the formation of single-helical coiled molecules. By changing the duration of treatment with alkali, it is possible to determine the time required for irreversible denaturation. This time depends on the temperature and viscosity of the medium*. These results are in agreement with the hypothesis that denaturation takes place through uncoiling of the double-helical molecule. The velocity of this process must diminish with an increase in the length of the molecule.

More complete information about the mechanism of the process can be obtained from kinetic data for thermal denaturation: a DNA solution is heated for a very short time, and the changes in optical density [168-174] or in the polarographic characteristics [175-177] are then recorded. It has been found that the kinetic picture of the conversion varies with the relationship between the temperature T_1 of the original DNA solution, the melting temperature T_m of that DNA, and the final temperature T_2. If T_1 is 20-30°C below T_m and if T_2 is not very high (5-10°C above T_m), changes in the properties of the molecule begin to be detectable after a definite induction period [172]. This induction period disappears with an increase in T_1 or T_2. If the DNA molecule has a certain number of dehelicized segments at T_1, then if the value of T_2 is high enough, a stage of extremely rapid increase in the degree of dehelicization is observed, characterized by an instantaneous increase in optical density, and this is followed by a rapid but measureable further increase in optical density. The magnitude of the initial jump increases with an increase in T_1 and T_2, and if the value of T_2 is high enough, the observable secondary structure disappears completely at the first, instantaneous stage. The time of completion of the

---

* See also [363]. The data for the dependence of the denaturation velocity on molecular weight are contradictory [166, 167, 363].

process following the instantaneous stage (provided that T_2 is high enough*) is approximately proportional to the square of the molecular weight. The presence of an instantaneous stage of denaturation has also been observed in experiments on the acid denaturation of DNA and the relative contribution of this stage to the process as a whole was increased with a decrease in the final pH value [178].

Instantaneous denaturation evidently takes place by the same mechanism as denaturation of circular DNA: interaction between the bases is disturbed without untwisting of the chains. The last stage is evidently a process of separation of the polynucleotide chains.

It is more difficult to interpret the presence of an induction period when the initial temperatures are low; perhaps it is associated with the need for active centres, without which untwisting of the chains cannot effectively take place [172].

VII. Processes leading to restoration of the double-helical structure (renaturation)

The restoration of the collapsed secondary structure is known as renaturation. The study of restoration of the secondary structure, when disturbed by one of the methods described above, is extremely important, no less so than investigation of denaturation itself. During renaturation of denatured DNA under optimal conditions (which we shall discuss below), the final result is the at least partial restoration of the original (native) structure of the DNA. This conclusion can be deduced from the following data:

1. Renatured DNA possesses optical properties (hypochromism, optical rotation, circular dichroism) corresponding to the native preparations [109].

2. The hydrodynamic characteristics of completely renatured DNA differ only very slightly from those for native double-stranded DNA, but they differ very considerably from the characteristics of the denatured molecule [109, 198].

3. X-ray structural analysis of renatured DNA gives the typical diffraction picture for the B form of DNA [109].

4. Electron-microscopic investigations show that (in the case of undegraded preparations) after renaturation molecules corresponding in length and diameter to the native preparations are formed [109, 183, 184].

It is concluded from these data that the renatured molecule corresponds in its physical properties to the native molecule, but it cannot be concluded from them that it is in fact the true native form. At the present time, such information can be obtained only by biological assessment of the renatured

---

* At low values of T_2 yet another stage is observed, but this will not be considered here because its nature is not yet clear.

DNA. Already, numerous investigations have shown that the biological properties are restored after renaturation. In one of these investigations,
restoration of the transforming activity of "transforming" DNAs was demonstrated (the denatured molecules are inactive). Restoration of biological
activity takes place during the process of renaturation, and the rate of its
restoration coincides with the rate of renaturation (as estimated spectrophotometrically [152]).

Methods of separating the complementary chains of DNA have recently
been developed [185-195]. During renaturation of single strands of the DNA
of phage λ, each of which alone possesses no biological activity (not more
than 10% of the initial activity), restoration of the biological function by about
half is observed [188]. It is important to note that only complementary chains
will interact with the formation of a double-helical molecule; chains of the
same type cannot undergo renaturation.

As a result of renaturation, the immunochemical properties characteristic of double-helical molecules also are restored [109].

If degraded specimens of DNA are used for renaturation, appreciable
differences are observed between the physical properties of the renatured
and original molecules [198]. For example, in electron micrographs of
renatured molecules, the ends appear coiled, while the middle of the molecule has the characteristic appearance of double-helical complexes [359].
The single-helical ends of these molecules can form double-helical segments
with complementary single-helical ends, as a result of which aggregates of
high molecular weight are formed [198]. Molecules differing from the native
form are also obtained by renaturation of denatured DNA with cyclic re-
arrangements, such as in the case of DNA of the T-even phages. Again,
aggregates consisting of a large number of molecules can be formed [198].

All this is evidence of the random combination of the complementary
chains during renaturation, as a result of which chains of different lengths
can be united.

The random character of renaturation has been conclusively proved by
experiments with DNA molecules labelled with heavy isotopes of nitrogen and
hydrogen. A double-helical molecule, both chains of which contain heavy
isotopes, has a definitely greater buoyant density than completely unlabelled
DNA. If an equimolar mixture of DNA, labelled in this way, and unlabelled
DNA is denatured, its subsequent renaturation leads to the appearance of
three zones corresponding in buoyant density to completely labelled (heavy),
completely unlabelled, and intermediate DNA, i.e., double-helical DNA with
one labelled and one unlabelled complementary chain*. The ratio between the
quantities of DNA in these zones is $1:1:2$, reflecting the random character
of renaturation [109, 196].

---

* The experiments just described provided the first adequate proof of separation of DNA chains by
denaturation.

1. Factors influencing the renaturation process

Two types of factors influence renaturation: factors dependent on the properties of DNA itself, i.e., structural, and factors associated with the conditions of renaturation.

Structural factors. This first group of factors naturally includes the molecular weight and nucleotide composition of the DNA, as well as a special characteristic, the complexity of the DNA. This last factor is measured by the number of nucleotide pairs in nonrepeating DNA sequences [184, 365]. In the case of DNA from viruses and bacteria, its complexity is simply the number of nucleotide pairs in the intact undegraded molecule, since the DNA of these organisms does not contain repeating sequences of sufficient length [365]. However, the DNA of higher organisms evidently contains very large numbers of repeating sequences, each with 200–300 nucleotide units [365].

Investigations of the kinetics of DNA renaturation under different conditions and using different methods have shown that this process is a second order reaction, embodying the following principles determined by the properties of the DNA [109, 184, 198, 199, 366].

For any given DNA a decrease in its molecular weight by cleavage is accompanied [184] by a decrease in the velocity constant of renaturation k_2, so that it remains permanently proportional to the square root of the length of the molecule L (length in this case is defined as the mean number of nucleotides in the single-stranded polynucleotide): $k_2 \approx \sqrt{L}$.

A study of the velocity of renaturation of fragments of equal length obtained from DNAs of different complexity shows that the velocity of their renaturation is inversely proportional to their complexity [184, 197, 366]. This provides, in principle, a method of determining the complexity of a molecule, or in the case of virus and bacterial DNAs, the molecular weight [184], on the basis of the kinetics of renaturation of degraded DNAs from the known length (which can be determined, for example, from the sedimentation constant) of the single-stranded fragments. If the fragments undergoing renaturation are of identical length, the DNAs can thus be arranged in the following order of velocity of renaturation:

DNA of phages and viruses > DNA of bacteria ≫ DNA of animals
(the slow component
is discussed below).

The observed velocity constants of renaturation for denatured DNAs of different origin are compared in Table 4.13. The general expression for the second order velocity constant is thus

$$k_2 \approx \frac{\sqrt{L}}{N}$$

where L is the mean length of the single-stranded polynucleotide and N the complexity of the native DNA molecule.

TABLE 4.13. Dependence of Velocity of Renaturation on Mean Molecular Weight of Denatured DNAs and on Complexity of Native DNA [184]

Source of DNA	Content of guanine·cytosine pairs, %	Sedimentation constant of denatured DNAs at pH 13 $S_{20,W}$	k_2, litres /mole ·sec	Molecular weight of native DNA (from k_2)
Ascites carcinoma of rat	42	6.8	0.0026	$(2.5\pm0.5)\cdot10^{12}$
Escherichia coli.	50	12.3	6.0	$(2.5\pm0.5)\cdot10^{9}$
Phage T4.	34	28.5 / 8.0	190 / 39 }	$(1.3\pm0.10)\cdot10^{8}$
Phage No. 1	64	30 / 7.9	1490 / 330 }	$(3.3\pm0.15)\cdot10^{7}$
Phage T7.	49	34.7 / 7.5	1790 / 320 }	$(2.5\pm0.1)\cdot10^{7}$
Virus SV 40	4J	6.6	1080	$(3.3\pm0.1)\cdot10^{6}$

The velocity of renaturation depends on the nucleotide composition of the DNA, and it increases slightly with an increase in the content of guanine ·cytosine pairs [184, 366]. If the original native DNA contains repeating sequences, on degradation to fragments roughly corresponding in length to one sequence, the concentration of these fragments will exceed the concentration of each fragment represented only once in the DNA by the same number of times as the repeating sequence is present in the DNA. Accordingly, renaturation of denatured DNAs from higher organisms is complex in character and consists of fast stages, during which the more numerous repeating sequences undergo renaturation, and slow stages, for sequences represented only once. The kinetics of renaturation can thus be used to detect repeating sequences [365, 369].

The theoretical examination of the kinetics of renaturation is based on the following propositions.

The first stage, which determines the velocity of the process, is formation of the "nucleus," consisting of one or several base pairs corresponding to pairs present in the native molecule. The formation of this "nucleus" (or "nuclei") may take place at one or several definite sites of the denatured molecule. After formation of the nucleus, the double helix is formed rapidly.

The velocity constant of renaturation is determined by the formula:

$$k_2 = \beta^3 k_N L/N$$

where β is the ratio between the number of nucleotides capable of being components of the developing nucleus, and the total number of nucleotides in the native molecule; k_N is the velocity constant of formation of one of the nuclei.

A possible explanation of discrepancies between theoretical predictions and the experimental data for the relationship between velocity of renaturation and length of the molecule is that, because of the coiled character of the denatured molecules, not all potential nuclei are formed equally quickly, for some of them are concealed inside the coil.

In some chains of denatured DNA, helical segments can be formed by interaction between complementary bases located at different points of the chain [359, 367, 368]. The number and stability of these segments will naturally increase with a decrease in temperature and an increase in ionic strength of the medium [367, 198]. The formation of segments of this type may affect the formation of nuclei, thus decreasing the rate of renaturation, and may also influence the subsequent helicization. If two chains with stable helical segments interact, only partial renaturation can take place. Segments capable of forming new nuclei with other chains can thus be left in the interacting chains, so ultimately it is possible for branching aggregates to be formed from many chains [367]. Renaturation must therefore be carried out under conditions in which the formation of internal double-helical segments is reduced to the minimum, yet the double-stranded structure still remains sufficiently strong.

Effect of external conditions. The curve of velocity constant of renaturation of DNA versus temperature has a maximum* at a temperature of about $T_m - 25°C$ [109, 184]; the precise position of this maximum depends on other conditions, especially on the ionic strength of the medium [368].

Within the pH interval 5-9, the velocity of renaturation is independent of pH; at lower and higher pH values the velocity of renaturation is reduced because of ionization of the bases.

Depending on the temperature at which the renaturation takes place, the ionic strength of the medium varies in its effect on the velocity of the process [368]. At sufficiently high temperatures (60-80°C), when there are virtually no intramolecular double-helical segments in the chains of denatured DNA, the velocity of renaturation rises steadily with an increase in ionic strength [366, 368]. If the conditions are such that the helical segments are stable (at 25-36°C), the dependence on ionic strength is complex in character. With an increase in salt concentration the velocity of renaturation at first increases because of screening of the charges on the phosphate groups and a decrease in repulsion between the chains. Later, when the degree of intramolecular helicization is about equal to half its possible maximum, the velocity of renaturation passes through a maximum and then declines [368]. At an ionic strength corresponding to the maximum of intramolecular helicization in the chains, the reaction velocity falls to a minimum, and rises again with a further increase in salt concentration [368]. At temperatures used for renaturation (50-60°C), the optimal salt concentration is about 0.4 M. At high ionic strengths of the medium, the actual order of the reaction may be changed.

* See, however, [366].

Finally, the velocity constant of renaturation is inversely proportional to the viscosity of the medium [184, 198, 199] if the difference between T_m of DNA and the renaturation temperature remains constant.

For practical purposes, the following conditions of renaturation are therefore optimal: a temperature 25°C below the value of T_m for the particular DNA; ionic strength 0.4; pH of the solution within the interval 5-9.

The reasons for the observed dependence of the velocity of renaturation on temperature and viscosity are at present the subject of discussion [184, 366].

Interaction between denatured DNAs and the complementary RNA chains evidently takes place by the same mechanism and is governed by the same rules [200, 364, 370-372].

The formation of a stable double-helical structure by interaction between complementary chains has been used very widely in molecular biology to determine genetic relationships between microorganisms, to determine the origin of various types of RNA in the cytoplasm of cells and, finally, to locate certain types of mutations in the DNA chain. In all these investigations, the investigator is interested only in the formation of mixed, hybrid molecules by the combination of two complementary DNA chains belonging to different organisms, or of the particular DNA and RNA which he is studying, and not in restoration of the structure of the original denatured DNA. The term "hybridization" is used to describe renaturation with the formation of hybrid molecules [109, 201-211].

Methods of renaturation in aqueous solutions as described above necessitate the use of fairly high temperatures (of the order of $T_m - 25°C$), and this leads to degradation of the nucleic acids during the renaturation process. If renaturation is carried out in the presence of substances lowering the stability of the double helix, such as organic solvents [373-375] or urea [376], and if a suitable concentration of the denaturing agent and suitable ionic strength are chosen, highly specific renaturation can be carried out rapidly and at a comparatively low temperature.

It can be concluded from investigations of the denaturation and renaturation of DNA that, under certain conditions, the double-helical Watson and Crick structure of DNA is more stable than the structure of its separated single-stranded components.

2. Intramolecular interactions in DNA

The behaviour of DNA in solution is determined by interaction between the bases, the influence of the solvent, and electrostatic effects of the phosphate groups. (The types of interactions between bases and the interplanar interactions between hydrogen-bonded base pairs have already been examined.) Regarding the influence of the solvent, it may be observed that as the structure of DNA is very compact, and the space between the bases is

impermeable to water molecules, a hydration membrane is formed around
the outer surface of the DNA double helix. Since the area of the outer surface
of the double-helical DNA molecule, which can be regarded approximately as
a cylinder with radius 10 Å, is much less than the surface area of the single-
stranded molecules formed by denaturation, according to Sinanoglu's theory (see
page 211), the double-helical state must be much more advantageous than the single-helical, from the point of view of solvent effects on the free energy of the system.

Besides interactions between bases and solvent effects, another important
factor influencing the stability of the double-helical molecule is interaction between charges on the phosphate groups. In native DNA these interactions take
place between charges on neighbouring phosphate groups in the same chain and
also between charges on groups in opposite chains of the double-helical complex. In the denatured state, only interactions between phosphate groups within the same chain remain. The distance between neighbouring phosphate
groups in one chain of the Watson and Crick helix (7 Å) corresponds almost
to the maximum possible distance between them. In the denatured state,
neighbouring phosphate groups are also the maximum possible distance apart,
evidently because of repulsion of negative charges. Interaction between neighbouring phosphate groups in the same chain both in the native and denatured
state must therefore be approximately the same, and it can have no significant
effect on differences in the stability of these two states. However, interaction
between distant phosphate groups belonging to the same chain may differ in
the native and denatured states because of differences in the conformations of
the double-helical and single-helical molecules. If this effect is disregarded,
it must be assumed that the main difference between the two states is interaction between phosphate groups belonging to opposite chains of the double-
helical structure, which is absent in the single-stranded polynucleotide. On
this basis it is possible to calculate the electrostatic potential created at the
site of any phosphate group of one chain by all charges on phosphate groups
in the other chain, and also to estimate the value of the electrostatic free
energy due to this interaction [147, 212-214].

In the presence of salts, because of a change in screening of the charges
on phosphate groups by the ionic atmosphere, the electrostatic free energy is
modified. Since the electrostatic free energy is the difference between the
free energies of interaction between charges on phosphate groups in the native
and denatured states of the DNA, a change in this value must affect the melting temperature of the double-helical structure. In fact, a linear relationship is observed between the values of the electrostatic free energy calculated
for different values of ionic strength and the corresponding values of T_m. An
increase in the screening of phosphate groups, reducing the free energy of
interaction between these groups, increases the stability of the double-helical
state, thereby increasing T_m.

Hence, whereas effects associated with interaction between bases and
interaction between polynucleotides and the solvent stabilize the double-helical
structure of the molecule, electrostatic interactions between charges on the
phosphate groups destabilize it.

An important problem arising in connection with denaturation of DNA is that of the causes of cooperativeness of the transition into the denatured state. A solution to this problem can be found by application of modern theories of the helix — coil transition for DNA, in which this process is considered from the point of view of statistical physics [146, 215-222, 377-380]. Cooperativeness of melting in DNA is due to interplanar interactions between base pairs. If ε represents the change in free energy during interplanar association of two isolated, but hydrogen-bonded, base pairs, the mutual arrangement of which corresponds to the Watson — Crick model, the cooperativeness of the transition can be characterized by the cooperativity factor σ:

$$\sigma = e^{-\frac{\varepsilon}{RT}}$$

where R is the universal gas constant, and T the absolute temperature in °K.

In the case of synthetic homopolynucleotides, the width of the transition interval is related to the cooperativity factor by the equation:

$$\Delta T_m = 12.4R \, \frac{T_m^2}{\Delta H} \, \sigma^{2/3}$$

i. e., it is proportional to $\sigma^{2/3}$ (here ΔH represents the enthalpy of the helix — coil transition, and the remaining symbols have their previous values).

As was mentioned above, an intermediate stage in the melting of DNA is the formation of defects in the double helix, so that in the intermediate state alternate helical and nonhelical segments are formed. The mean length of a helical segment ν at the melting point is also connected with the cooperativity factor in the case of double-helical synthetic homopolynucleotides with a random distribution of defects appearing in the double helix along the chain during melting. This relationship is expressed by the equation:

$$\nu \approx \frac{1}{\sigma^{2/3}}$$

For DNAs, which differ slightly in the mechanism of their melting from homopolynucleotides, defects arise in the double helix initially in sites rich in adenine·thymine pairs. For the width of the transition interval of DNA:

$$\Delta T_m \approx \frac{1}{[\ln \sigma]} \approx \frac{1}{\varepsilon}$$

It thus follows from the theoretical examination of melting of DNA that, with an increase in the free energy of interplanar interaction between base pairs, the width of the transition interval is reduced and the number of nucleotides in helical segments at the transition point is increased.

The free energy of interplanar interaction between base pairs can be determined from experimentally obtained values for the enthalpy of helix — coil transition and the width of the transition interval. Assuming that $\Delta H = 8$ kcal /mole [147, 155], the change in the free energy ε during association will be

approximately 7 kcal/mole for double helices composed of synthetic homopolynucleotide chains.   The calculated values of ΔT_m are in good agreement
with experimentally obtained values of the transition interval.   The conclusion that, with a decrease in the molecular weight of DNA the value of T_m must
fall and ΔT_m must rise, is also confirmed experimentally.

To complete this examination of double-helical molecules, a few words
are necessary on the effect of the macromolecular structure on reactivity of
the bases composing these molecules relative to various chemical agents.
The general rule here is evidently a sharp decrease in reaction velocities
compared with those of the monomers.

The reaction with formaldehyde, widely used to investigate the structure of nucleic acids, can serve as an example.   Although this compound
reacts readily with the bases in nucleosides and nucleotides, it hardly reacts at all with DNA [233].   Similar effects are observed for reagents such as
carbodiimide [224, 225] and semicarbazide [226].   This problem is examined
in more detail for particular reagents in the second half of the book.   It is
difficult at present to give a precise specification of all the factors responsible
for this diminution of reactivity, but it is quite certain that an important role
in this effect is played by steric factors associated with packing of the bases
within the double helix, and also by energetic factors when the reaction takes
place at an atom participating in a system of hydrogen bonds.

It is important to note in this connection that DNA in solution is not a
static structure, fixed once and for all, but a dynamic structure, continuously
undergoing local denaturation and renaturation [227].   This is evidently the
reason why chemical reactions with groups of bases located within the structure of the double helix can take place.   This applies, in particular, to the
exchange of DNA protons which participate in hydrogen bonding for deuterium
and tritium [97, 98, 228, 341], and also to the reaction with formaldehyde
[381].   The possibility of local denaturation and, consequently, the velocity
of the chemical reaction diminish with an increase in the degree of cooperativeness of the double-helical molecule.

VIII. Single-stranded polynucleotides

Natural polynucleotides possess a single-helical conformation only
when in a denatured state.   They are either double-helical, like DNA, or
they include double-helical segments alternating with single-helical, like
RNA.   To understand the properties of single-stranded RNAs, it is therefore
essential to have model compounds consisting of practically single-helical
molecules, on the basis of whose properties conclusions can be drawn regarding the properties of the single-helical segments of RNA.   Suitable model
compounds are the synthetic homopolynucleotides: polyadenylic and polycytidylic acids in neutral and alkaline media* and polyuridylic acid† [100].

* In an acid medium, these polynucleotides form double-helical complexes [100].
† The structure of polyguanylic acid has not yet been adequately explained [100].

Although in their hydrodynamic properties, polyadenylic and polycytidylic acids in neutral and alkaline media are statistical coils [229-232], the presence of a hypochromic effect [57, 60, 232, 233], of circular dichroism [71, 234, 235], and of optical rotatory dispersion [62, 236, 237] indicate that interactions take place between the bases in these polynucleotides. The curves of optical rotatory dispersion and circular dichroism are very similar to the corresponding curves for dinucleoside monophosphates (see page 200), and this suggests that no hydrogen bonds are formed between the bases in the homopolymer. Theoretical curves of optical rotatory dispersion for polyadenylic and polycytidylic acids [238], calculated from the properties of dinucleoside diphosphates, give results in agreement with experimental observations*. The most direct evidence in favour of the single-helical structure of these polynucleotides is the analogy between their properties and those of polynucleotides in which the formation of hydrogen bonds is impossible because of substitution of the corresponding hydrogen atoms by alkyl radicals [71, 239, 240], and also the results of kinetic investigations of the reaction with formaldehyde [241].

Differences between the optical properties of homopolynucleotides and their isolated monomer components are due to interplanar interactions between neighbouring bases, leading to interference of the chromophore groups of the bases with one another. This effect is observed at the dinucleoside monophosphate level (see page 200). With a rise of temperature, a gradual change, which is noncooperative in character, takes place in the optical properties of the homopolynucleotides [57, 70]. As an example, the change in strength of rotation of the positive circular dichroism band of oligoadenylates of different lengths with elevation of the temperature is shown in Fig. 4.20.

Analysis of the curves leads to the conclusion that the values of T_m, and also the changes in enthalpy, entropy, and free energy of the oligoadenylates, during the transition from a state in which the bases interact to a state in which (judging from the optical properties) they do not interact, remain approximately the same regardless of the length of the chain [57, 70].

These results are confirmed by data for the hypochromic effect and optical rotatory dispersion [57, 60, 77, 242]. Although, just as in the case of ApA (see page 209), the thermodynamic parameters differ when determined by different methods, the general conclusion that they are independent of chain length remains unchanged. This means that the free energy of attachment of one base to an existing sequence of orderly bases is equal to the free energy of interaction between the two bases in the absence of any preliminary orderliness. Consequently, the equilibrium constant between interacting and

---

*In the case of deoxypolynucleotides, the calculated values do not agree with the experimental data [340]. This is evidently due to what is known as the terminal effect: nucleoside components at the ends of the chain and those inside the chain interact to a different degree with water. The reasons for the differences between ribo- and deoxyribonucleotide oligomers and polymers are not clear [76, 340].

It is interesting to note that 2'-O-methylribonucleotide oligomers and polymers are closer in their properties to ribo- than to deoxyribo-derivatives.

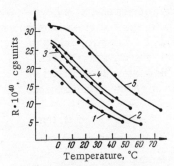

Fig. 4.20. Changes in strength of rotation of positive band in solutions of oligoadenylates of different lengths with an increase in temperature at pH 7.4 [70]: 1) $(A)_3$; 2) $(A)_5$; 3) $(A)_7$; 4) $(A)_{12}$; 5) $(A)_n$ $(n > 50)$.

noninteracting bases in a dinucleotide is equal to the corresponding equilibrium constant in the polymer. This is the typical situation for noncooperative interactions, and it differs substantially from cooperative ones (see page 240). From the standpoint of macromolecular structure this means that at any given moment of time only short sequences of stacked bases can exist in single-helical polynucleotides, with regions of noninteracting components between them. Moreover, the orderly regions are continuously being broken down and reformed at random along the polynucleotide chain. This is the first and principal distinguishing feature of the secondary structure of single-helical polynucleotides; in double-helical polynucleotides interaction between neighbouring base pairs is much more stable.

The mean length of the orderly regions in single-stranded polynucleotides increases with a decrease of temperature. At 20°C, for example, the helical regions contain about 2/3 of the total number of bases, and at 0°C as much as 7/8 of the total number. This leads to a change in the conformation of the whole molecule, which is manifested, for example, as a rapid increase in viscosity [100].

The second distinguishing feature of single-helical polynucleotides is the independence of their melting temperature, determined from changes in optical characteristics, of the ionic strength of the medium [57], at values of $\mu = 0.01$ or above. This is evidently the result of the fact that the distance between neighbouring phosphate groups is the maximum possible both when the corresponding bases are stacked and also when no interaction exists between them (this corresponds to the situation within one chain of the double-helical molecule: see above). The transition from an orderly to a disorderly state cannot depend on electrostatic interactions. Finally, the reactivity of single-helical polynucleotides is much greater than that of double-helical molecules. This is because of the greater accessibility of the reactive groups in single-helical molecules.

The properties of single-helical polynucleotides thus differ essentially from the properties of double-helical molecules. In RNA molecules, in which all four types of bases are present in one chain, both double-helical and single-helical regions exist, as will be shown below, and as a result of this, the properties of these compounds are largely intermediate in character between the two cases examined above. This is reflected particularly clearly in their reactivity toward chemical agents.

The properties of ribonucleic acids will be examined relative to two classes of these compounds for some of which the primary structure is now known, namely: transfer RNAs and 5S ribosomal RNAs. The properties of ribonucleic acids of higher molecular weight are in many ways analogous, but their secondary structure cannot at present be discussed at the level of concrete models, because the sequence of the bases in their polynucleotide chain is still unknown [243, 388].

1. The secondary structure of tRNA

It was stated in Chapter 1 that the tRNAs are evidently the shortest of the natural polynucleotides at present known. They also have a higher content of minor components than the other polynucleotides. On the basis of these two facts, a primary structure has been established for a number of tRNAs. This, in turn, has enabled concrete models of their secondary structure to be built. However, even before the nucleotide sequence was established, certain conclusions were drawn regarding general features of the secondary structure of tRNA [244, 245].

Form of the tRNA molecule in solution. Temperature transitions. With elevation of the temperature of tRNA solutions, the optical density at 260 nm increases [246-249] and the optical rotation decreases [250]. Melting of the macrostructure in the presence of monovalent cations is only weakly cooperative in character, and the width of the ΔT_m interval is about 40°C. The value of T_m for various tRNAs is dependent on the ionic strength of the medium. For total tRNA from yeast in 0.001 M phosphate buffer (pH 6.7), for example, $T_m = 50$°C, while in 1 M NaCl it is about 65°C (Fig. 4.21). The observed hyperchromic effect amounts to 23-28% and is completely reversible, disappearing on cooling [246, 247] (even during rapid cooling [149], when the double-helical polynucleotides are not renatured). In the presence of bivalent ions the character of melting of tRNA changes: T_m rises considerably and the melting interval is narrowed. For example, T_m of total yeast tRNA is raised to 72°C in the presence of $3 \cdot 10^{-3}$ M magnesium chloride, and 75% of the total hypochromic effect is observed in an interval of 10°C. The values of T_m and its changes following addition of magnesium ions differ for different individual tRNAs, but the general tendency is for T_m to rise with an increase in the content of guanine and cytosine and with an increase in the ionic strength of the medium, as well as for the cooperativeness of the transition to be strengthened in the presence of magnesium ions.

The tRNAs differ significantly from single-helical homopolynucleotides in the marked dependence of their T_m on the ionic strength of the solution. This fact suggests that the tRNA molecule contains double-helical segments with bases linked together as in the Watson and Crick structure. This view is confirmed by the difference in reactivity of bases incorporated in tRNA relative to certain chemical agents. After a rapid reaction between any reagent and the tRNA molecule, a much slower stage is observed. Analysis shows that at the end of the fast stage only some of the potentially reactive bases

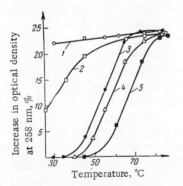

Fig. 4.21. Temperature change in
optical density of tRNA solutions
for different values of ionic strength
of medium [246]: 1) solution of
tRNA in water; 2) in 6 M urea solution; 3) in 0.001 M phosphate buffer,
pH 6.7; 4) in 0.1 M NaCl solution;
5) in 1 M NaCl solution.

have taken part in the reaction, the actual proportion depending on the experimental conditions. On the assumption that the rapidly reacting bases are located in the single-helical segments, and those reacting more slowly in the double-helical segments, this method can be used to estimate the degree of helicization of the complete polynucleotide molecule.

It has been concluded from the results of a study of interaction between tRNA and formaldehyde [251-255] in the presence of Mg^{++} and a study of the ability of hydrogen atoms of tRNAs to be exchanged with ^3H_2O [228] that approximately 70-75% of bases in the molecule participate in the formation of complementary pairs.

Differences in the activity of bases incorporated in tRNA have also been found following its modification with perphthalic acid [257], which reacts in the presence of Mg^{++} and oxidizes only 10% of the total number of adenine residues in the tRNA molecule. Acrylonitrile, in a 0.5 M solution of sodium chloride, converts 2.2% of the pseudouridine contained in tRNA and thereafter virtually ceases to react with the bases [258]. Hydroxylamine, in an alkaline medium [259], also acts selectively by modifying only 25% of the total number of uracil residues in tRNA at a low temperature (about 10°C). If, however, the temperature of the reaction is increased to 40°C, practically all the uracil nucleotides are converted (see page 410). Double-helical and single-helical segments of the tRNA chain likewise differ in their sensitivity to RNases, which split double-helical polyribonucleotides much more slowly than single-helical [256].

Investigations using synthetic polynucleotides and, in particular, experiments to study complex formation between polyuridylic acid and a copolymer of adenylic and uridylic acids, have led to the conclusion that complexes possessing defects in the double helix can exist. These defects consist of nonhelical segments in which noncomplementary bases lie opposite each other [260]. The application of these concepts to a hypothetical RNA possessing a random base sequence led to the conclusion that for a random sequence of 90 nucleotides it is possible to build a structure with 60% of its bases linked by hydrogen bonds [260]. This corresponds approximately to the estimate obtained by chemical methods. In the light of these concepts of the helical structure of RNA with defects, the degree of helicization can also be determined from measurements of the change in optical density of tRNA solutions at different wavelengths with an increase of temperature. The increase in

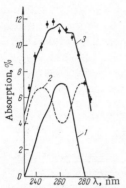

Fig. 4.22. Denaturation spectra of polynucleotides obtained by subtracting spectrum at 0°C from spectrum at 100°C [262]: 1) double-helical (poly-A)·(poly-U) complex; 2) double-helical (poly-G)·(poly-C) complex; 3) tRNA from yeast.

optical density during melting of the secondary structure is evidently connected with three principal effects: the breakdown of pairs of the guanine ·cytosine type, the breakdown of pairs of the adenine·uracil type, and disturbance of interplanar interactions between bases in nonhelical areas. As a first approximation it can be assumed that all three effects make their independent contribution to the change in properties of tRNA during melting of the secondary structure. Changes taking place in the UV-spectrum of tRNA as a result of break-down of base pairs can be assessed from data for melting of synthetic double-helical polynucleotide (poly-A)·(poly-U) and (poly-G)·(poly-C) complexes. A characteristic feature of the changes is a differ-ence between the extinctions of denatured and helical forms at different wavelengths [261, 262]: what is known as the denaturation spectrum (Fig. 4.22).

It is easy to see that at 255 and 270 nm the contribution of breakdown of the adenine·uracil and guanine·cytosine pairs during denaturation to the change in spectrum is identical, so that during denaturation the change in optical density of tRNA at these points cannot depend on the nucleotide composition of the RNA.

The hypochromic effect at 255 nm due to breakdown of the base pairs can be estimated from determinations of melting of double-helical (poly-A)·(poly-U) and (poly-G)·(poly-C) complexes. The hypochromic effect with a change in temperature of the solutions from 15 to 85°C is about equal for both base pairs at 255 nm, and amounts to 30% [263].

Assessment of the hypochromic effect due to melting of the single-helical regions is more complex. In the case of RNA treated with formal-dehyde [263], when the formation of base pairs is no longer possible, the hypochromic effect is 4-6% (Table 4.14). This figure can be taken as an estimate of the hypochromic effect due to single-helical segments of RNA.

If x denotes the proportion of nucleotides in double-helical segments, h_t the percentage hypochromism of native tRNA, h_d the percentage hypo-chromism due to melting of double-helical regions, and h_s that due to melt-ing of single-helical regions, then assuming that h_d and h_s are independent, the equation

$$h_t = x h_d + (1 - x) h_s$$

can be written.

TABLE 4.14. Hypochromic Effects for Native RNA and RNA
Treated with Formaldehyde [263]

Polynucleotide	Hypochromism, %	
	native RNA	RNA treated with formaldehyde
tRNA from Escherichia coli.	19,5	4,5
tRNA from yeast.	21	6
RNA from phage R17	23	4,5
RNA from tobacco mosaic virus (TMV) .	21	4

Using the values given in Table 4.14 for h_t and h_s, and assuming that h_d is 30%, the value of x can be calculated. For tRNA this value is of the order of 59-63%. If melting of single-helical areas is disregarded, the corresponding values rise to 65-70%. These very approximate estimates nevertheless give the same value for the degree of helicization as chemical methods.

Preliminary x-ray structural data [389-392] obtained on crystalline specimens of tRNA, the preparation of which has recently become possible [389-392], indicate that double-helical regions exist in crystalline forms of tRNA.

The general conclusion from all these various methods is thus that tRNA shows a high degree of helicization*. Many ways of organization of the RNA molecule so that it contains double-helical segments can be postulated a priori [260]. Some possible arrangements of the molecule of alanine tRNA from yeast are shown in Fig. 4.23 (for a discussion of this problem, see page 249).

Some concrete evidence regarding the secondary structure of tRNA is given by determination of cleavage of the molecule by phosphodiesterases in the presence of magnesium ions [246-266]. As a result, only the 3'-terminal trinucleotide pCpCpA, common to all molecules of tRNA, is detached; no further cleavage takes place. This fact evidently indicates that this particular 3'-terminal trinucleotide does not participate in formation of macrostructure of tRNA. Unlike other polyribonucleotides, tRNA is not split completely by polynucleotide phosphorylase in the presence of Mg^{++}, and this also evidently indicates a high degree of stability of the tRNA structure under these conditions [244]. The concrete nature of the macrostructure of tRNA became of much greater importance after the nucleotide sequence of a number of individual tRNA molecules had been established [267-276, 290].

Models of the secondary structure of tRNA. On the principle of the need to have the greatest possible number of paired bases, three probable structures have been suggested for the first individual tRNA to have its primary structure completely decoded: alanine tRNA [267] (Fig. 4.23).

* Approximate estimates of the content of guanine•cytosine and adenine•uracil pairs in the double-helical segments can also be obtained from the denaturation spectra of RNA if the small contribution of single-helical segments to the spectra is disregarded [261].

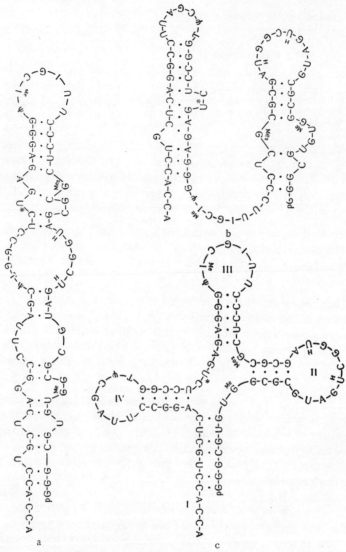

Fig. 4.23. Possible schemes of the secondary structure of alanine tRNA from yeast [267]: a) single-hairpin model; b) double-hairpin model; c) "clover-leaf" model (Roman numbers denote separate loops of the structure).

Two of these structures correspond to structures of the "hairpin" (a) and branched (b) models previously suggested [260] on the basis of a random sequence. The conformation (c), described as "clover leaf" differs from structures suggested previously (although it is not impossible that this type of structure could be obtained for a certain random nucleotide sequence). An unexpected discovery was that a "clover leaf" type of structure can be built for all the tRNA molecules so far decoded.

There are several arguments in support of the validity of the clover leaf structure.

1. The results of degradation of serine tRNA from yeast [268] by pancreatic RNase and by guanyl-RNase agree more closely with the clover leaf model of secondary structure than with the hairpin type. In addition, more of the bases are hydrogen bonded in the clover leaf model than in the hairpin type.

2. Chemical modification of alanine yeast tRNA by the action of bromosuccinimide, nitrous acid [277], or water-soluble carbodiimide [278] mainly affects segments corresponding to loops in the clover leaf model. Absence of modification in the "universal" pGpTpΨpC oligonucleotide, occurring in the loop, during the action of both N-bromosuccinimide and carbodiimide is noteworthy. The same effect is observed when total preparations of tRNA are modified by acrylonitrile [279], indicating that this oligonucleotide is equally difficult of access in all tRNAs. It is difficult at present to decide whether this differentiation between the properties of nucleotides lying in loop segments is due to incorrectness of the suggested secondary structure or to the existence of a tertiary structure in tRNA (see below). Another possible argument, this time not experimental in character, could be that the clover leaf type of structure enables the anticodon trinucleotide, the oligonucleotides with dihydrouridine residues, and the "universal" pGpTpΨpC nucleotide sequence to be located in identical loop segments of different tRNAs. The possibility cannot be ruled out that the universal secondary structure with a universal arrangement of homologous sequences is directly linked with the identical role of different tRNAs in protein biosynthesis.

It is concluded from analysis of the physicochemical properties of halves of molecules obtained by cleavage of the molecules of alanine and valine tRNAs from yeast by guanyl-RNase T_1 and pancreatic RNase, under conditions of maximum stabilization of the secondary tRNA structure [280, 281], that these tRNAs under the particular conditions used have a double-hairpin configuration as in structure b in Fig. 4.23. In addition, calculations have shown that the double-hairpin model corresponds better to the structure containing the maximum number of hydrogen-bonded base pairs in the case of alanine tRNA from yeast [282].

However, for some tRNAs, the double-hairpin model cannot be reconciled with data for association of halves of the molecule with the formation of a complex corresponding closely in its physicochemical properties with intact tRNA [283, 383-385]. Moreover, such complexes have acceptor activity close to that of intact tRNA [283]. Complex formation between the halves evidently takes place through hydrogen bonding, in agreement with the single-hairpin model or the clover leaf type (in which the halves are linked together), but not with the double-hairpin model, in which all hydrogen bonds are concentrated inside the halves.

I*

Optical rotatory dispersion. Some additional information on the secondary structure of tRNA can be obtained by the use of optical rotatory dispersion data [238, 284]. Making use of the noncooperative character of
interplanar interaction between the bases, the optical rotatory dispersion of
an RNA not possessing any other secondary structure than one determined
by interplanar interactions can be calculated from the optical rotatory dispersion data of the 16 possible dinucleoside monophosphates (meaning
ordinary nucleosides) [56, 236] and mononucleotides. Calculations of this
type, undertaken for polyadenylic, polycytidylic, and polyuridylic acids
give satisfactory agreement with the experimental data obtained under conditions when these polynucleotides exist in a single-helical conformation
[238]. However, similar calculations for tRNA give a picture which differs
sharply from the experimental observations. This is evidently because of
the large contribution made by base pairs to the secondary structure of
tRNA.

When a model of the secondary structure of tRNA has been constructed,
with the specified number of complementary adenine·uracil and guanine·cyto-
sine pairs, the optical rotatory dispersion corresponding to this model can
be calculated if the changes introduced into the dispersion pattern of the
single-helical molecule by base pairing are known. The magnitude of these
changes can be estimated approximately by comparing the curves of optical
rotatory dispersion of double-helical (poly-A)·(poly-U) and (poly-G)·(poly-C)
complexes and single-helical poly-A, poly-U, poly-G, and poly-C molecules
[285, 286]. Generally speaking, a comparison of this type enables the most
likely types of formation of the secondary structure to be selected in principle. However, such calculations, undertaken for the three types of secondary structure suggested for alanine tRNA from yeast (see page 248), give
practically the same curve. This is evidently because of the great inaccuracy
of the approximations made. The best agreement with the experimental data
for optical rotatory dispersion is given by a model of alanine tRNA consisting
of a slightly modified (compared with that suggested by Holley [267]) clover
leaf with the greatest possible number of hydrogen-bonded pairs [238].
Nevertheless, perfectly satisfactory results are also obtained when the experimental curves are compared with theoretical ones obtained by the use of the
original clover leaf models for alanine and tyrosine tRNAs from yeast [284].

The existing data on optical rotatory dispersion thus give further confirmation of the formation of hydrogen-bonded segments in the tRNA molecule*, but they do not enable the correct choice of type of secondary structure to be made.

---

* Comparison of theoretical and experimental data can be a useful method of detecting double-
helical segments in molecules of single-stranded polynucleotides, because experimental curves
for single-stranded molecules with double-helical segments are shifted strongly toward the short-
wave compared with curves calculated for single-helical polynucleotides.

IR-Spectra. Interplanar interactions between bases in polynucleotides have little effect on IR-spectra, whereas the formation of hydrogen-bonded pairs leads to the appearance [96, 97, 386, 387, 393-395] of characteristic peaks in the region 1500-1700 cm^{-1}. Seven absorption maxima, due to absorption by adenine·uracil and guanine·cytosine pairs and also by nucleotide components not participating in pairing, are found in the IR-spectrum of formylmethionine tRNA from E. coli in D_2O at 33°C in the region 1450-1750 cm^{-1} [386]. With a rise of temperature, absorption at 1686 cm^{-1} is reduced, and that at 1651 cm^{-1} is increased, because of rupture of base pairs. The melting profile in the IR-spectrum is the same as at 260 nm in the UV-region. The maximum at 1620 cm^{-1}, corresponding to absorption of the adenylyl component in the polynucleotide, is not changed until 97°C; this means that the adenylyl components participate to a slight degree only in the formation of hydrogen-bonded pairs. Starting from values of the molar extinction coefficients of free nucleotide components and of hydrogen-bonded components at different wavelengths [determined from data for nucleotides and for (poly-A)·(poly-U) and (poly-G)·(poly-C) complexes respectively], IR-spectra of tRNA corresponding to different pairing schemes can be calculated [386]. The best agreement with the experimental spectrum of formylmethionine tRNA from E. coli was obtained by the use of a clover leaf type of model [274]. Similar comparison of the calculated and experimental data for a fragment corresponding to the anticodon loop of this tRNA [387] also shows agreement for the pairing scheme corresponding to the clover leaf model [274].

Finally, the experimental melting profile of tRNA can be compared with that calculated theoretically, just as in the case of DNA (see page 240) [396]. This comparison has been made for several individual tRNAs, and in the case of serine tRNA from yeast, good agreement between the experimental data and values calculated on the basis of the clover leaf model has been demonstrated. For alanine and tyrosine tRNAs from yeast, however, considerable discrepancies have been found between the calculated and experimental values. Of course, calculations of this type can be regarded as only very approximate, because they disregard interplanar interactions in unpaired regions and also the effect of cations and of the tertiary structure on the thermal conversions of tRNA [396].

This material can be summarized by saying that at present there is no convincing evidence in support of the existence of any suggested model for the secondary structure of tRNA. The clover leaf model, the most popular at the present time, likewise has no solid experimental basis.

Investigations of tRNA by the methods of birefringence, electron microscopy, fluorescence polarization [251], and low-angle x-ray scatter [287, 288] have led to the conclusion that tRNA molecules are rod-shaped, and measure 100 Å in length and 20 Å in width. This conclusion is also confirmed by the study of the hydrodynamic properties of tRNA molecules [289]. In the case of any branched scheme of secondary structure, some form of superstructure is obviously necessary to give the molecule its rod-shaped character.

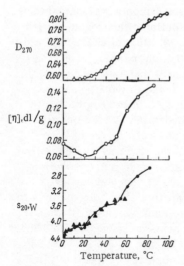

Fig. 4.24. Changes in some physical parameters of tRNA from yeast in a solution of 0.2 M NaCl + 0.01 M sodium phosphate + 0.0005 M Na₂EDTA, pH 6.85, as a function of temperature (effects observed when temperature raised from 0 to 20°C and associated with aggregation of tRNA) [291].

2. The tertiary structure of tRNA

It has already been mentioned that the pGpTpΨpC tetranucleotide is universal for all individual tRNAs so far investigated; it evidently lies in the loop region and is not modified by chemical action. This may be the result of screening of the corresponding nucleotides by the formation of the tertiary structure of the polymer molecule. As a result of the action of nucleases on tRNAs under conditions where the secondary structure of the molecule is stabilized, of all the possible sites of enzyme attack it is the loop containing the hypothetical anticodon which is split first [283, 290]. As a result, the tRNA molecule is split into two halves. This evidently means that the loop with the anticodon is under conditions which differ somewhat from those of the other nonhelical segments, and it suggests the existence of a special folding of the molecule over and above that due to the secondary structure.

Further confirmation of this conclusion was found by studying thermal denaturation of tRNA. If the medium does not contain magnesium ions, before the optical density of the tRNA solution has changed significantly, the hydrodynamic characteristics of the molecule (the characteristic viscosity and sedimentation coefficient, for example) have already undergone considerable changes [289, 291] (Fig. 4.24). Since the optical properties are determined chiefly by interplanar interactions between the bases (and base pairs) of the chain, while the hydrodynamic properties are determined by the volume and shape of the molecule, these results mean that a slight change in the degree of the interplanar interactions (i. e., in the secondary structure of the molecule) is associated with considerable changes in the spatial arrangement of individual parts of the tRNA molecule, i. e., in its tertiary structure.

The biphasic character of the transition of tRNA during denaturation*
is also revealed by investigations of the fluorescence polarization of the dye

---

* The existence of at least three phase transitions during the thermal denaturation of alanine tRNA from yeast has recently been demonstrated by differential spectrophotometry [397]. However, it is not yet clear whether they are connected with the tertiary structure or with independent melting of individual helical segments of the clover leaf.

acriflavine, when bound to the adenosine residue on the 3'-end of the tRNA molecule during its denaturation [292, 293].

Evidence in support of the existence of a tertiary structure in a number of individual tRNAs from E. coli is the biphasic character [400] of the change in absorption in the region 335 nm, characteristic of 4-thiouridine. Since, as a rule, one tRNA molecule contains one 4-thiouridine residue, the biphasic character of melting can hardly be the result of independent melting of individual helical segments, but is more likely to be the result of successive collapse of the tertiary and secondary structures of the molecule [400].

It is interesting to note that in the presence of magnesium ions, the hydrodynamic and optical properties of tRNA change synchronously, with appreciable increases in the value of T_m and in the degree of cooperativity of the transition [291] (an increase in the concentration of monovalent cations leads to an increase in T_m, but not in the degree of cooperativity). This effect can also be ascribed to the existence of a tertiary structure if it is accepted that the tertiary structure, when stabilized by magnesium ions, protects individual helical regions of the polynucleotide against separate melting, characteristic of the unstabilized tRNA molecule (in the absence of magnesium ions); under these circumstances the tertiary structure collapses simultaneously with the secondary.

The physical data examined above do not rule out the possibility that all the effects observed are due simply to slight changes in secondary structure. Nevertheless, in conjunction with the chemical modification data and the enzymic cleavage of stabilized tRNA molecules these effects must evidently be ascribed, with a far greater measure of probability, to changes in the tertiary structure.

One possible result of the ability of tRNA to form a tertiary structure is the recently discovered existence of a series of individual transfer RNAs [289, 291, 294-301] in two forms, differing in conformation and capable of conversion from one to the other, only one of which corresponds to the biologically active (native) form. These two forms differ appreciably from each other in their hydrodynamic properties, the biologically inactive ("denatured") form being less compact than the native, and differing only slightly from it in its hypochromic effect. Estimates of the differences between the number of base pairs in the active and inactive forms [291, 297] give the values 2-5. The conditions of interconversion of these forms are such that the presence of an appreciable energy barrier between them can be assumed. The native form can be converted into the "denatured" by incubation with chelating agents such as trilon B, even at 0°C [291]. However, the reverse change on the addition of magnesium ions at room temperature takes place slowly, and can be accelerated if the solution of "denatured" tRNA is heated to about 60°C and then cooled. This may mean that the transition from the "denatured" to the native form requires the collapse of a certain structure (possibly the tertiary).

Similar transitions take place with a change in pH. This does not, of course, rule out the possibility that during denaturation there is simply a considerable reorganization of the secondary structure, leading to the formation of a structure with close to the native number of base pairs, but differing from it essentially in the arrangement of the helical and loop regions.

It may be postulated that the condition for biological activity of tRNA is a tertiary structure properly formed above the secondary structure. Changes in this tertiary structure, possibly resulting from changes (of slight degree) in secondary structure, lead to loss of biological activity. The conditions under which restoration of the correct secondary and tertiary structures is possible also lead to restoration of biological activity.

The hypothesis regarding the existence of a tertiary structure of tRNA also raises the question of concrete methods of formation of this structure. The available evidence is scanty. As has already been mentioned, the loop with the anticodon is probably screened to a lesser degree by the tertiary structure than the loop containing the "universal" pGpTpΨpC sequence. This tetranucleotide segment evidently plays some type of special role in the formation of the tertiary structure, as follows from comparison of the properties of native tRNA and tRNA in which the pseudouridine residue in the tetranucleotide has been cyanoethylated by the action of acrylonitrile [264]. Cyanoethylated tRNA can be split much more readily by pancreatic ribonuclease in the presence of magnesium ions than native tRNA. During the action of snake venom phosphodiesterase on cyanoethylated tRNA in the presence of Mg^{++}, the cleavage does not stop after detachment of the terminal trinucleotide pCpCpA sequence, as is observed with the native tRNA. Analysis of the melting curves shows that after cyanoethylation the number of adenine·uracil and guanine·cytosine pairs formed (compared with the native preparation) shows little change; in both cases the hyperchromic effects are low. Judging from these findings, cyanoethylation has little effect on secondary structure, but causes considerable changes in tertiary structure. This conclusion is also confirmed by the fact that melting of the macrostructure of cyanoethylated tRNA in the presence of Mg^{++} is a less cooperative process than melting of native tRNA.

Suggestions have been expressed regarding the possibility of maintaining the tertiary structure by the formation of complementary pairs between bases from different loops, which is possible if the loops lie one above the other [290, 398]. In the case of tyrosine tRNA from yeast, for example, if the side loops are placed one above the other, two such guanine·cytosine and two adenine·uracil pairs can be formed.

Stacking of loops in some tRNAs has been shown to be possible [398] through the formation of hydrogen bonds between pΨpC dinucleotides in the pTpΨpC loop and pApG of the 3'-terminal trinucleotide pCpCpA. In the case of serine tRNA from yeast, bonding is also possible between the pUpU sequence in the additional loop with the pApA fragment in the dihydrouridine

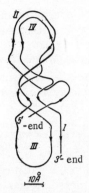

Fig. 4.25. Hypothetical scheme of the tertiary structure of alanine tRNA from yeast (the secondary structure corresponds to a model of the clover leaf type). For nomenclature of loops, see Fig. 4.23c [288].

loop. This structure is supported by data for the oxidation of tRNA by monoperphthalic acid at different temperatures [398], although the results of IR-spectroscopy [386] do not agree with the number of base pairs specified in this model.

Attempts have been made to build stereochemical models of the various tertiary structures of alanine tRNA from yeast on the basis of the known nucleotide sequence and of the clover leaf type of secondary structure of the molecule suggested by Holley. For each model thus obtained, the intensities of low-angle x-ray scatter have been calculated and compared with those obtained experimentally [288]. Better agreement was obtained for the model shown schematically in Fig. 4.25 (although at angles exceeding 60 mrad appreciable discrepancies are observed). This conformation evidently is in agreement with the available data. In fact, the loop with the anticodon is less completely screened than the two other loops, and this may explain the higher reactivity of this segment. The details described above naturally do not rule out the possibility of existence of other [399] models*.

3. The secondary and tertiary structures of 5S RNA

The primary structure of the molecule has now been established for two individual 5S RNAs [302-304]. The sedimentation coefficient of 5S RNA is relatively independent of the magnesium ion concentration (to 10^{-2} M) and the sodium ion concentration (to 0.5 M) [305], thus suggesting that the 3-dimensional structure of the 5S RNA molecule possesses high rigidity.

It is concluded from the marked increase in molar extinction of specimens of 5S RNA from E. coli after treatment with formaldehyde [305] that its molecule contains many base pairs. This conclusion is confirmed by the optical rotatory dispersion data (experimental data are compared with calculated values for a known nucleotide sequence, on the assumption that this sequence does not form helical segments, on page 250) [306, 307]. Further evidence in support of this conclusion is given by differences in the resistance of oligonucleotide segments of the molecule to the action of nucleases, and also by differences in the reactivity of identical bases in the polynucleotide sequence of 5S RNA [308].

---

* As a result of several low-angle x-ray scatter studies, the calculated parameters of the length and diameter of the molecule suggest that a hairpin conformation is the most likely one for tRNA (see, for example, [287]).

The number of base pairs in the 5S RNA molecule can be estimated in several ways. It has been concluded from differences in the hypochromism of native and formaldehyde-treated 5S RNA from E. coli at 255 nm [305, 306] (see page 246) that the double-helical segments account for between 63% [305] and 67-72% [306] of the total number of nucleotides in the molecule. Estimation of the relative number of complementary adenine·uracil and guanine·cytosine base pairs [306] from the denaturation spectra (see page 246), gives a figure of 60-70% for guanine·cytosine pairs. During oxidation of the adenine rings of the 5S RNA molecule from E. coli with monoperphthalic acid [308] at 20°C, only 43.7% of the bases, i.e., 23 adenine residues, are reactive. This agrees with the number of adenine·uracil pairs determined from the denaturation spectra [197]. Hence, for 5S RNA just as for tRNA, the molecule shows a high degree of helicization.

Several models have now been suggested for the secondary structure of 5S RNA [302, 303, 307, 309]. The model proposed by Sanger for 5S RNA from E. coli contains 23 base pairs. This corresponds to a degree of helicity of the polynucleotide of about 38%, which does not agree with the data given above. Other models give closely similar values for the number of base pairs and agree with the observed physicochemical characteristics of 5S RNA. At present it is impossible to give preference to any one of them. All these models resemble the clover leaf type of structure proposed for tRNA. This type of structure can be assigned to both 5S RNAs which have so far been decoded [309]. It is too early, however, to decide to what extent it is common to all.

For it to have the compact shape corresponding to its observed hydrodynamic properties, the molecule of 5S RNA must probably also possess a tertiary structure. Recent work has revealed different forms of 5S RNA [310] mutually interconvertible under conditions similar to those of interconversion of tRNA "conformers" [310, 401]. The denatured form of 5S RNA does not possess biological activity. It is not yet clear to what extent this is evidence of a close analogy between the conformations of 5S RNA and tRNA, but it seems that the mutual interconversion of different forms of 5S RNA, just as in the case of tRNA, can most probably be explained by modification of the tertiary structure of the molecule associated with only minor changes in its secondary structure. On this example of the two types of RNA molecules possessing a single-stranded structure, it is thus evident that such molecules in solution form double-helical segments separated by nonhelical regions; the same bases, present in the same molecule, can in this way possess different properties.

A similar situation is evidently characteristic of RNAs of higher molecular weight, such as 16S and 23S ribosomal RNAs and virus RNAs. There is evidence in support of this hypothesis [243]; the results of cleavage of high-molecular weight RNAs by nucleases show, in particular, that these molecules contain long oligonucleotides resistant to enzyme action [311-314]. The use of the same methods as with tRNA and 5S RNA has demonstrated a high degree of helicization of these high-polymer molecules, but naturally this is still very far from the construction of concrete models of their secondary structure. Even when the primary structure of high-polymer RNAs has been established, this problem will still be much more complex than in the case of RNAs of low molecular weight, because they must evidently have many more alternative ways of formation of their various double-helical segments.

It may well be, however, that there is no particular reason for establishment of the precise secondary structure of high-polymer RNAs, because these RNAs exist in biological systems, as a rule, as complexes with proteins. The secondary structure of the RNA in these complexes may differ completely from the structure of their "pure" nucleic acid component, so that knowledge of the latter will not be very informative about the biological functions of the RNA. Moreover, the possibility is not ruled out that high-polymer RNAs do not possess a single secondary structure, but exist in solution as an assortment (perhaps large) of molecules with different schemes of base pairing.

For the chemist the most important consequence of the existence of secondary and tertiary structures of polymer molecules of the nucleic acids is the difference between reactivity not only of the bases in double-helical and single-helical molecules and segments of molecules, but also of the bases as monomer units and as components of polynucleotide chains.

Bibliography

1. J. D. Watson and F. H. C. Crick, Nature, 171:156, 740 (1953).
2. R. M. Hamlin, R. C. Lord, and A. Rich, Science, 148:1734 (1965).
3. D. W. Green, F. S. Mathews, and A. Rich, J. Biol. Chem., 237:3573 (1962).
4. Y. Kyogoku, R. C. Lord, and A. Rich, Science, 154:518 (1966).
5. J. Pitha, R. Jones, and P. Pithova, Canad. J. Chem., 44:1045 (1966).
6. J. A. Pople, W. G. Schneider, and H. J. Bernstein, in: High-Resolution Nuclear Magnetic Resonance, McGraw-Hill, New York – Toronto – London (1959).
7. R. R. Shoup, H. T. Miles, and E. D. Becker, Biochem. Biophys. Res. Comm., 23:194 (1966).
8. L. Katz and S. Penman, J. Mol. Biol., 15:220 (1966).
9. K. H. Scheit, Angew. Chem., 79:90 (1967).
10. L. Pauling and R. Corey, Arch. Biochem. Biophys., 65:164 (1956).
11. Yu. G. Baklagina, Molekul. Biol., 2:635 (1968).
12. K. Hoogsteen, Acta Cryst., 12:822 (1959); 16:907 (1963).
13. F. S. Mathews and A. Rich, J. Mol. Biol., 8:89 (1964).
14. Yu. G. Baklagina, M. V. Vol'kenshtein, and Yu. D. Kondrashev, Biofizika, 10:165 (1965); Zh. Strukt. Khimii, 7:399 (1966); Dokl. Akad. Nauk SSSR, 169: 229 (1966).
15. K. Tomita, L. Katz, and A. Rich, J. Mol. Biol., 30:545 (1967).
16. A. E. V. Haschemeyer and H. M. Sobell, Proc. Nat. Acad. Sci. USA, 50:872 (1963); Acta Cryst., 18:525 (1965).
17. L. Katz, R. Tomita, and A. Rich, J. Mol. Biol., 13:340 (1965); Acta Cryst., 21:754 (1966).
18. E. J. O'Brien, J. Mol. Biol., 7:107 (1963).
19. E. J. O'Brien, Acta Cryst., 23:92 (1967).
20. H. M. Sobell, K. Tomita, and A. Rich, Proc. Nat. Acad. Sci. USA, 49:885 (1965).
21. A. E. V. Haschemeyer and H. M. Sobell, Nature, 202:969 (1964); Acta Cryst., 19:125 (1965).
22. E. J. O'Brien, J. Mol. Biol., 22:377 (1966).
23. K. Hoogsteen, Acta Cryst., 16:28 (1963).
24. F. S. Mathews and A. Rich, Nature, 201:179 (1964).
25. R. F. Stewart and L. H. Jensen, J. Chem. Phys., 40:2071 (1964).
26. J. Iball and H. R. Wilson, Nature, 198:1193 (1963).
27. B. Pullman, P. Claverie, and J. Caillett, Proc. Nat. Acad. Sci. USA, 55:904 (1966).
28. Y. Kyogoku, R. C. Lord, and A. Rich, J. Am. Chem. Soc., 89:496 (1967).
29. Y. Kyogoku, R. C. Lord, and A. Rich, Proc. Nat. Acad. Sci. USA, 57:250 (1967).
30. J. S. Binford and D. M. Holloway, J. Mol. Biol., 31:91 (1968).
31. E. Kuchler and J. Derkosch, Z. Naturforsch, 21b:209 (1966).
32. J. H. Miller and H. M. Sobell, J. Mol. Biol., 24:345 (1967).
33. H. de Voe and I. Tinoco (Jr.), J. Mol. Biol., 4:500 (1962).
34. H. A. Nash and D. F. Bradley, Biopolymers, 3:261 (1965).
35. H. A. Nash and D. F. Bradley, J. Chem. Phys., 45:1380 (1966).
36. R. C. Lord and G. J. Thomas, Biochim. Biophys. Acta, 142:1 (1967).
37 P. O. P. T'so, I. S. Melvin, and A. C. Olson, J. Am. Chem. Soc., 85:1289 (1963).
38. A. D. Broom, M. P. Schweizer, and P. O. P. Ts'o, J. Am. Chem. Soc., 89: 3612 (1967).
39. P. O. P. Ts'o and S. I. Chan, J. Am. Chem. Soc., 86:4176 (1964).
40. T. N. Solie and J. A. Schellman, J. Mol. Biol., 33:61 (1968).

41.   G. P. Rossetti and K. E. Van Holde, Biochem. Biophys. Res. Comm. , 26:717 (1967).

42.   S. J. Gill, M. Downing, and G. F. Sheats, Biochemistry, 6:272 (1967).

43.   E. L. Farquhar, M. Downing, and S. J. Gill, Biochemistry, 7:1224 (1968).

44.   M. Gellert , M. N. Lipsett, and D. K. Davies, Proc. Nat. Acad. Sci. USA, 48:2013 (1962).

45.   P. K. Sarkar and J. T. Yang, Biochem. Biophys. Res. Comm. , 20:346 (1965).

46.   R. W. Ravindranathan and H. T. Miles, Biochim. Biophys. Acta, 94:603 (1965).

47.   S. J. Chan, M. P. Schweizer, P. O. P. Ts'o, and G. K. Helmkamp, J. Am. Chem. Soc. , 86:4182 (1964).

48.   G. K. Helmkamp and N. S. Kondo, Biochim. Biophys. Acta, 145:27 (1967).

49.   M. P. Schweizer, S. J. Chan, and P. O. P. Ts'o, J. Am. Chem. Soc. , 87:5241 (1965).

50.   G. K. Helmkamp and N. S. Kondo, Biochim. Biophys. Acta, 157:242 (1968).

51.   O. Jardetzky, Biopolymers Symposia, (1):501 (1964).

52.   H. De Voe, J. Chem. Phys. , 43:3199 (1965).

53.   I. Tinoco (Jr.), in: Molecular Biophysics, B. Pullman and A. Weissbluth (editors), Academic Press, New York (1965), p. 269.

54.   I. Tinoco (Jr.), J. Am. Chem. Soc. , 82:4785 (1960).

55.   A. M. Michelson, The Chemistry of Nucleosides and Nucleotides, Academic Press, London – New York (1963).

56.   H. M. Warshaw and I. Tinoco (Jr.), J. Mol. Biol. , 20:29 (1966); 13:54 (1965).

57.   G. Felsenfeld and M. Leng, J. Mol. Biol. , 15:455 (1966).

58.   W. M. Stanley and R. M. Bock, Anal. Biochem. , 13:43 (1965).

59.   M. Warshaw, C. A. Bush, and I. Tinoco, Biochem. Biophys. Res. Comm. , 18:633 (1965).

60.   J. Applequist and V. Damle, J. Am. Chem. Soc. , 88:3895 (1966).

61.   R. C. Davis and I. Tinoco, Biopolymers, 6:223 (1968).

62.   D. N. Holcomb and I. Tinoco, Biopolymers, 3:121 (1965).

63.   Y. Inoue, S. Aoyagi, and K. Nakanishi, J. Am. Chem. Soc. , 89:5701 (1967).

64.   Y. Inoue, S. Aoyagi, and K. Nakanishi, Tetrahedron Letters, 3575 (1967).

65.   J. M. Vournakis, H. A. Scheraga, G. W. Rushizky, and H. A. Sober, Biopolymers, 4:33 (1966).

66.   S. Aoyagi and Y. Inoue, J. Biol. Chem. , 243:514 (1968).

67.   C. A. Bush and I. Tinoco, J. Mol. Biol. , 23:601 (1967).

68.   J. Brahms, J. C. Maurizot, and A. M. Michelson, J. Mol. Biol. , 25:465 (1967).

69.   J. Brahms, J. C. Maurizot, and A. M. Michelson, J. Mol. Biol. , 25:481 (1967).

70.   J. Brahms, A. M. Michelson, and K. E. Van Holde, J. Mol. Biol. , 15:467 (1966).

71.   K. E. Van Holde, J. Brahms, and A. M. Michelson, J. Mol. Biol. , 12:726 (1965).

72.   Y. Inoue and S. Aoyagi, Biochem. Biophys. Res. Comm. , 28:973 (1967).

73.   K. H. Scheit, F. Cramer, and A. Franke, Biochim. Biophys. Acta, 145:21 (1967).

74.   F. E. Hruska and S. S. Danyluk, Biochim. Biophys. Acta, 157:238 (1968).

75.   S. J. Chan, B. W. Bangerter, and H. H. Peter, Proc. Nat. Acad. Sci. USA, 55:720 (1966).

76.   P. O. P. Ts'o, S. A. Rappaport, and F. J. Bollum, Biochemistry, 5:4153 (1966).

77.   D. Poland, J. N. Vournakis, and H. A. Scheraga, Biopolymers, 4:223 (1966).

78.   H. Simpkins and E. G. Richards, Biochemistry, 6:2513 (1967).

79.   D. Glaubiger, D. A. Lloyd, and I. Tinoco, Biopolymers, 6:4091 (1968).

80. P. Claverie, B. Pullman, and J. Caillet, J. Theoret. Biol., 12:419 (1966).
81. O. Sinanoglu and S. Abdulnur, Photochem. Photobiol., 3:333 (1964).
82. O. Sinanoglu and S. Abdulnur, Fed. Proc., 24:(2), Part 3, S12 (1965).
83. A. Wacker and E. Lodemann, Angew. Chem., 77:133 (1965).
84. F. H. C. Crick and J. D. Watson, Proc. Roy. Soc., A223:80 (1954).
85. D. M. Brown and A. R. Todd, J. Chem. Soc., 52 (1952).
86. C. R. Dekker, A. M. Michelson, and A. R. Todd, J. Chem. Soc., 947 (1953).
87. D. M. Brown and B. Lithgoe, J. Chem. Soc., 1990 (1950).
88. B. Lythgoe, H. Smith, and A. R. Todd, J. Chem. Soc., 355 (1947); J. Davoll, B. Lythgoe, and A. R. Todd, J. Chem. Soc., 833 (1944); V. M. Clark, A. R. Todd, and J. Zussman, J. Chem. Soc., 2952 (1951); S. Furberg, Acta Chem. Scand., 4:751 (1950).
89. W. Cochran, Acta Cryst., 4:81 (1951).
90. J. M. Broomhead, Acta Cryst., 1:324 (1948).
91. J. M. Broomhead, Acta Cryst., 4:92 (1951).
92. S. Furberg, Acta Cryst., 3:325 (1950).
93. J. Zussman, Acta Cryst., 6:504 (1953).
94. J. D. Watson and F. H. C. Crick, Nature, 171:964 (1953).
95. J. M. Gulland and D. O. Jordan, Symp. Soc. Exp. Biol., I. Nucleic Acids, 56, Cambridge Univ, Press (1946).
96. H. T. Miles, Proc. Nat. Acad. Sci. USA, 51:1104 (1964).
97. H. T. Miles, Proc. Nat. Acad. Sci. USA, 47:791 (1961).
98. M. P. Printz and P. H. von Hippel, Proc. Nat. Acad. Sci. USA, 53:364 (1965).
99. P. H. von Hippel and M. P. Printz, Fed. Proc., 24:1458 (1965).
100. A. M. Michelson, J. Massoulie, and W. Guschlbauer, Prog. Nucl. Acid Res., 6:83 (1967).
101. S. Arnott, M. H. F. Wilkins, L. D. Hamilton, and R. Langridge, J. Mol. Biol., 11:391 (1965).
102. M. Fenghelman, R. Langridge, W. E. Seeds, A. R. Stockes, H. R. Wilson, C. W. Hooper, M. F. H. Wilkins, R. K. Barkley, and L. D. Hamilton, Nature, 175:834 (1955).
103. R. Langridge, H. R. Wilson, C. W. Hooper, M. H. F. Wilkins, and L. D. Hamilton, J. Mol. Biol., 2:19 (1960); R. Langridge, D. A. Marvin, W. E. Seeds, H. R. Wilson, C. W. Hooper, M. H. F. Wilkins, and L. D. Hamilton, J. Mol. Biol., 2:38 (1960).
104. W. Fuller, M. H. F. Wilkins, H. R. Wilson, and L. D. Hamilton, J. Mol. Biol., 12:60 (1965).
105. P. J. Cooper and L. D. Hamilton, J. Mol. Biol., 16:562 (1966); L. D. Hamilton, Nature, 218:633 (1968).
106. D. A. Marvin, M. Spencer, M. H. F. Wilkins, and L. D. Hamilton, J. Mol. Biol., 3:547 (1961).
107. V. Luzzati, A. Nikolaieff, and F. Masson, J. Mol. Biol., 3:185 (1961).
108. J. Josse and J. Eigner, Ann. Rev. Biochem., 35:789 (1966).
109. J. Marmur, R. Rownd, and C. L. Schildkraut, in: Progress in Nucleic Acid Research, Vol. 1, W. E. Cohn and J. N. Davidson (editors), Academic Press, New York — London (1963).
110. D. M. Crothers and B. H. Zimm, J. Mol. Biol., 12:525 (1965).
111. J. Vinograd and J. Lebowitz, J. Gen. Physiol., 49:103, Part 2 (1966); J. A. Cohen, Science, 158:343 (1967).
112. H. S. Janz, P. D. Baces, P. H. Pouwel, E. E. J. Van Bruggen, and H. Oldenziel, J. Mol. Biol., 32:159 (1968).
113. L. Shapiro, L. Grosman, J. Marmur, and A. V. Kleinschmidt, J. Mol. Biol., 33:097 (1968).

114. V. C. Bode and L. A. MacHattie, J. Mol. Biol., 32:673 (1968).

115. F. T. Hickson, T. F. Roth, and D. K. Helinski, Proc. Nat. Acad. Sci. USA, 58, 1731 (1967).

116. L. Pikó, D. G. Blair, A. Tyler, and J. Vinograd, Proc. Nat. Acad. Sci. USA, 59:838 (1968).

117. D. R. Wolstenholme and J. B. Dawid, Chromosoma, 20:445 (1967).

118. J. B. Dawid and D. R. Wolstenholme, J. Mol. Biol., 28:233 (1967).

119. E. F. J. Van Bruggen, C. M. Runner, G. J. C. M. Ruttenberg, A. M. Kroon, and F. M. A. H. Schurmans-Stekhoven, Biochim. et Biophys. Acta, 161:402 (1968).

120. E. T. Young and R. L. Sinsheimer, J. Mol. Biol., 30:165 (1967).

121. M. Gellert, Proc. Nat. Acad. Sci. USA, 57:148 (1967).

122. J. Vinograd, J. Leibowitz, and R. Watson, J. Mol. Biol., 33:173 (1968).

123. D. Glaubiger and J. E. Hearst, Biopolymers, 5:591 (1967).

124. W. Bauer and J. Vinograd, J. Mol. Biol., 33:141 (1968).

125. C. A. Thomas and L. A. MacHattie, Ann. Rev. Biochem., 36:485 (1967).

126. G. J. C. M. Ruttenberg, E. M. Smit, P. Borst, and E. F. J. Van Bruggen, Biochim. Biophys. Acta, 157:429 (1968).

127. L. V. Crawford and M. J. Waring, J. Mol. Biol., 25:23 (1967); J. Gen. Virol., 1:387 (1967).

128. H. Bujard, J. Mol. Biol., 33:503 (1968).

129. D. Lang, H. Bujard, B. Wolff, and D. Russell, J. Mol. Biol., 23:163 (1967).

130. S. Arnott, K. Hutchinson, M. Spencer, M. H. F. Wilkins, W. Fuller, and R. Langridge, Nature, 211:227 (1966).

131. S. Arnott, M. H. F. Wilkins, W. Fuller, and R. Langridge, J. Mol. Biol., 27:525 (1967).

132. S. Arnott, M. H. F. Wilkins, W. Fuller, and R. Langridge, J. Mol. Biol., 27:535 (1967).

133. S. Arnott, M. H. F. Wilkins, W. Fuller, J. H. Venable, and R. Langridge, J. Mol. Biol., 27:549 (1967).

134. T. Samejima, H. Hashizume, K. Imahori, Y. Fujii, and K. Miura, J. Mol. Biol., 34:39 (1968).

135. K. I. Miura, Y. Kimura, and N. Suzuki, Virology, 28:571 (1966).

136. T. Sato, Y. Kyogoku, S. Higuchi, I. Mitsui, Y. Itaka, M. Tsuboi, and K. I. Miura, J. Mol. Biol., 16:180 (1966).

137. L. M. Blach, R. Markham, and J. Neth, Plant. Path., 69:215 (1965).

138. P. J. Gomatos and J. Tamm, Proc. Nat. Acad. Sci. USA, 49:707 (1963).

139. R. Langridge and P. J. Gomatos, Science, 141:694 (1963).

140. K. Tomita and A. Rich, Nature, 201:1160 (1964).

141. H. Hashizume and K. Imahori, J. Biochem. Japan, 61:738 (1967).

142. J. Kraut, Ann. Rev. Biochem., 34:247 (1965).

143. D. R. Davies, Ann. Rev. Biochem., 36:646 (1967).

144. R. Langridge and J. Marmur, Science, 143:1450 (1964).

145. A. M. Michelson, in: The Chemistry of Nucleosides and Nucleotides, Academic Press, London — New York (1963).

146. M. D. Frank-Kamenetskii, in: Physical Methods of Investigation of Proteins and Nucleic Acids [in Russian], Yu. S. Lazurkin (editor), Nauka (1967), p. 130.

147. G. Folsenfeld and H. T. Miles, Ann. Rev. Biochem., 36:407 (1967).

148. M. F. Singer, L. A. Heppel, G. W. Rushizky, and H. A. Sober, Biochim. Biophys. Acta, 61:474 (1962).

149. P. Doty, J. Polymer. Sci., 55:1 (1961).

150. M. N. Lipsett, L. A. Heppel, and D. F. Bradley, J. Biol. Chem., 236:857 (1961).

151. M. N. Lipsett, L. A. Heppel, and D. F. Bradley, Biochim. Biophys. Acta, 41:175 (1960).

152. K. Hamaguchi and E. P. Geiduschek, J. Am. Chem. Soc., 84:1329 (1962); D. K. Robinson and M. E. Grant, J. Biol. Chem., 241:4030 (1966); C. E. Emanuell, Biochim. Biophys. Acta, 42:91 (1960).

153. A. M. Michelson, in: The Chemistry of Nucleosides and Nucleotides, Academic Press, London – New York (1963).

154. M. T. Record, Biopolymers, 5:993 (1967).

155. L. G. Bunville, E. P. Geiduschek, M. A. Rawitscher, and J. M. Sturtevant, Biopolymers, 3:213 (1965).

156. B. Weila and S. J. Gill, Biochim. Biophys. Acta, 112:179 (1966).

157. H. J. Lin and E. Chargaff, Biochim. Biophys. Acta, 145:398 (1967).

158. A. M. Michelson, in: The Chemistry of Nucleosides and Nucleotides, Academic Press, London – New York (1963).

159. P. H. Pouwels, C. M. Knijnenburg, J. Van Rotterdam, J. A. Cohen, and H. S. Jansz, J. Mol. Biol., 32:169 (1968).

160. B. M. Alberts, J. Mol. Biol., 32:405 (1968); B. M. Alberts and P. Doty, J. Mol. Biol., 32:379 (1968); E. P. Geiduschek, Proc. Nat. Acad. Sci. USA, 47:950 (1961).

161. J. Vinograd, J. Leibowitz, R. Radloff, R. Watson, and P. Laipis, Proc. Nat. Acad. Sci. USA, 53:1104 (1965).

162. E. A. C. Follett and L. V. Crawford, J. Mol. Biol., 28:455 (1967); 34:565 (1968).

163. L. V. Crawford, J. Mol. Biol., 13:362 (1965).

164. M. Ageno, E. Dore, and K. Frontali, Atti. Acad. Naz. Lincei Rend. Cl. Sci., Fis. Mat. e Natur., 40:346 (1966).

165. M. Ageno, E. Dore, and K. Frontali, Atti. Acad. Naz. Lincei Rend. Cl. Sci., Fis, Mat. e Natur., 40:540 (1966).

166. P. F. Davison, J. Mol. Biol., 22:97 (1966).

167. P. F. Davison, Biopolymers, 5:715 (1967).

168. H. C. Spatz and R. L. Baldwin, J. Mol. Biol., 11:213 (1965).

169. S. Tawashima and E. A. Arnolds, Biochim. Biophys. Acta, 94:546 (1965).

170. H. Spatz and D. M. Crothers, J. Mol. Biol., 42:191 (1969).

171. Yu. M. Evdokimov and Ya. M. Varshavskii, Dokl. Akad. Nauk SSSR, 170:1205 (1966).

172. Yu. M. Evdokimov, K. G. Knorre, and Ya. M. Varshavskii, Molekul. Biol., 3:163 (1969).

173. S. A. Rice and P. Doty, J. Am. Chem. Soc., 79:3939 (1957).

174. A. R. Peacocke and L. O. Walker, J. Mol. Biol., 5:560 (1962).

175. H. Berg, H. Bär, and F. A. Gollmick, Biopolymers, 5:61 (1967).

176. H. Berg, Yu. M. Evodokimov, H. Bär, and Ya. M. Varshavskii, Molekul. Biol., 2:830 (1968).

177. Yu. M. Evdokimov, H. Berg, H. Bär, and Ya. M. Varshavskii, Molekul. Biol. (in press).

178. L. G. Bunville, E. P. Geiduschek, M. A. Rawitscher, and J. M. Sturtevant, Biopolymers, 3:213 (1965).

179. E. A. C. Follett and L. V. Crawford, J. Mol. Biol., 28:461 (1967).

180. R. B. Inman, J. Mol. Biol., 18:464 (1966).

181. R. B. Inman, J. Mol. Biol., 28:103 (1967).

182. A. V. Shugalii, M. D. Frank-Kamenetskii, and Yu. S. Lazurkin, Molekul. Biol., 3:133 (1969).

183. C. A. Thomas and L. A. MacHattie, Proc. Nat. Acad. Sci. USA, 52:1297 (1964).

184. J. C. Wettmur and N. Davidson, J. Mol. Biol., 31:349 (1968).

185. R. Muriel, Proc. Nat. Acad. Sci. USA, 59:200 (1968).

186. J. D. Karkas, K. Rudner, and E. Chargaff, Proc. Nat. Acad. Sci. USA, 60:915 (1968).

187. R. Rudner, J. Karkas, and E. Chargaff, Proc. Nat. Acad. Sci. USA, 60:921 (1968).

188. W. Doerfler and D. S. Hogness, J. Mol. Biol., 33:635 (1968).

189. S. Aurisicchio, in: Procedures in Nucleic Acid Research, G. L. Cantoni and D. R. Davies (editors), Harper and Row, New York – London (1966), p. 562.

190. R. Wu and A. D. Kaiser, Proc. Nat. Acad. Sci. USA, 57:170 (1967).

191. P. Sherdrick and W. Szybalski, J. Mol. Biol., 29:217 (1967).

192. G. Corneo, E. Ginelli, and E. Polli, J. Mol. Biol., 33:331 (1968).

193. A. Guha and W. Szybalski, Virology, 34:608 (1968).

194. M. D. Chilton, Science, 157:817 (1967).

195. S. Cordes, H. T. Epstein, and J. Marmur, Nature, 191:1097 (1961).

196. C. L. Schildkraut, J. Marmur, and P. Doty, J. Mol. Biol., 3:595 (1961).

197. M. Waring and R. J. Britten, Science, 154:791 (1966).

198. J. A. Subirana and P. Doty, Biopolymers, 4:171 (1966); J. E. Subirana, Biopolymers, 4:189 (1966).

199. K. J. Thrower and A. R. Peacocke, Biochim. Biophys. Acta, 119:652 (1966).

200. A. P. Nygaard and B. D. Hall, J. Mol. Biol., 9:125 (1964).

201. R. W. Davis and N. Davidson, Proc. Nat. Acad. Sci. USA, 60:243 (1968).

202. R. L. C. Brimacombe and K. S. Kirby, Biochim. Biophys. Acta, 157:362 (1968).

203. M. Arca, E. D. Di Mauro, L. Frontali, G. Tecce, Europ. J. Biochem., 5:466 (1968).

204. A. Landy, J. Abelson, H. M. Goodman, and J. D. Smith, J. Mol. Biol., 29:457 (1967).

205. M. Ageno, E. Dore, C. Frontali, M. Area, L. Frontali, and G. Tecce, J. Mol. Biol., 15:555 (1966).

206. T. Zehavi-Willner and D. G. Comb, J. Mol. Biol., 16:250 (1966).

207. S. O. Warnaar and J. A. Cohen, Biochem. Biophys. Res. Comm., 24:554 (1966).

208. E. H. McConcey and D. T. Dubin, J. Mol. Biol., 15:102 (1966).

209. N. Murata and W. Szybalski, J. Gen. Appl. Microbiol., 14:57 (1968).

210. E. Winocour, Virology, 31:15 (1967).

211. D. J. Brenner, M. A. Martin, and B. H. Hoyer, J. Bacteriol., 94:486 (1967).

212. C. Schildkraut and S. Lifson, Biopolymers, 3:195 (1965).

213. L. Kotin, J. Mol. Biol., 7:309 (1963).

214. M. T. Record, Biopolymers, 5:975, 993 (1967).

215. B. H. Zimm and J. K. Bragg, J. Chem. Phys., 31:526 (1959).

216. D. M. Crothers and B. H. Zimm, J. Mol. Biol., 9:1 (1964).

217. A. A. Vedenov, A. M. Dykhne, A. D. Frank-Kamenetskii, and M. D. Frank-Kamenetskii, Molekul. Biol., 1:313 (1967).

218. M. D. Frank-Kamenetskii and A. D. Frank-Kamenetskii, Molekul. Biol., 3:375 (1969).

219. D. M. Crothers, Biopolymers, 6:1391 (1968).

220. E. W. Montroll and N. S. Goel, Biopolymers, 4:855 (1966).

221. T. R. Fink and D. M. Crothers, Biopolymers, 6:783 (1968).

222. N. R. Kallenbach and D. M. Crothers, Proc. Nat. Acad. Sci. USA, 56:1018 (1966).

223. H. Fraenkel-Conrat, Biochim. Biophys. Acta, 15:307 (1954).

224. V. F. Drevich, D. G. Knorre, É. G. Malygin, and R. I. Salganik, Molekul. Biol., 1:299 (1967).

225. V. F. Drevich, R. J. Salganic, D. G. Knorre, and E. G. Malygin, Biochim. Biophys. Acta, 123:207 (1966).

226. H. Hayatsu and T. Ukita, Biochim. Biophys. Acta, 123:458 (1966).

227. M. Eigen, J. Chim. Phys., 65:53 (1968).

228. S. W. Englander and J. J. Englander. J. Proc. Nat. Acad. Sci. USA, 53:370 (1965).

229. J. R. Fresco and P. Doty, J. Am. Chem. Soc., 79:3928 (1957).

230. R. Steiner and R. Beers, Biochim. Biophys. Acta, 26:336 (1957).

231. C. L. Stevens and G. Felsenfeld, Biopolymers, 2:293 (1964).

232. D. Barszcz and D. Shugar, Acta Biochim. Polon., 11:481 (1964).

233. G. K. Helmcamp and P. O. P. Ts'o, Biochim. Biophys. Acta, 55:601 (1962).

234. J. Brahms, Nature, 202:797 (1964).

235. J. Brahms, J. Am. Chem. Soc., 85:3298 (1963).

236. H. M. Warshaw and I. Tinoco (Jr.), J. Mol. Biol., 13:54 (1965).

237. G. D. Fasman, C. Lindblow, and L. Grossman, Biochemistry, 3:1015 (1964); A. Adler, L. Grossman, and G. D. Fasman, Proc. Nat. Acad. Sci. USA, 57: 423 (1967).

238. C. R. Cantor, S. K. Jaskunas, and J. Tinoco, J. Mol. Biol., 20:39 (1966).

239. R. L. C. Brimacombe, Biochim. Biophys. Acta, 142:24 (1967).

240. B. E. Griffin, W. J. Haslam, and C. B. Reese, J. Mol. Biol., 10:353 (1964).

241. C. L. Stevens and A. Rosenfeld, Biochemistry, 5:2714 (1966).

242. R. M. Epand and H. A. Scheraga, J. Am. Chem. Soc., 89:3888 (1967).

243. A. S. Spirin, in: Progress in Nucleic Acid Research, Vol. 1, W. E. Cohn and J. N. Davidson (editors), Academic Press, London – New York (1963).

244. G. L. Brown, Progr. Nucl. Acid Res., 2:259 (1963).

245. K.-I. Miura, Prog. Nucl. Acid Res., 6:39 (1967).

246. A. Tissiers, J. Mol. Biol., 1:365 (1960).

247. R. A. Cox and U. Z. Littauer, J. Mol. Biol., 2:166 (1960).

248. G. L. Brown and G. Zubay, J. Mol. Biol., 2:287 (1960).

249. E. J. Ofengand, M. Dieckman, and P. Berg, J. Biol. Chem., 236:1741 (1961).

250. G. D. Fasman, C. Lindblow, and F. Seaman, J. Mol. Biol., 12:630 (1965).

251. L. L. Kiselev, Uspekhi Sovr. Biol., 58:177 (1964).

252. J. T. Penniston and P. Doty, Bioploymers, 1:145 (1963).

253. G. Zubay and R. Marciello, Biochem. Biophys. Res. Comm., 11:79 (1963).

254. L. L. Kiselev, L. Yu. Frolova, O. F. Borisova, and M. K. Kukhanova, Bio-khimiya, 29:116 (1964).

255. R. Marciello and C. Zubay, Biochem. Biophys. Res. Comm., 14:272 (1964).

256. M. Hayashi and S. Spiegelman, Proc. Nat. Acad. Sci. USA, 47:1564 (1961).

257. H. Seidel and F. Cramer, Biochim. Biophys. Acta, 108:367 (1965).

258. M. Yoshida and T. Ukita, J. Biochem. Japan, 58:191 (1965).

259. N. K. Kochetkov, É. I. Budovskii, and V. P. Demushkin, Dokl. Akad. Nauk SSSR, 168:102 (1966).

260. J. R. Fresco, N. M. Alberts, and P. Doty, Nature, 188:98 (1960).

261. J. R. Fresco, L. C. Klotz, and E. G. Richards, Cold Spring Harbor Symposia, 28:83 (1963).

262. J. R. Fresco, in: Informational Macromolecules, H. J. Vogel, V. Bryson, and H. O. Lampen (editors), Academic Press, New York (1963).

263. H. Boedtker, Biochemistry, 6:2718 (1967).

264. D. B. Millar, Biochim. Biophys. Acta, 174:32 (1969).

265. K. S. McCully and G. L. Cantoni, J. Mol. Biol., 5:497 (1962).

266. G. Zubay and M. Takanami, Biochem. Biophys. Res. Comm., 15:207 (1964).

267. R. W. Holley, J. Apgar, G. A. Everett, T. J. Madison, M. Marquisse, S. H. Merril, J. K. Penswic, and A. Zamir, Science, 147:1462 (1965).

268. H. G. Zachau, D. Dütting, H. Feldman, F. Melchers, and W. Karau, Cold Spring Harbor Symposia Quant. Biol., 31:417 (1966).

269. H. Zachau, D. Dütting and H. Feldman, Angew. Chem., 78:392 (1966).

270. A. A. Baev, T. V. Venkstern, A. I. Mirzabekov, A. I. Krutilina, L. Li, and V. D. Aksel'rod, Molekul. Biol., 1:754 (1967).

271. U. L. RajBhandary and S. H. Chang, J. Biol. Chem., 243:598 (1967).

272. S. Takemura, T. Mirutani, and J. Mijazahi, J. Biochem. Japan, 63:277 (1968).

273. H. M. Goodman, J. Abelson, A. Landy, S. Brenner, and J. D. Smith, Nature, 217:1019 (1968).

274. S. K. Dube, K. A. Marcker, B. F. C. Clark, and S. Cory, Nature, 218:232 (1968).

275. S. Cory, K. A. Marcker, S. K. Dube, and B. F. C. Clark, Nature, 220:1039 (1968).

276. M. Staehelin, H. Rogg, B. C. Baguley, T. Ginsburg, and W. Wehrli, Nature, 219:1363 (1968).

277. J. A. Nelson, S. C. Ristow, and R. W. Holley, Biochim. Biophys. Acta, 149: 590 (1967).

278. S. W. Brostoff and W. M. Ingram, Science, 158:666 (1967).

279. M. Yoshida and T. Ukita, Biochim. Biophys. Acta, 123:214 (1966).

280. A. Armstrong, H. Hagopian, V. M. Ingram, and E. K. Wagner, Biochemistry, 5:3027 (1966).

281. E. K. Wagner and V. M. Ingram, Biochemistry, 5:3019 (1966).

282. V. G. Tumanyan, in: Nucleic Acids [in Russian], V. I. Orekhovich (editor), Meditsina (1966), pp. 72-77.

283. A. A. Baev, I. Fodor, A. D. Mirzabekov, V. D. Aksel'rod, and L. Ya. Kazarinova, Molekul. Biol., 1:859 (1967).

284. J. N. Vournakis and H. A. Scheraga, Biochemistry, 5:2997 (1966).

285. P. K. Sarkar and J. T. Yang, J. Biol. Chem., 240:2088 (1965).

286. P. K. Sarkar and J. T. Yang, Biochemistry, 4:1238 (1965).

287. V. G. Tumanyan, N. G. Esipova, and L. L. Kiselev, Dokl. Akad. Nauk SSSR, 168:211 (1966).

288. J. A. Lake and W. W. Beeman, J. Mol. Biol., 31:115 (1968).

289. D. D. Henley, T. Lindahl, and J. R. Fresco, Proc. Nat. Acad. Sci. USA, 55:191 (1966).

290. J. T. Madison, G. A. Everett, and H. Kung, Science, 153:531 (1966).

291. J. R. Fresco, A. Adams, R. Ascione, D. Henley, and T. Lindahl, Cold Spring Harbor Symposia Quant. Biol., 31:527 (1966).

292. D. B. Millar and R. F. Steiner, Biochemistry, 5:2289 (1966).

293. D. B. Millar and M. C. McKenzie, Biochem. Biophys. Res. Comm., 23:724 (1966).

294. T. Lindahl, A. Adams, M. Geroch, and J. R. Fresco, Proc. Nat. Acad. Sci. USA, 57:178 (1967).

295. T. Lindahl, A. Adams, and J. R. Fresco, Proc. Nat. Acad. Sci. USA, 55:941 (1966).

296. W. Gartland and N. Sueoka, Proc. Nat. Acad. Sci. USA, 55:948 (1966).

297. T. Ishida and N. Sueoka, Proc. Nat. Acad. Sci. USA, 58:1080 (1967).

298. T. Lindahl and A. Adams, Science, 152:512 (1966).

299. N. Sueoka, T. Kano-Sueoka, and W. J. Gartland, Cold Spring Harbor Symposia Quant. Biol., 31:571 (1966).

300. K. Muench, Cold Spring Harbor Symposia Quant. Biol., 31:539 (1966).

301. A. Adams, T. Lindahl, and J. R. Fresco, Proc. Nat. Acad. Sci. USA, 57:1684 (1967).

302. G. G. Brownlee, F. Sanger, and B. G. Barrell, Nature, 215:735 (1967).
303. G. G. Brownlee, F. Sanger, and B. G. Barrell, J. Mol. Biol., 34:379 (1968).
304. B. G. Forget and N. Weissman, Science, 158:1695 (1967).
305. H. Boedtker and D. G. Kelling, Biochem. Biophys. Res. Comm., 29:758 (1968).
306. C. R. Cantor, Proc. Nat. Acad. Sci. USA, 59:478 (1968).
307. C. R. Cantor, Nature, 216:513 (1967).
308. F. Cramer and V. A. Erdmann, Nature, 218:92 (1968).
309. J. D. Raacke, Biochem. Biophys. Res. Comm., 31:528 (1968).
310. M. Aubert, J. F. Scott, M. Reynier, and R. Monier, Proc. Nat. Acad. Sci.
     USA, 61:292 (1968).
311. P. McPhie, J. Hounsell, and W. B. Gratzer, Biochemistry, 5:988 (1966).
312. H. Gould, Biochemistry, 5:1103 (1966); Biochim. Biophys. Acta, 123:441 (1966).
313. H. Gould, J. Mol. Biol., 29:307 (1967).
314. J. H. Strauss and R. L. Sinsheimer, J. Mol. Biol., 34:453 (1968).
315. J. C. Wang, D. Baumgarten, and B. M. Olivera, Proc. Nat. Acad. Sci. USA,
     58:1852 (1967).
316. B. Hudson, W. B. Upholt, J. Devinny, and J. Vinograd, Proc. Nat. Acad. Sci.
     USA, 62:813 (1969).
317. Y. Kyogoku, R. C. Lord, and A. Rich, Biochim. Biophys. Acta, 179:10 (1969).
318. P. O. P. Ts'o, N. S. Kondo, M. P. Schweizer, and D. P. Hollis, Biochemistry,
     8:997 (1969).
319. S. I. Chan and J. H. Nelson, J. Am. Chem. Soc., 91:168 (1969).
320. C. C. McDonald, W. D. Phillips, and J. Lazar, J. Am. Chem. Soc., 89:4166
     (1967).
321. I. C. P. Smith, B. J. Blackburn, and T. Yamane, Canad. J. Chem., 47:513
     (1969).
322. K. H. Sheit and W. Saenger, FEBS Letters, 2:305 (1969).
323. A. J. Adler, L. Grossman and C. D. Fasman, Biochemistry, 7:3836 (1968).
324. J. C. Maurizot, W. J. Wechter, J. Brahms, and C. Sadron, Nature, 219:377
     (1968).
325. J. C. Maurizot, J. Brahms and F. Eckstein, Nature, 222:559 (1969).
326. A. J. Adler, L. Grossman and C. D. Fasman, Proc. Nat. Acad. Sci. USA, 57:
     423 (1967).
327. J. Brahms, J. C. Maurizot, and J. Pilet, Biochim. Biophys. Acta, 186:110
     (1969).
328. J. Brahms, A. Aubertin, G. Ditheimer, and M. Grunberg-Manago, Biochemistry
     8:3269 (1969).
329. Y. Inoue and K. Satoh, Biochem. J., 113:843 (1969).
330. A. Opschoor, P. H. Pouwels, C. M. Knijnenberg, and J. B. T. Aten, J. Mol.
     Biol., 37:13 (1968).
331. J. C. Wang, J. Mol. Biol., 43:45 (1969).
332. J. C. Wang, J. Mol. Biol., 43:263 (1969).
333. H. Eisenberg and G. Cohen, J. Mol. Biol., 37:355 (1968).
334. V. Luzzati, F. Masson, A. Mathis, and P. Saludjan, Biopolymers, 5:491 (1967).
335. M. M. K. Nass, Science, 165:25 (1969).
336. M. F. Bourguignon and P. Bourgaux, Biochim. Biophys. Acta, 169:476 (1968).
337. W. Fuller and M. J. Waring, Ber. Bunsenges. Physik. Chem., 68:805 (1964).
338. M. J. Waring, Biochem. J., 109:28 (1968).
339. S. Arnott, W. Fuller, A. Hodson, and J. Prutton, Nature, 220:561 (1968).
340. C. A. Bush and H. A. Scheraga, Biopolymers, 7:395 (1969).
341. M. P. Printz and P. H. Von Hippel, Biochemistry, 7:3194 (1968).
342. B. McConnel and P. H. Von Hippel, Fed. Proc., 27:802 (1968).
343. J. Marmur and P. Doty, J. Mol. Biol., 5:109 (1962).

344. L. G. Silvestri and L. R. Hill, J. Bacteriol. 90:136 (1965).

345. J. Bonachek, M. Kokur, and T. Martinec, J. Gen. Microbiol., 46:369 (1966).

346. R. J. Owen, L. R. Hill, and S. R. Lapage, Biopolymers, 7:503 (1969).

347. D. W. Gruenwedel and C.-H. Hsu, Biopolymers, 7:557 (1969).

348. P. Doty, H. Boedtker, J. R. Fresco, R. Haselkorn, and M. Litt, Proc. Nat. Acad. Sci. USA, 45:482 (1959).

349. G. Löber and C. H. Zimmer, Biochem. Biophys. Res. Comm., 31:641 (1968).

350. A. M. Michelson and F. Pochon, Biochim. Biophys. Acta, 174:604 (1969).

351. G. Luck and Ch. Zimmer, Biochim. Biophys. Acta, 169:466 (1968).

352. C. Zimmer, G. Luck, H. Venner, and J. Fric, Biopolymers, 6:563 (1968).

353. Y. Courtois, P. Fromageot, and W. Gushlbauer, Europ. J. Biochem., 6:493 (1968).

354. G. Luck, Ch. Zimmer, and G. Shatzke, Biochim. Biophys. Acta, 169:548 (1968).

355. C. Zimmer, Biochim. Biophys. Acta, 161:584 (1968).

356. R. E. Chapman and J. M. Sturtewant, Biopolymers, 7:527 (1969).

357. S. Z. Basu, Naturforsch, 24b:511 (1969).

358. P. Bartl and M. Boublik, Biochim. Biophys. Acta, 103:678 (1965).

359. B. Bagchi, D. N. Misra, S. Basu, and N. N. Das Gupta, Biochim. Biophys. Acta, 182:551 (1969).

360. V. P. Kushner and N. V. Zakharova, Molekul. Biol., 3:384 (1969).

361. H. S. Jansz, P. D. Baas, P. H. Pouwels, E. F. J. van Bruggen, and H. J. Oldenziel, J. Mol. Biol., 32:159 (1968).

362. P. H. Pouwels, J. van Rotterdam, and J. A. Cohen, J. Mol. Biol., 40:379 (1969).

363. H. R. Massie and B. H. Zimm, Biopolymers, 7:475 (1969).

364. B. J. McCarthy, Bacter. Rev., 31:215 (1967).

365. R. J. Britten and D. E. Kohne, Science, 161:529 (1968).

366. K. J. Thrower and A. R. Peacocke, Biochem. J., 109:543 (1969).

367. F. W. Studier, J. Mol. Biol., 41:189 (1969).

368. F. W. Studier, J. Mol. Biol., 41:199 (1969).

369. D. D. Wood and D. J. L. Luck, J. Mol. Biol., 41:211 (1969).

370. J. O. Bishop, Biochem. J., 108:35 (1968).

371. J. O. Bishop, Biochem. J., 113:805 (1969).

372. M. Melli and J. O. Bishop, J. Mol. Biol., 40:117 (1969).

373. J. Bonner, G. Kung, and I. Bekhor, Biochemistry, 6:3650 (1967).

374. B. L. McConaughy, C. D. Laird, and B. J. McCarthy, Biochemistry, 8:3289 (1969).

375. J. Legault-Demare, B. Dessaux, T. Heyman, S. Seror, and G. P. Ress, Biochem. Biophys. Res. Comm., 28:550 (1967).

376. I. Bekhor, J. Bonner, and G. K. Dahmus, Proc. Nat. Acad. Sci. USA, 62:271 (1969).

377. L. C. Klotz, Biopolymers, 7:265 (1969).

378. G. W. Lehman and J. P. McTague, J. Chem. Phys., 49:3170 (1969).

379. A. D. Frank-Kamenetskii and M. D. Frank-Kamenetskii, Molekul. Biol., 2:778 (1968).

380. N. S. Goel and S. C. Maitra, J. Theoret. Biol., 23:87 (1969).

381. I. Hiroyashi and P. Doty, Repts. Progr. Polym. Phys. Japan, 11:533 (1968).

382. A. M. Bobst, P. A. Cerrutti, and F. Rottman, J. Am. Chem. Soc., 91:1246 (1969).

383. A. D. Mirzabekov, L. Ya. Kazarinova, D. Latity, and A. A. Baev, FEBS Letters, 3:268 (1969).

384. N. Imura, G. B. Weiss, and R. W. Chambers, Nature, 222:1147 (1969).

385. N. Imura, H. Schwamm, and R. W. Chambers, Proc. Nat. Acad. Sci. USA, 62:1203 (1969).

386. M. Tsuboi, S. Higuchi, Y. Kyogoku, and S. Nishimura, Biochim. Biophys. Acta, 195:23 (1969).

387. K. Morikawa, M. Tsuboi, Y. Kyogoku, T. Seno, and S. Nishimura, Nature, 223:537 (1969).

388. R. A. Cox, Quart. Rev., 82:499 (1968).

389. B. F. Clark, B. P. Doctor, K. C. Holmes, A. Klug, K. A. Marcker, S. J. Morris, and H. H. Paradies, Nature, 219:1222 (1968).

390. B. S. Vold, Biochem. Biophys. Res. Comm., 35:222 (1969).

391. J. R. Fresco, R. D. Blake, and R. Langridge, Nature, 220:1285 (1968).

392. B. P. Doctor, W. Fuller, and N. L. Webb, Nature, 221:58 (1969).

393. G. J. Thomas (Jr.), Biopolymers, 7:325 (1969).

394. R. I. Cotter and W. B. Gratzer, Nature, 221:154 (1969).

395. G. J. Thomas (Jr.) and M. Spencer, Biochim. Biophys. Acta, 179:360 (1969).

396. N. R. Kallenbach, J. Mol. Biol., 37:445 (1968).

397. D. Riesner, R. Römer, and G. Maass, Biochem. Biophys. Res. Comm., 35:369 (1969).

398. F. Cramer, H. Doepner, F. Haar, E. Schlimme, and H. Seidel, Proc. Nat. Acad. Sci. USA, 61:1384 (1969).

399. G. Melcher, FEBS Letters, 3:185 (1969).

400. T. Seno, M. Kobayashi, and S. Nishimura, Biochim. Biophys. Acta, 174:71 (1969).

401. J. F. Scott, R. Monier, M. Aubert, and M. Reynier, Biochem. Biophys. Res. Comm., 33:794 (1968).

Index

The index for Parts A and B are consolidated at the
end of Part B. Starting on page 619.